REGULATION OF MACROMOLECULAR SYNTHESIS BY LOW MOLECULAR WEIGHT MEDIATORS

Academic Press Rapid Manuscript Reproduction

Proceedings of the

*Regulation of
Macromolecular Synthesis
by Low Molecular Weight Mediators*

Workshop
Held at Hamburg-Blankenese
May 29–31, 1979

REGULATION OF MACROMOLECULAR SYNTHESIS BY LOW MOLECULAR WEIGHT MEDIATORS

edited by

GEBHARD KOCH
Physiologisch-Chemisches Institut
der Universität Hamburg
Abteilung Molekularbiologie
Hamburg, West Germany

DIETMAR RICHTER
Physiologisch-Chemisches Institut
der Universität Hamburg
Abteilung Zellbiochemie
Hamburg, West Germany

1979

ACADEMIC PRESS

A Subsidiary of Harcourt Brace Jovanovich, Publishers
New York   London   Toronto   Sydney   San Francisco

COPYRIGHT © 1979, BY ACADEMIC PRESS, INC.
ALL RIGHTS RESERVED.
NO PART OF THIS PUBLICATION MAY BE REPRODUCED OR
TRANSMITTED IN ANY FORM OR BY ANY MEANS, ELECTRONIC
OR MECHANICAL, INCLUDING PHOTOCOPY, RECORDING, OR ANY
INFORMATION STORAGE AND RETRIEVAL SYSTEM, WITHOUT
PERMISSION IN WRITING FROM THE PUBLISHER.

ACADEMIC PRESS, INC.
111 Fifth Avenue, New York, New York 10003

United Kingdom Edition published by
ACADEMIC PRESS, INC. (LONDON) LTD.
24/28 Oval Road, London NW1 7DX

LIBRARY OF CONGRESS CATALOGING IN PUBLICATION DATA

Regulation of Macromolecular Synthesis by Low Molecular Weight
Mediators Workshop, Blankenese, Ger., 1979.
   Regulation of macromolecular weight mediators by low molecular weight mediators.

   Includes index.
   1. Nucleotides--Physiological effect--Congresses.
2. Macromolecules--Metabolism--Congresses.   3. Metabolic
regulation--Congresses.   4. Genetic translation--Congresses.
I. Koch, Gebhard. II. Richter, Dietmar. III. Title.

QP625.N89R44   1979   574.8'732   79-26279

ISBN   0-12-417580-5

PRINTED IN THE UNITED STATES OF AMERICA

79 80 81 82   9 8 7 6 5 4 3 2 1

Fritz Lipmann

CONTENTS

Contributors   xi
Foreword   xvii
Preface   xix

Opening Remarks:
Oligonucleotides and Growth Regulation   P. Zamecnik   1

I. FUNCTIONS AND METABOLISM OF GUANOSINE 3',5'-BIS(DIPHOSPHATE) [ppGpp]

1. Stringent Control in Bacterial Systems

Stringent Control of Translational Accuracy   J. A. Gallant and D. Foley   5

Guanosine 3',5'-Bispyrophosphate Is a Dispensable Metabolite   J. A. Engel, J. Sylvester, and M. Cashel   25

Studies on the Coordination of tRNA-Charging and Polypeptide Synthesis Activity in *Escherichia coli*   W. Piepersberg, D. Geyl, P. Buckel and A. Böck   39

Deletion of *relA* and *relX* in *Escherichia coli* Has No Effect on Basal or Carbon-Downshift ppGpp Synthesis   A. G. Atherly   53

2. Effect of ppGpp on RNA Polymerase and Transcription in E. coli

Molecular Interactions in the Initiation of rRNA Synthesis in Prokaryotes   M. Gruber, J. Hamming, B. A. Oostra, and G. AB   67

The Regulation of RNA Polymerase by Guanosine 5'-Disphosphate 3'-Diphosphate   *P. G. Debenham, R. Buckland, and A. A. Travers*   77

3. Metabolism of ppGpp In Cell-Free Bacterial Systems

Synthesis and Degradation of the Pleiotropic Effector Guanosine 3',-5'Bis(disphosphate) in Bacteria   *D. Richter*   85

Biosynthesis of Guanosine Tetraphosphate in *Bacillus brevis*   *J. Sy*   95

4. Failure to Detect ppGpp In Eukaryotes

A Radioimmunoassay for ppGpp   *R. H. Silverman, A. G. Atherly, G. Glaser, and M. Cashel*   107

Guanosine 3',5'-Bis (diphosphate) Search in Eukaryotes   *R. Silverman, A. Atherly, and D. Richter*   115

II. PURINE NUCLEOTIDES AND SPORULATION

Initiation of Bacterial and Yeast Sporulation by Partial Deprivation of Guanine Nucleotides   *E. Freese, J. M. Lopez, and E. B. Freese*   127

Regulation of Sporulation by Highly Phosphorylated Nucleotides in *Bacillus subtilis*   *H. J. Rhaese, R. Groscurth, R. Vetter, and H. Gilbert*   145

III. HIGHLY PHOSPHORYLATED NUCLEOTIDES IN EUKARYOTES

1. Isolation and Characterization of Novel Dinucleotides

Studies on the Biosynthesis and Function of Dinucleoside Polyphosphates in *Artemia* Embryos   *A. H. Warner*   161

Historical Background on Adenosine $5',5'''\text{-}P^1,P^4$-Tetraphosphate (Ap4A) and Current Developments   *P. C. Zamecnik and E. Rapaport*   179

The Accumulation of Highly Phosphorylated Compounds in *Drosophila melanogaster* Cells after Heat Shock   *A. A. Travers*   185

HS3—A Bizzare Dinucleoside Polyphosphate as Possible 193
Pleiotypic Regulator of Eukaryotes   H. B. LéJohn,
G. R. Klassen, and S. Han Goh

2. Effects of Dinucleotide and Adenosine Nucleotide Pools in Replication of DNA and Protein Synthesis

Diadenosine Tetraphosphate ($A_{P4}A$)—A Ligand of DNA 209
Polymerase α and Trigger of Replication   F. Grummt,
G. Waltl, H.-M. Jantzen, K. Hamprecht, U. Huebscher,
and C. C. Kuenzle

Elevated Nuclear ATP Pools and ATP/ADP Ratios Mediate 223
Adenosine Toxicity in Fibroblasts   E. Rapaport, and
S. K. Svihovec

Relation of Protein Synthesis to the Content of Adenosine 233
Polyphosphates   P. Plesner and K. Kristiansen

IV. ALTERATION OF TRANSLATIONAL MECHANISMS

Nucleosidetriphosphate Mediated Discrimination of Gene 249
Expression in T1-Infected *E. coli*   M. Schweiger and
E. F. Wagner

Compartmentalization of Phosphate Donor Pool for Ribosomal 263
Protein S6 before and after Stimulation of Quiescent
Mouse Fibroblast Cells with Serum, and Nondependence
of the Phosphorylation of cAMP Pools   J. Gordon,
L. Jimenez de Asua, M. Siegmann, A.-M. Kübler, and
G. Thomas

Membrane Mediated Amplification of Translational Control in 273
Eukaryotes: A Pleiotropic Effect   G. Koch, P. Bilello, and
J. Kruppa

Mode of Action of Hemin-Controlled Translational Inhibitor 291
C. de Haro and S. Ochoa

V. 2',5'-OLIGO A AND INTERFERON

Induction, Purification, and Properties of 2',5'-Oligoadenylate 303
Synthetase   L. A. Ball and C. N. White

Mechanism of Action of 2-5A in Intact Cells 319
A. G. Hovanessian, J. N. Wood, E. Meurs, and L. Montagnier

2',5'-Oligo(A): A Mediator of Viral RNA Cleavage in Interferon-   329
Treated Cells?   *T. W. Nilsen, P. A. Maroney, and C. Baglioni*

Studies on Interferon Action: Synthesis, Degradation, and   341
Biological Activity of (2',5')-Oligoisoadenylate   *M. Revel,*
A. Kimchi, A. Schmidt, L. Shulman, and Y. Chernajovsky

Concluding Remarks
Low Molecular Weight Mediators of Macromolecular Synthesis—   361
An Overview   *A. A. Travers*

Index   *365*

CONTRIBUTORS

Numbers in parentheses indicate the pages on which the authors' contributions begin.

G. AB (67), *Biochemisch Laboratorium, The University, Groningen, The Netherlands*

ALAN G. ATHERLY (53, 107, 115), *Department of Genetics, Iowa State University, Ames, Iowa*

CORRADO BAGLIONI (329), *Department of Biological Sciences, State University of New York at Albany, Albany, New York*

L. ANDREW BALL (303), *University of Wisconsin, Madison, Wisconsin*

PATRICIA BILELLO (273), *Physiologisch-Chemisches Institut der Universität Hamburg, Abteilung Molekularbiologie, Hamburg, West Germany*

AUGUST BÖCK (39), *Lehrstuhl für Mikrobiologie, Universität Regensburg, West Germany*

PETER BUCKEL (39), *Lehrstuhl für Mikrobiologie, Universität Regensburg, West Germany*

R. BUCKLAND (77), *MRC Laboratory of Molecular Biology, Hills Road, Cambridge, CB 2 2QH, England*

MICHAEL CASHEL (25, 107), *Laboratory of Molecular Genetics, National Institute of Child Health and Development, National Institutes of Health, Bethesda, Maryland*

YUTI CHERNAJOVSKY (341), *Department of Virology, Weizmann Institute of Science, Rehovot, Israel*

P. G. DEBENHAM (77), *MRC Laboratory of Molecular Biology, Hills Road, Cambridge, CB2 2QH, England*

CÉSAR DE HARO (291), *Centro di Biologia Molecular, Universidad Autónoma de Madrid, Madrid, Spain*

JO A. ENGEL (25), *Laboratory of Molecular Genetics, National Institute of Child Health and Human Development, National Institutes of Health, Bethedsa, Maryland*

DIANE FOLEY (5), *Department of Genetics, University of Washington, Seattle, Washington*

ELISABETH B. FREESE (127), *Laboratory of Molecular Biology, National Institute of Child Health and Human Development, National Institutes of Health, Bethesda, Maryland*

ERNST FREESE (127), *Laboratory of Molecular Biology, National Institute of Child Health and Human Development, National Institutes of Health, Bethesda, Maryland*

JONATHAN A. GALLANT (5), *Department of Genetics, University of Washington, Seattle, Washington*

DIETER GEYL (39), *Lehrstuhl für Mikrobiologie, Universität Regensburg, West Germany*

HANNELORE GILBERT (145), *Institut für Mikrobiologie, Universität Frankfurt, Frankfurt am Main, West Germany*

GAD GLASER (107), *Laboratory of Molecular Genetics, National Institute of Child Health and Development, National Institutes of Health, Bethesda, Maryland*

JULIAN GORDON (263), *Friedrich Miescher-Institut, P. O. Box 273, CH-4002 Basel, Switzerland*

REINHARD GROSCURTH (145), *Institut für Mikrobiologie, Universität Frankfurt, Frankfurt am Main, West Germany*

M. GRUBER (67), *Biochemisch Laboratorium, The University, Groningen, The Netherlands*

FRIEDRICH GRUMMT (209), *Max-Planck-Institut für Biochemie, D-8033 Martinsried bei Müchen, West Germany*

J. HAMMING (67), *Biochemisch Laboratorium, The University, Groningen, The Netherlands*

KLAUS HAMPRECHT (209), *Max-Planck-Institut für Biochemie, D-8033 Martinsried bei München, West Germany*

SWEE HAN GOH (193), *Department of Microbiology, University of Manitoba, Winnipeg, Manitoba R3T 2N2, Canada*

Contributors

ARA G. HOVANESSIAN (319), *Unité d'Oncologie Virale, Département de Virologie, Institut Pasteur, Paris, France*

ULRICH HUEBSCHER (209), *Institut für Pharmakalogie, Veterinärmedizinische Fakultat der Universität Zürich, CH-8057 Zürich, Switzerland*

HANS-MICHAEL JANTZEN (209), *Max-Planck-Institut für Biochemie, D-8033 Martinsried bei München, West Germany*

LUIS JIMENEZ DE ASUA (263), *Friedrich Miescher-Institut, P. O. Box 273, CH-4002 Basel, Switzerland*

ADI KIMCHI (341), *Department of Virology, Weizmann Institute of Science, Rehovot, Israel*

GLEN R. KLASSEN (193), *Department of Microbiology, University of Manitoba, Winnipeg, Manitoba, R3T 2N2, Canada*

GEBHARD KOCH (273), *Physiologisch-Chemishes Institut der Universität Hamburg, Abteilung Molekularbiologie, Hamburg, West Germany*

KARSTEN KRISTIANSEN (233), *Department of Chemistry, Carlsberg Laboratory, Copenhagen, Denmark; and Department of Molecular Biology, University of Odense, Odense, Denmark*

JOACHIM KRUPPA (273), *Physiologisch-Chemisches Institut der Universität Hamburg, Abteilung Molekularbiologie, Hamburg, West Germany*

ANNE-MARIE KÜBLER (263), *Friedrich Miescher-Institut, P. O. Box 273, CH-4002 Basel, Switzerland*

CLIVE C. KUENZLE (209), *Institut für Pharmakalogie, Veterinärmedizinische Fakultat der Universität Zürich, CH-8057 Zürich, Switzerland*

YEHUDA LAPIDOT (341), *Department of Biological Chemistry, Hebrew University, Jerusalem, Israel*

HERB LÉ JOHN (193), *Department of Microbiology, University of Manitoba, Winnipeg, Manitoba, R3T 2N2, Canada*

JUAN M. LOPEZ (127), *Laboratory of Molecular Biology, National Institute of Child Health and Human Development, National Institutes of Health, Bethesda, Maryland*

PATRICIA A. MARONEY (329), *Department of Biological Sciences, State University of New York at Albany, Albany, New York*

ELIANE MEURS (319), *Unité d'Oncologie Virale, Département de Virologie, Institut Pasteur, Paris, France*

LUC MONTAGNIER (319), *Unité d'Oncologie Virale, Département de Virologie, Institut Pasteur, Paris, France*

TIMOTHY W. NILSEN (329), *Department of Biological Sciences, State University of New York at Albany, Albany, New York*

SEVERO OCHOA (291), *Roche Institute of Molecular Biology, Nutley, New Jersey*

B. A. OOSTRA (67), *Biochemisch Laboratorium, The University, Groningen, The Netherlands*

WOLFGANG PIEPERSBERG (39), *Lehrstuhl für Mikrobiologie, Universität Regensburg, West Germany*

PAUL PLESNER (233), *Department of Chemistry, Carlsberg Laboratory, Copenhagen, Denmark; and Department of Molecular Biology, University of Odense, Odense, Denmark*

ELIEZER RAPAPORT (179, 223), *Huntington Laboratories, Massachusetts General Hospital, Boston, Massachusetts*

SARAH RAPOPORT (341), *Department of Biological Chemistry, Hebrew University, Jerusalem, Israel*

MICHEL REVEL (341), *Department of Virology, Weizmann Institute of Science, Rehovot, Israel*

HANS J. RHAESE (145), *Institut für Mikrobiologie, Universität Frankfurt, Frankfurt am Main, West Germany*

DIETMAR RICHTER (85, 115), *Physiologisch-Chemisches Institut der Universität Hamburg, Abteilung Zellbiochemie, Hamburg, West Germany*

AZRIEL SCHMIDT (341), *Department of Virology, Weizmann Institute of Science, Rehovot, Israel*

MANFRED SCHWEIGER (249), *Institut für Biochemie, Universität Innsbruck, Innsbruck, Austria*

LESTER SHULMAN (341), *Department of Virology, Weizmann Institute of Science, Rehovot, Israel*

MICHEL SIEGMANN (263), *Friedrich Miescher-Institut, P. O. Box 273, CH-4002 Basel, Switzerland*

ROBERT SILVERMAN (107, 115), *Department of Genetics, Iowa State University, Ames, Iowa*

SANDRA K. SVIHOVEC (223), *The John Collins Warren Laboratories, Massachusetts General Hospital, Boston, Massachusetts 02114*

JOSE SY (95), *The Rockefeller University, New York, New York*

JAMES SYLVESTER (25), *Section on Molecular Regulation, Laboratory of Molecular Genetics, National Institute of Child Health and Human Development, National Institutes of Health, Bethesda, Maryland*

GEORGE THOMAS (263), *Friedrich Miescher-Institut, P. O. Box 273, CH-4002 Basel, Switzerland*

A. A. TRAVERS (77, 185, 361), *MRC Laboratory of Molecular Biology, Hills Road, Cambridge, CB2 2QH, England*

ROMAN VETTER (145), *Institut für Mikrobiologie, Universität Frankfurt, Frankfurt am Main, West Germany*

ERWIN F. WAGNER (249), *Institut für Biochemie, Universität Innsbruck, Innsbruck, Austria*

GERT WALTL (209), *Max-Planck-Institut für Biochemie, D-8033 Martinsried bei München, West Germany*

A. H. WARNER (161), *Depatment of Biology, University of Windsor, Windsor, Ontario, Canada*

CAROL N. WHITE (303), *Microbiology Section, University of Connecticut, Storrs, Connecticut*

JOHN N. WOOD (319), *Unité d'Oncologie Virale, Département de Virologie, Institut Pasteur, Paris, France*

PAUL C. ZAMECNIK (1, 179), *Huntington Laboratories, Massachusetts General Hospital, Boston, Massachusetts*

FOREWORD

This volume contains the proceedings of the Workshop on Regulation of Macromolecular Synthesis by Low Molecular Weight Mediators organized by Gebhard Koch and Dietmar Richter. The meeting was held in Hamburg, May 29-31, 1979.

The low molecular weight mediators discussed at the workshop are mostly nucleotides of unusual structure. Their occurrence was recognized as early as 1963 in the case of GppppG and 1966 in the case of AppppA. However, a biological role for nucleotides of uncommon structure was first indicated in 1969 through the discovery of the "magic spots" ppGpp and pppGpp and the recognition of their involvement in the stringent response of bacterial cells to amino acid starvation. This consists of a marked decrease in the synthesis of rRNA and tRNA, mediated at the transcriptional level, and a restriction of other biosynthetic and metabolic processes. The widespread occurrence of ppGpp in prokaryotes contrasts with its apparent absence from eukaryotes, particularly from higher organisms. Although the protein-synthesizing machinery of eukaryotic organelles, e.g., mitochondria, chloroplasts, is related to that of prokaryotes, there are no clear indications that ppGpp or similar compounds are made in these organelles.

The occurrence of other unusual nucleotides, such as GppppG in *Artemia* and AppppA in a variety of organisms, has raised questions as to their function and significance. No regulatory functions can as yet be ascribed to GppppG but there are indications that AppppA may function as a pleiotypic activator, and an intriguing role for this compound as an activator of DNA synthesis has recently been suggested. The latest addition, and a most exciting one, to the list of unusual nucleotides is that of pppA(2')p(5')A(2')p(5')A, a potent translational inhibitor active at subnanomolar concentrations.

The discovery of nucleotides of unusual structure has always been unexpected and there is no reason to believe that the list is complete; one might probably bet that it is not. I think we may expect more surprises as the field unfolds further. This volume contains reports from most of the groups responsible for the basic discoveries in the field and for studies on the distribution, mechanism of biosynthesis, metabolism, and biological significance of unusual nucleotides. It also contains a report on the mode of action of a translational inhibitor that is controlled by heme, a non-nucleotide low molecular weight mediator.

The volume is dedicated to Fritz Lipmann on the occasion of his eightieth anniversary. Lipmann and his students have made important contributions to our knowledge of ppGpp biosynthesis and function, and this volume contains papers by some of Lipmann's former students. I am happy for the opportunity to add my personal tribute of admiration and esteem for one of the greatest biochemists of all times.

Severo Ochoa
Roche Institute of Molecular Biology
Nutley, New Jersey

PREFACE

The elucidation of the important role of low molecular weight mediators in the regulation of macromolecular synthesis is presently the aim of many research teams. Investigators working with prokaryotes have analyzed the effect of guanosine tetra- and pentaphosphates on RNA and protein synthesis and have also studied the metabolism of these compounds. Several research groups working with eukaryotic systems have begun to unravel the involvement of unique dinucleotides in the regulation of DNA, RNA, and protein synthesis. An exciting new direction of research was opened by the discovery of 2',5'-oligoadenylic acid in interferon-treated cells. Therefore, we thought the time had come for a fruitful exchange of research findings and experiences obtained with various prokaryotic and eukaryotic systems.

Representatives from different fields were invited to the first conference on macromolecular synthesis in Hamburg in May 1979. The proceedings of the meeting are hereby presented to a wider audience.

This meeting was made possible through financial support by Stiftung Volkswagenwerk, Hamburgische Wissenschaftliche Stiftung and Behörde für Wissenschaft und Kunst der Freien und Hansestadt Hamburg.

**REGULATION OF MACROMOLECULAR SYNTHESIS
BY LOW MOLECULAR WEIGHT MEDIATORS**

OLIGONUCLEOTIDES AND GROWTH REGULATION

Paul Zamecnik

Huntington Laboratories of Harvard University
Massachusetts General Hospital
Boston, Massachusetts

INTRODUCTORY REMARKS *

During the past decade, the speed of development of knowledge of the physical characteristics and metabolic behavior of macromolecules has diverted attention from the existance of a small but growing group of unusual nucleotides which play regulatory roles in living cells. Cyclic AMP may be regarded as the first of this number to be recognized, and contrary to the statement above, it has received enormous scrutiny. Setting aside this model regulator, however, a number of other di- and oligonucleotides have surfaced on the scientific scene during recent years, and have to date led modest lives, outside the limelight of general attention. Let me enumerate several such molecules, while leaving others for discussion later in this meeting: Gp_4G, discovered in 1963 by Finamore and Warner to serve as an energy source in rehydrated brine shrimp cysts (Finamore & Warner, 1963), Ap_4A, uncovered in 1966 in our laboratories, formed in the back reaction of the first step in protein synthesis (Zamecnik et al., 1966; Randerath et al., 1966) and now implicated in cell proliferation (Rapaport & Zamecnik, 1976) and DNA synthesis (Grummt, 1978); ppGpp, found by Cashel and Gallant in 1969 to play a role in the stringent response of bacteria (Cashel & Gallant, 1969); and pppA2'p5'A2'p5'A, reported by Kerr (Kerr

---

* *Supported by ERDA contract 79-EV02403, ACS Grant, BC 278, and NCI Virus Contract NO177001.*

& Brown, 1978) and by Ball (Ball & White, 1978) to be an intermediate formed in enhanced amounts in interferon system.

In a separate category are the oligonucleotides regarded as primers for DNA polymerase (Bouche, Rowen & Kornberg, 1978), some isolated by Richardson and colleyues (Masamune & Richardson, 1977) from commercial GTP preparations.

Finally, one may include, in this general grouping of regulatory nucleotides synthetic ribo- and deoxyribonucleotides which have been shown to exert regulatory effects on cellular events. In particular may be mentioned rUGG, synthesized by Miller et al. (1977) and found to inhibit protein synthesis by hybridizing with the terminal - CCA of transfer RNAs. Also, quite lately a report has come from our laboratories of the use of a synthetic tridecamer deoxynucleotide complementary to the terminal reiterated sequences of Rous sarcoma virus 35S RNA to inhibit viral replication and cell transformation (Zamecnik & Stephenson, 1978; Stephenson & Zamecnik, 1978).

The development of high pressure liquid chromatography (Chen, Brown & Rosie, 1977; Plesner et al., 1979) has also presented to the investigator of nucleotides a number of unknown peaks, presumably due to nucleotides, which remain to be identified. One harbors the suspicion that in the 10^{-7} - 10^{-9} molar nucleotide range, which now becomes more possible to investigate in tissue culture systems, a new family of such cellular regulators may be waiting. I am impressed by the thoughtful review of the late Gordon Tomkins (Tomkins, 1975) in which he assigned to the nucleotide family the role of intracellular regulators, in general, because of their membrane metabolic lability and speed of interaction in cellular events. He consigned to the role as intercellular regulators peptide and steroid hormones, more stable metabolically, and exerting their effects in a more leisurely way.

REFERENCES

Ball, L.A. & White, C.N. (1978), pppA2'p5'A2'p5'A: An inhibitor of protein synthesis with an enzyme faction from interferon-treated cells. Proc. Natl. Acad. Sci. USA 75, 1167.
Bouche, J.-P., Rowen, L. & Kornberg, A. (1978), The RNA primer synthesized by primase to initiate phage T4 DNA replication. J. Biol. Chem. 253, 765.

Cashel, M. & Gallant, T.J. (1969), Two compounds implicated in the function of the RC gene of Escherichia coli. *Nature* 221, 838.

Chen, S.-C., Brown, R.P. & Rosie, D.M. (1977), Extraction procedures for use prior to HPLC nucleotide analysis using microparticle chemically bonding packings. *J. Chromatog. Sci.* 15, 218.

Finamore, F.J. & Warner, A.H. (1963), The occurrence of $P^1 P^4$-diguanosine 5'-tetraphosphate in brine shrimp eggs. *J. Biol. Chem.* 238, 344.

Grummt, F. (1978), Diadenosine 5'5'''-P^1,P^4-tetraphosphate triggers initiation of in vitro DNA replication in baby hamster kidney cells. *Proc. Natl. Acad. Sci. USA* 75, 371.

Kerr, I.M. & Brown, R.E. (1978), pppA2'p5'A2'p5'A: An inhibitor of protein synthesis synthesized with an enzyme fraction from interferon-treated cells. *Proc. Natl. Acad. Sci. USA* 75, 256.

Masamune, Y. & Richardson, C.C. (1977), Priming of deoxyribonucleic acid synthesis on phage fd and ØX174 templates by oligoribonucleotides contaminating nucleoside 5'-triphosphates. *J. Biol. Chem.* 252, 8498.

Miller, P.S., Braiterman, L.T. & Ts'o P.O.P. (1977), Effects of a trinucleotide ethyl phosphotriester, $G^m p(Et) G^m p(Et)U$, on mammalian cells in culture. *Biochemistry* 16, 1988.

Plesner, P., Stephenson, M.L., Zamecnik, P.C. & Bucher, N.L.R. (1979), Diadenosine tetraphosphate (Ap_4A), an activator of gene function. *Alfred Benzon Sympos.* 13, Munksgaard, Copenhagen (in press).

Randerath, K., Janeway, C.I., Stephenson, M.L. & Zamecnik, P.C. (1966), Isolation and characterization of dinucleoside tetra- and triphosphates formed in the presence of lysyl-sRNA synthetase. *Biochem. Biophys. Res. Comm.* 24, 98.

Rapaport, E. & Zamecnik, P.C. (1976). Presence of Ap_4A in mammalian cells in levels varying widely with proliferative activity of the tissue. A possible "pleiotypic activator". *Proc. Natl. Acad. Sci. USA* 73, 3984.

Stephenson, M.L. & Zamecnik, P.C. (1978), Inhibition of Rous sarcoma viral RNA translation by a specific oligodeoxyribonucleotide. *Proc. Natl. Acad. Sci. USA* 75, 285.

Tomkins, G. (1975), The metabolic code. *Science* 189, 760.

Zamecnik, P.C., Stephenson, M.L., Janeway, C.L. & Randerath, K. (1966), Enzymatic synthesis of diadenosine tetraphosphate and diadenosine triphosphate with a purified lysyl-sRNA synthetase. *Biochem. Biophys. Res. Comm.* 24, 91.

Zamecnik, P.C. & Stephenson, M.L. (1978), Inhibition of Rous sarcoma virus replication and transformation by a specific oligodeoxynucleotide. *Proc. Natl. Acad. Sci. USA* 75, 280.

Part I
FUNCTIONS AND METABOLISM OF GUANOSINE 3',5'-BIS(DIPHOSPHATE) [ppGpp]

Section 1
Stringent Control in Bacterial Systems

Section 2
Effect of ppGpp on RNA Polymerase and Transcription in *E. coli*

Section 3
Metabolism of ppGpp in Cell-Free Bacterial Systems

Section 4
Failure to Detect ppGpp in Eukaryotes

REGULATION OF MACROMOLECULAR SYNTHESIS BY LOW MOLECULAR WEIGHT MEDIATORS

STRINGENT CONTROL OF TRANSLATIONAL ACCURACY

Jonathan A. Gallant and Diane Foley

Department of Genetics
University of Washington
Seattle, Washington

We are accustomed to thinking of control mechanisms in biology as agencies which adjust the *rates* of processes without affecting their underlying specificities. Under certain circumstances, however, it is likely that the discrimination specificity of protein synthesis is actively adjusted. This conclusion follows from the fact that wild type bacteria do not show an increase in translational errors during limitation for one aminoacyl-tRNA species, a situation in which a passive system would have to show such an increase.

Consider our old friend the messenger RNA codon UUU. It calls for PHE-tRNA^{phe}. But the discrimination specificity of the translation apparatus is not absolute. *In vitro* polyU directs LEU into protein as well, at rates which are not only detectable, but sometimes positively alarming. Other non-cognate tRNA's whose normal codons differ from UUU at but one position are also bound by polyU programmed ribosomes; and these errors are exacerbated when the ribosomes are "detuned" by abnormal conditions or the presence of aminoglycoside antibiotics such as streptomycin (reviewed by Woese, 1966). *In vivo*, errors of amino acid substitution occur at lower but detectable frequencies: the few estimates we have indicate a normal error frequency in the vicinity of 3×10^{-4} (Loftfield, 1963; Loftfield and Vanderjagt, 1972; Edelmann and Gallant, 1977a), a value which can be increased by more than an order of magnitude during growth in streptomycin (Edelmann and Gallant, 1977b).

Now, consider what happens when cells are starved for phenylalanine, or the phenylalanyl-tRNA synthetase is specifically inhibited. The level of PHE-tRNA^{phe} in the cell drops, but the levels of the other non-cognate aminoacyl-tRNA's which UUU can recognize do not. The ratio of incorrect to correct substrates appearing in the product of a selection reaction must be a monotonically increasing function of the ratio of incorrect to correct substrates available for selection. Therefore, when the level of PHE-tRNA^{phe} is reduced relative to the others, then the error frequency at UUU codons would have to increase. The same is true, of course, during limitation for any individual aminoacyl-tRNA species.

But there is no sign of increased translational error in wild type cells under such conditions. This paradox admits of only two explanations. Either the normal error frequency is so low that even a large increase remains below the limits of detection; or the intrinsic discrimination specificity of translation is *increased* under the conditions discussed.

The former interpretation can be ruled out in two ways. First, if the normal error frequency is of the order of 3×10^{-4}, then simple calculations show that an increase of a factor of ten or more *should* be detectable, and indeed it is in the case of growth in streptomycin. More to the point, an increased error frequency during limited aminoacyl-tRNA generation can be demonstrated in *rel* mutants. This demonstration proves, at one blow, that sensitivity of detection is not at issue, and therefore that wild type cells do increase the discrimination specificity of translation; it further demonstrates, of course, that the *rel* system is involved in this adjustment.

In section (A) below, I will review the evidence for *rel* mistranslation, including some recent unpublished work which distinguishes between certain elementary biochemical hypotheses. In section (B), I will describe some rather unexpected new findings about the types of mistranslation to which *rel* mutants are prone. In section (C), I will offer some speculative ideas on the control of discrimination specificity in the hope of stimulating further experimentation on the mechanism which enables rel^+ cells to prevent the occurrence of translational errors in the face of imbalances of tRNA aminoacylation.

(A) The first clear evidence of *rel* mistranslation came from experiments on beta-galactosidase synthesis performed about nine years ago. We compared the rates of synthesis of active enzyme, measured by the conventional assay of ONPG hydrolysis, and of the *lacZ* protomer, isolated by gel electrophoresis, in a $rel^+/-$ congenic pair subjected to amino acid limitation. We found that the presence of a rel^- mutation reduced the synthesis of *lacZ* protomer by a factor of three, but reduced the synthesis of active enzyme by a factor of ten (Hall and Gallant, 1972). The *rel* defect in beta-galactosidase production thus comprises two distinct effects, one pertaining to the quantity of *lacZ* protein product produced, and the other having to do with the protein's enzymatic activity.

The first effect is undoubtedly of transcriptional origin: several groups have shown that ppGpp stimulates *lacZ* transcription by about a factor of three *in vitro* (Yang et al., 1974; DeCrombrugghe et al., 1971; Aboud and Pastan, 1973), and we find a comparable deficit of *lac* messenger RNA, measured both by rifampicin capacity and by direct RNA:DNA hybridization

(Foley, Dennis, and Gallant, manuscript in preparation).

The second effect, the reduced enzymatic activity per unit *lacZ* monomer, strongly suggests the occurrence of translational errors. This interpretation was supported by the abnormal thermolability of the beta-galactosidase formed in amino acid starved *rel⁻* cells, a phenomenon first observed in preliminary experiments by Alan Atherly, Chris Smith, and Aaron Novick, and subsequently confirmed by us (Hall and Gallant, 1972) and by Fiil *et al.* (1977).

More specific evidence of amino acid substitutions in translation comes from two quite different experiments. Flagellin, the protomeric subunit of bacterial flagellae, contains no cysteine residues, and it is possible to isolate this protein at a sufficiently high purity to detect the trace quantities of ^{35}S-labelled cysteine present throught errors in translation (Edelmann and Gallant, 1977a and b). The basal level -- about six molecules of cysteine per 10,000 molecules of flagellin -- is in good agreement with what one would expect based on Loftfield's classic measurements of mistranslation at specific sites (Loftfield, 1963; Loftfield and Vanderjagt, 1972). TABLE 1, taken from Edelmann and Gallant (1977a), shows that partial arginine starvation significantly increases the frequency of ^{35}S-cysteine incorporation into flagellin in a *relA⁻* mutant, but not in its congenic *relA⁺* sibling. The CYS codons UGU and UGC are related to two of the six ARG codons by a single transition mistake in the first base, a type of error which has been found to be relatively common *in vitro* (Davies *et al.*, 1965; Davies *et al.*, 1966). The effect is amino acid specific, and therefore probably codon specific; partial starvation for ILE, whose three codons cannot be confounded with the CYS codons by single base errors, does not increase cysteine incorporation into flagellin.

The second type of experiment concerns the isoionic points of a large number of proteins resolved by two-dimensional electrophoresis. O'Farrell (1978) discovered that nearly all the proteins made by *relA* mutant cells during histidine starvation exhibited a striking pattern of electrophoretic heterogeneity: most protein spots were accompanied by a collection of satellite spots of the same molecular weight, but differing from the normal species by integral reductions in positive charge. This effect was also amino acid specific: a somewhat similar pattern was seen during starvation for arginine, but not for isoleucine or proline (O'Farrell, 1978).

This remarkable phenomenon was confirmed by Parker *et al.* (1979) in a *relC* mutant. Parker has recently made an important further observation: starvation of a *relA* mutant for asparagine leads to a symmetrically related pattern of electrophoretic heterogeneity, in which normal proteins are accom-

TABLE 1. $^{35}S/^{3}H$ Ratios in Purified Flagellin from an Isogenic relA+/- Pair of Strains Labeled under the Following Growth Conditions

	C91 relA+	C92 relA
+ARG	0.65 ± 0.11 (4)	0.50 ± 0.09 (4)
-ARG	0.66 ± 0.18 (2)	2.5 ± 0.86 (3)
+ARG, +Spc[a]	0.47 ± 0.05 (2)	
+ARG, +Sm[b]	3.85 ± 0.81 (3)	3.86 ± 0.58 (6)
-ARG, +Sm[b]	3.18 ± 0.17 (2)	10.27 ± 0.36 (6)
-ILEU, +Sm[b]		4.1 ± 1.06 (3)
+ARG, +Neo[c]		8.8 ± 0.19 (2)

[a] 15 μg/ml spectinomycin.
[b] 2 μg/ml streptomycin.
[c] 2 μg/ml neomycin.

Values are mean ± SEM with the number of replicate experiments in parentheses. ^{35}S cpm values were corrected to a normalized ^{35}S-sulphate specific activity of 2 μCi/nm. See Edelmann and Gallant (1977) for Methods.

panied by satellite spots differing by integral *increases* in positive charge (Parker, personal communication).

The amino acid specificity of these effects leaves little doubt that mistranslation is involved. They make no sense in terms of less interesting hypotheses (such as post-translational modification of proteins) but make perfect sense in terms of single base errors of decoding. Histidine is positively charged, and single base decoding mistakes would bring neutral amino acids into positions normally filled by histidine, to produce just the kind of electrophoretic heterogeneity observed. Asparagine is neutral, but decoding mistakes in the third position would bring in positively charged lysine, producing the kind of electrophoretic heterogeneity Parker has observed. I emphasize the amino acid (or codon) specificities of these phenomena because they place very severe constraints on the hypotheses one must entertain, and, in my view, exclude all hypotheses other than decoding errors. In section (B) below, we will present further evidence for mistranslation of specific codons in *rel* mutants. First, however, let me consider two elementary biochemical hypotheses as to the mechanism. The *relA* gene product is associated with the ribosome, and its activity responds to a signal from decacylated tRNA binding at the A site. One might imagine, therefore, that *rel* mistranslation is due to aberrant behavior by ribosomes associated with a mutant *relA* factor. But the work of Pedersen and Kjeldgaard (1978) tends to rule this hypothesis out of consideration. They observed that one *relA* mutant strain (which exhibits mistranslation) contains fewer than one *relA* gene product per thousand ribosomes. It is doubtful that the major aberrations discussed here are due to only a few of *E. coli*'s complement of 10,000 to 20,000 ribosomes.

In that case, what is the origin of *rel* mistranslation? Since *rel* mutants are unable to accumulate ppGpp during limited aminoacylation, one possibility is that high levels of ppGpp exercise a corrective effect on translational accuracy.

Two different kinds of experiment support this hypothesis. First, *rel* mutants can be tricked into accumulating ppGpp by a pathway which evidently is independent of the stringent factor. This occurs following carbon/energy source downshift. We find that the translational defect disappears when a *relA* mutant is taken through such a downshift prior to being subjected to amino acid limitation: that is, the differential rate of beta-galactsidase synthesis is increased to nearly that of the control wild type strain (FIGURE 1), and the enzyme activity per *lacZ* monomer is normal (TABLE 2). Fiil *et al*. (1977) have performed a comparable experiment, using a genetic trick to increase the ppGpp level in relaxed mutants. They found that a second site mutation in $spoT^-$, which inactivates the ppGpp degrading enzyme, permits leaky *relA* and *relC* mutants to accumulate near normal levels of ppGpp during amino acid limitation; in such double mutant strains, the differential rate of beta-galactosidase synthesis, and the thermolability of the enzyme become nearly normal.

A second experiment is to assess the effect of ppGpp on mistranslation *in vitro*. We find that concentrations of the nucleotide which partially inhibit total protein synthesis in a crude polyU programmed system bring about a preferential reduction in the incorporation of an incorrect amino acid (LEU) as opposed to the correct one (PHE). The effect is quite small, about a 40% reduction on average in the misreading coefficient, but it is consistently observed (TABLE 3). We should add that Just Justesen, working in N.-O. Kjeldgaard's laboratory in Aarhus, has independently observed a similar phenomenon (personal communication).

TABLE 4 presents some control experiments indicating that ppppG and ppG, which do not inhibit total protein synthesis,

TABLE 2. Specific activity of β-galactosidase.

	enzyme units per microgram β-galactosidase	amount of β-galactosidase monomers percent of total protein
C91, rel$^+$ downshifted and amino-acid starved	0.68	0.8%
	0.58	0.9%
	0.68	0.8%
	$\bar{x} = 0.65$	$\bar{x} = 0.83$
	$s = 0.058$	$s = 0.058$
C92, rel$^-$ downshifted and amino-acid starved	0.63	0.6%
	0.77	0.5%
	0.45	0.9%
	$\bar{x} = 0.62$	$\bar{x} = 0.67$
	$s = 0.16$	$s = 0.21$

The amount of label in β-galactosidase was measured by double labelling and gel electrophoresis, as described by Hall and Gallant (1972). The raw counts in the β-galactosidase area from the induced extract were corrected for background as measured by migration of counts to the same location in the uninduced extract. The corrected counts were converted to μg using the specific activity of the radioactive amino acids, the number of residues per β-gal monomer, and the molecular weight of the β-gal monomer. Enzymes were assayed as described previously (Hall and Gallant, 1972). One unit of enzyme is defined as the amount giving a ΔOD per minute of one, at 420 nm. Total protein synthesis was measured by TCA precipitating and counting a 100 λ aliquot of labelled cells.

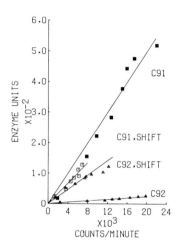

FIGURE 1. *Differential rate of beta-galactosidase synthesis. Cultures of C91 (relA⁺) and C92 (relA⁻), growing either in excess glucose or 50 minutes after a glucose to acetate downshift, were subjected to isoleucine limitation by the addition of valine (400 µg/ml). Beta-galactosidase was induced (IPTG, 2mM) in the presence of excess cAMP (5 mM), and aliquots were labelled with ^{3}H-leucine. Samples were removed at intervals for assay of beta-galactosidase activity and TCA-precipitable leucine. The data are presented as a differential plot.*

have no such effect on misreading; while fusidic acid, which inhibits EF-G, does reduce misreading. Evidently, the enhancement of discrimination specificity depends upon inhibition of at least one of the partial reactions of polypeptide synthesis. In section C, we will suggest that Jacques Ninio's kinetic analysis of discrimination specificity provides a simple and elegant explanation of this phenomenon.

(B) In thinking about the mechanism of *rel* mistranslation, one tends to assume that the mistakes are confined to codons calling for the aminoacyl-tRNA species in short supply. But this is not so. Phenotypic suppression of nonsense codons, which of course don't call for any normal aminoacyl-tRNA species, also occurs in relaxed mutants during certain conditions of limited aminoacylation. FIGURE 2 shows an example. Here we have limited lysine activation (codons AAA and AAG) and observed phenotypic suppression of a UAA nonsense codon in the *phoA* messenger for alkaline phosphatase. The increasing enzyme specific activity shown in the middle panel of FIGURE 2 corresponds to a 5-6 fold increase in the differential rate of enzyme synthesis. Note that the other two panels provide the essential controls. The left-hand panel shows that there is no such increase in enzyme synthesis in a congenic *phoA⁺*

TABLE 3. *Effect of ppGpp on fidelity of polyU translation.*

Experiment	Q (control)	Q (+ ppGpp)	Q ratio
#1	1.35	0.68 (0.9 mM)	0.50
#2	0.57	0.34 (0.9 mM)	0.60
#3	0.35	0.15 (0.9 mM)	0.43
		0.15 (1.8 mM)	0.43
#4	0.8	0.62 (0.9 mM)	0.78
#5	0.55	0.41 (0.9 mM)	0.75
#6 (streptomycin)	16	10.9 (0.9 mM)	0.68
#7 (streptomycin)	16.4	11.6 (0.9 mM)	0.71
#8 (streptomycin)	15.6	8.2 (0.9 mM)	0.53

Cells were subjected to 30 minutes of inhibition by rifampicin (20 µg/ml) to eliminate endogenous messenger RNA, chilled, and crude S-30 extracts were prepared essentially according to Halsetine et al. (1972). Various different S-30 preps were then assayed for mistranslation of polyU (200 µg/ml) with a mixture of ^{3}H-phenylalanine and ^{14}C-leucine (in experiments 1-5) or ^{14}C-isoleucine (in experiments 6-8). In experiments 6-8, isoleucine mistakes were stimulated by the presence of streptomycin (10 µg/ml). The magnesium concentration was 19 mM, and incubation was for 16 minutes at 37°. TCA precipitable radioactivity was assayed by precipitation with 10% TCA, heating for 30 minutes at 90°, filtration, and liquid-scintillation spectrometry of the precipitated material. Q = µµmoles ^{14}C-amino acid/µµmoles ^{3}H-amino acid incorporated.

strain, proving that the increase seen in the $phoA^{UAA}$ reflects read-through of the ochre codon. The right-hand panel shows that this is not observed in a congenic $relA^+$ strain carrying the same $phoA^{UAA}$ allele.

If this read-through phenomenon reflects a general aspect of translation, it should occur in translating other messengers as well. This is the case. TABLES 5 and 6 show that a similar enhanced read-through occurs with a UAG mutant in *lacZ* the gene for beta-galactosidase. Note that the congenic $lacZ^+$ strain shows a two-fold *reduction* in active beta-galactosidase

TABLE 4. *Effects of this and that on the fidelity of polyU translation. Methods as in TABLE 3.*

	INCORPORATION (corrected for blank without polyU)				
REACTION	μμmoles PHE	μμmoles LEU	Both	% Control	Q
control	4.8	5.1	9.9	≡ 100	1.06
+ ppGpp (0.9 mM)	4.2	2.65	6.85	69	0.63
+ ppG (1.0 mM)	5.8	6.3	12.1	122	1.09
+ ppppG (1.0 mM)	6.2	6.6	12.8	129	1.06
+ Fusidic acid (4 γ/ml)	3.7	2.05	5.75	58	0.55
+ Fusidic acid (8 γ/ml)	2.45	1.6	4.05	41	0.65

synthesis, due presumably to a combination of the transcriptional effect noted above and to mistranslation at positions other than the UAG codon; the two-fold *increase* in enzyme synthesis in the UAG mutant is thus all the more striking, and implies a four-fold increase in read-through. TABLE 6 confirms that the read-through is specific to the $relA^-$ genotype, as in the case of phosphatase.

We do not understand the mechanism of this phenomenon, but we have some clues. First, it is allele specific. Six other nonsense mutants have been tested (three in *lacZ* and three in *phoA*) and none of them show phenotypic suppression in response to limited lysine activation.

In order to assess the specificities involved in this phenomenon, we are in the process of testing for phenotypic suppression in eight nonsense mutants (four in *lacZ* and four in *phoA*) and eight frame-shift mutants in *lacZ*, subjecting all of them to limited synthesis or activation of 12 different amino acids. It would be a herculean task to fill in this 192 box matrix by the conventional methods illustrated above, especially since phenotypic suppression is observed only within a rather narrow range of aminoacyl-tRNA limitation. We

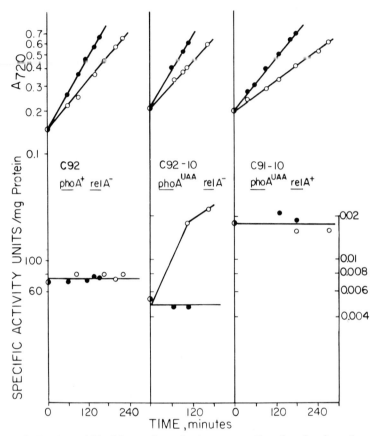

FIGURE 2. Alkaline phosphatase synthesis during lysyl-tRNA limitation. The three strains are congenic and carry a phoS⁻ mutation, making them constitutive for alkaline phosphatase synthesis. Exponential cultures of each growing in Tris-glucose minimal medium were divided into an untreated control (filled circles) and a subculture (open circles) subjected to inhibition by lysine-hydroxamate (300 μg/ml). The upper panels show growth, monitored by turbidity at 720 nm. The lower panels show alkaline phosphatase specific activity, measured in sonicates (suitably concentrated in the case of the ochre mutants). The left hand ordinate applies to the $phoA^+$ strain; the right hand ordinate applies to the two $phoA^{UAA}$ strains. Note the increase in phosphatase activity in the middle panel; similar results have been obtained in five independent experiments.

have therefore devised a simple Petri plate assay for phenotypic suppression. We plate the tester strains on agar con-

TABLE 5. Enzyme activities of YA625 and lac$^+$ derivative.

	Control	Plus lysine hydroxamate
YA625	.0057	.0144
	.0072	.0185
YA625 lac$^+$	12	5
	16	9

The strains were cultured with and without 300 μg per ml lysine hydroxamate. β-galactosidase activities are given in enzyme units per ml (corrected to culture) divided by the ΔOD_{720} during induction. Each value was measured twice.

taining indoxyl dyes which are very sensitive substrates for beta-galactosidase or alkaline phosphatase, giving a blue color reaction in the presence of the appropriate enzymatic activity (XG and XP, described in Miller's invaluable cookbook of molecular biology); after the soft agar hardens, we place small filter discs saturated with amino acid analogues which inhibit the biosynthesis or activation of specific natural amino acids. The inhibitor diffuses out, producing a circular zone of complete growth inhibition, surrounded by a zone of partial inhibition. At whatever radius is right for the degree of amino acid limitation that produces phenotypic suppression, we see a striking ring of blue color, reflecting enhanced hydrolysis of the chromogenic substrate.

PLATE 1 illustrates the method. It can be seen that O-methylthreonine, an inhibitor of isoleucine activation, produces phenotypic suppression of the *lacZ* UAG allele in the *rel*$^-$ strain but not in its congenic *rel*$^+$ partner. Tests of this sort with a variety of specific inhibitors have shown that there is nothing special about lysine: the responsive *lacZ* amber mutant is also suppressed during limitation for isoleucine, phenylalanine, and tyrosine, while the phosphatase ochre mutant also responds to limitation for tryptophan, phenylalanine and proline.

We hope that a survey of the amino acid specificities of read-through in a larger collection of nonsense mutants may, eventually, make some codological sense. Certain conclusions can already be drawn from our observations on the two suppressible nonsense mutants. First, as mentioned above, *rel* misreading is obviously not confined to codons calling for the aminoacyl tRNA in short supply. Second, read-through is almost certainly *not* a generalized response to the abnormal physiology of relaxed mutant cells. This conclusion follows from the specificity the phenomenon displays with respect to amino acid *and* nonsense allele suppressed. For example, limitation for proline or tryptophan suppresses the *phoA* ochre, but not the *lacZ* amber. These specificities strongly suggest that context effects in the message is what determines which amino acid limitations give rise to phenotypic suppression of a particular nonsense codon. That is, we postulate that some aberration of the translation process occurs at codons calling for the limiting amino acid, and this aberration then generates read-through of the nonsense codon, which is presumably located nearby in the message.

The only aspect of translation which is known to be propagated from one position to another in a message is the reading frame. Therefore, we suggest the following specific

TABLE 6. *Enzyme activities of YA624 and YA626.*

	Control	Plus lysine hydroxamate
YA624 (rel^+)	.0051	.0058
	.0043	.0042
	.0038	.0054
	.0056	.0052
YA626 (rel^-)	.0058	.0106
	.0068	.0110
	.0060	.0113
	.0062	.0112

Methods as in TABLE 5. The enzyme activity of YA624 is not significantly increased by the addition of lysine hydroxamate ($t = .854$, $.2 < p < .25$, in a two-sample t-test). The enzyme activity of YA626 is significantly increased ($t = 18.15$, $p < .0005$). Enzyme activities are given in enzyme units per ml (corrected to culture) divided by the ΔOD_{720} during induction. Each pair of entries (control and lysine hydroxamate) represent an independent experiment, with enzyme assays in duplicate.

mechanism. Misrecognition at sense codons near the nonsense codon frequently results in aberrant translocation, and consequent frame-shifting; translation then proceeds in a different reading frame, thus reading through the nonsense codon, and then shifts back into the correct frame, presumably at the next codon calling for the limiting amino acid.

This hypothesis makes two strong predictions. First, phenotypic suppression of frameshift mutations should also occur in rel^- mutants subjected to appropriate amino acid limitation; second, frameshifting in both directions must occur, and therefore phenotypic suppression should be observed with frameshift mutants of both + and - sign. Both of these predictions are fulfilled. PLATE 2 shows that inhibition of isoleucine activation suppresses two *lacZ* frameshift mutants of opposite sign. We have tested eight *lacZ* frameshift mutants, and six of them exhibit phenotypic suppression in response to limitation for one or another amino acid or aminoacyl-tRNA. In order to confirm that the suppression was *rel* specific, we constructed a congenic $rel^+/-$ pair carrying the frameshift mutation (#38) which exhibited the strongest response. PLATE 3 shows that the *rel* mutant exhibits greater phenotypic suppression than its rel^+ partner in response to partial starvation for phenylalanine and proline; interestingly, it also shows greater suppression

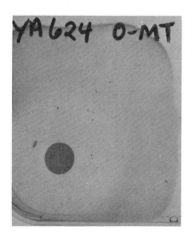 

PLATE 1. *Phenotypic suppression of a lacZ UAG mutant. Strains YA624 (relA$^+$) and YA626 (relA$^-$) are a congenic pair constructed from amber mutant YA625. They were plated in soft agar on plates containing medium 63 with glucose as the carbon source, with IPTG (2 mM), cAMP (5 mM) and XG (40 mg/L). After the agar hardened, filter discs containing 1 mg of O-methylthreonine were placed at appropriate positions, and the plates incubated at 37° for 3-5 days.*

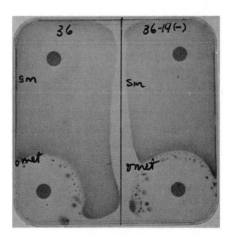

PLATE 2. *Phenotypic suppression of two lacZ frameshift mutants of opposite sign. The frameshift mutants, isolated and characterized by Austin Newton (1970), in a relA⁻ genetic background, were plated as in PLATE 1.*

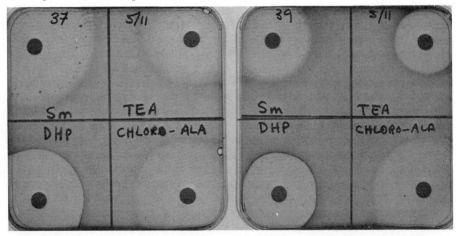

PLATE 3. *Congenic strains 37 (relA⁺) and 39 (relA⁻) were constructed from frameshift mutant #38. They were assayed for phenotypic suppression as described in PLATES 1 and 2. The filter discs contained, respectively, Sm (0.5 mg), thienylalanine (0.1 mg), chloroalanine (0.25 mg), and dehydroproline (0.25 mg).*

by streptomycin, which does not involve aminoacyl-tRNA limitation. This probably means that frameshifting is affected by the basal level of ppGpp in unstarved cells, which is considerably reduced in rel^- mutants.

Our provisional conclusion is that translational error in rel^- mutants pertains not only to missense errors (amino acid substitutions) but also to errors of translocation, resulting in phenotypic frameshift. There are theoretical reasons for supposing that these two kinds of error are in fact manifestations of the same basic phenomenon (see section C below). One puzzle is that nonsense read-through ought to involve two independent frameshift events, and therefore it should be very much less frequent than frameshift suppression. This does not seem to be the case. We find similar efficiencies of phenotypic suppression for those nonsense and frameshift alleles which respond to the rel effect. However, given the importance of context effects, it may be that our limited sample of suppressible alleles of either type is insufficient to assess *average* efficiencies of the process.

(C) Interpretation

The studies reviewed in part (A) demonstrate that imbalances in tRNA aminoacylation elicit an increase in missense errors, that is, amino acid substitutions, in rel mutant cells. In part (B), we showed that the same circumstance leads to phenotypic suppression of frameshift mutations, strongly suggesting that frameshifting, that is, errors of translocation, occurs as well. Finally, it should be recalled that amino acid limited rel cells also show a marked increase in the proportion of small polypeptides, and a corresponding reduction in the proportion of large proteins (Hall and Gallant, 1972; O'Farrell, 1978), suggesting the occurrence of premature chain termination.

Thus, we need to explain three different aberrations of the translation process: errors of aminoacyl-tRNA recognition, errors of translocation, and probably premature termination. One would like to be able to account for all three aberrations on a unitary basis. Charles Kurland has recently (1979) provided the theoretical grounds for doing so. Kurland's argument is that codon:anti-codon recognition and translocation are physically coupled processes. This conclusion follows inescapably from the well-documented case of the $sufD$ frameshift suppressor (Riddle and Roth, 1970 and 1972). The suppressor is a mutant form of the GLY-tRNA$_{CCC}$ which has four, rather than three, C residues in the anticodon loop (Riddle and Carbon, 1973). The fact that this mutant tRNA produces frameshift suppression proves that four base codon:anti-codon interaction leads, with appreciable frequency, to abnormal four-base translocation. Kurland goes on to show that coupling between codon:anti-codon interaction and translocation is implicit in the conformational selection model of tRNA recognition (Kurland, 1979; Kurland *et al.*, 1974).

The three kinds of errors characteristic of *rel* mistranslation are easily explained in this light. Incorrect aminoacyl-tRNA recognition is the primary event, and gives rise to amino acid substitution errors. These errors are in turn frequently associated with aberrant translocation. The latter process yields phenotypic suppression of frameshift mutations, and also elicits premature termination, because frameshifted ribosomes are likely to encounter an out-of-frame nonsense codon. QED.

The remaining problem of interpretation is to explain how ppGpp, a humble low molecular weight mediator, can affect a process as seemingly sophisticated as aminoacyl-tRNA selection by the ribosome. Here, the theoretical ideas of Jacques Ninio provide an elegantly simple explanation. Ninio's scheme of "kinetic amplification" (1974, 1975) considers the following situation: (a) a correct and an incorrect substrate bind in equilibrium fashion to an enzymatic site; (b) the rates of association are identical, but the incorrect substrates dissociates from the complex more rapidly -- in Ninio's terminology it has a shorter "sticking time"; (c) following association of either substrate, a rapid *irreversible* reaction occurs which alters the state of the complex and removes it from equilibrium with the free substrates.

Ninio showed that the discrimination between correct and incorrect substrates appearing in the products of the irreversible reaction increases as the rate of that reaction is slowed down. In essence, a delay in the occurrence of the second reaction allows time for the binding of the correct and incorrect substrates to come to equilibrium before being cast in stone through the irreversibility of the second reaction. Adjustment of this delay time would thus permit adjustment of the discrimination specificity. In a little-noted monograph, Ninio advanced the prescient suggestion that there might be "un compromis vitesse-précision, ajustable à differents nivaux à l'aide d'un élément secondaire" (Ninio, 1975). We suggest that ppGpp functions as just such an "élément secondaire", by delaying the irreversible partial reactions of protein synthesis, namely the hydrolysis of GTP mediated by $EF-T_u$ and EF-G. This hypothesis fits the effect of ppGpp on errors of polyU translation we reported in section (B) above, and also the similar effect of fusidic acid, an inhibitor of EF-G. We are currently examining the effects of a variety of ribosome inhibitors on mistranslation, in order to decide which of the partial reactions is most consequential in regard to the control of discrimination specificity.

The effects of ppGpp on translational specificity, and the theoretical explanation we have advanced, raise an interesting implication in this age of rapid jet travel, rapid media communication, rapid scientific publication, and rapid everything else. It is that, in some circumstances at least, slower can be better, or at least perfectly all right. The scientific literature itself is not without some embarrassing examples of the trade-off between speed and precision. Therefore, let me conclude by suggesting that the progress of science would not be irreparably damaged, and might even be improved, if we were to add a bit of ppGpp and slow down just a bit. They used to say in Germany: "wer langsam reit', kommt grad' so weit."

REFERENCES

Aboud, M. and Pastan, I. (1973). Activation of Transcription by Guanosine-5'-Diphosphate-3'-Diphosphate, Transfer RNA, and a Novel Protein from *E. coli*. J. Biol. Chem. *250:*2189-2195.

Davies, J., Gorini, L., and Davis, B. (1965). Misreading of RNA Codewords Induced by Aminoglycoside Antibiotics. J. Mol. Pharmacol. *1:*93-106.

Davies, J., Jones, D. S., and Khorana, H. G. (1966). A Further Study of Misreading of Codons Induced by Streptomycin and Neomycin. J. Mol. Biol. *18:*48-57.

de Crombrugghe, B., Chen, B., Gottesman, M., Pastan, I., Varums, H., Emmer, M. and Perlman, R. (1971). Regulation of *lac* mRNA Synthesis in a Soluble Cell-free System. Nature New Biol. *230:*37-40.

Edelmann, P. and Gallant J. (1977). Mistranslation in *E. coli*. Cell *10:*131-137.

Edelmann, P. and Gallant J. (1977). On the Translational Error Theory of Aging. Proc. Nat. Acad. Sci. USA *74:*3396-3398.

Fiil, N. P., Willumsen, B. M., Friesen, J. D. and von Meyenburg, K. (1977). Interaction of Alleles of the *relA*, *relC*, and *spoT* Genes in *E. coli*. Molec. Gen. Genet. *150:* 87-101.

Hall, B. G. and Gallant J. (1972). Defective Translation in RC⁻ Cells. Nature New Biol. *237:*131-135.

Haseltine, W. A., Block, R., Gilbert, W. and Weber, K. (1972). MSI and MSII Made on Ribosome in Idling Step of Protein Synthesis. Nature *238:*381384.

Kurland, C. G., Rigler, R., Ehrenberg, M. and Blomberg, C. (1975). Allosteric Mechanism for Codon-dependent tRNA Selection on Ribosomes. Proc. Nat. Acad. Sci. USA 72:4248-4251.

Kurland, C. G. (1979). Reading Frame Errors on Ribosomes. In: Nonsense Mutations and tRNA Suppressors, eds. Celis, J. E., and Smith, J. D. (Academic Press) pp. 95-108.

Loftfield, R. B. (1963). The Frequency of Errors in Protein Synthesis. Biochem. J. 89:2-85.

Loftfield, R. B. and Vanderjagt, D. (1972). The Frequency of Errors in Protein Biosynthesis. Biochem. J. 128:1353-1356.

Ninio, J. (1974). A Semi-quantitative Treatment of Missense and Nonsense Suppression in the *strA* and *ram* Ribosomal Mutants of *E. coli*. J. Mol. Biol. 84:297-313.

Ninio, J. (1975). Kinetic Amplification of Enzyme Discrimination. Biochimie 57:587-595.

Ninio, J. (1975). La Précision dans la Traduction Génétique. In: L'Evolution des Macromolécules Biologiques, C. Sadron (ed.), Editions du C.N.R.S., Paris.

O'Farrell, P. H. (1978). The Suppression of Defective Translation by ppGpp and its Role in the Stringent Response. Cell 14:545-557.

Parker, J., Pollard, J. W., Friesen, J. D. and Stanners, C. P. (1978). Stuttering: High-level Mistranslation in Animal and Bacterial Cells. Proc. Nat. Acad. Sci. USA 75:1091-1094.

Pedersen, F. S. and Kjeldgaard, N.-O. (1977). Analysis of the *relA* Gene Product of *Escherichia coli*. Eur. J. Biochem. 76: 91-97.

Riddle, D. L. and Roth, J. R. (1970). Suppressors of Frameshift Mutations in *Salmonella typhimurium*. J. Mol. Biol. 54: 131-144.

Riddle, D. L., and Roth, J. R. (1972). Frameshift Suppressors III: Effects of Suppressor Mutations on Transfer RNA. J. Mol. Biol. 66:495-507.

Riddle, D. L. and Carbon, J. (1973). Frameshift Suppression: a Nucleotide Addition in the Anticodon of a Glycine Transfer RNA. Nature New Biol. *242*:230-234.

Woese, C. R. (1967). The Genetic Code (Harper and Row, New York).

Yang, H., Zubay, G., Urm, E., Reiness, G., and Cashel, M. (1974). Effects of Guanosine Tetraphosphate, Guanosine Pentaphosphate, and Methylenyl-Guanosine Pentaphosphate on Gene Expression of *E. coli in vitro*. Proc. Nat. Acad. Sci. USA *71*:62-68.

**REGULATION OF MACROMOLECULAR SYNTHESIS
BY LOW MOLECULAR WEIGHT MEDIATORS**

GUANOSINE 3',5'-BISPYROPHOSPHATE IS A DISPENSABLE METABOLITE

Jo A. Engel[1], James Sylvester and Michael Cashel

Section on Molecular Regulation
Laboratory of Molecular Genetics
National Institute of Child Health and Human Development
National Institutes of Health
Bethesda, Maryland

A mutant of an E. coli relaxed (relA) strain has been isolated that is unable to accumulate detectable levels of guanosine 3',5'-bispyrophosphate (ppGpp). The mutant was obtained by nitrosoguanidine mutagenesis and screened for ppGpp production during carbon source starvation.

No absolute growth requirements are associated with the absence of ppGpp accumulation. The mutant strain does show a longer lag than its parent before growth resumption after a shift down from amino acid rich to amino acid poor media. We conclude that ppGpp mediates regulatory functions and is not essential for growth.

The mutant, called relS, has been localized to a chromosomal region distinct from known genes affecting ppGpp regulation during the stringent response and from genes known to encode ribosomal structural proteins. We suggest this is genetic evidence that there are probably two pathways for ppGpp production. One pathway can be altered by mutation of the relA, B, C or spoT genes while the other pathway is blocked by the relS mutation.

[1] Present address: Stanford University
                     School of Medicine
                     Stanford, California

INTRODUCTION

As far as we know all prokaryotic organisms are able to synthesize and accumulate at least low basal levels of guanosine 3',5'-bispyrophosphate (ppGpp). This generalization has led us to wonder what would happen if Escherichia coli were mutated so as to be completely unable to make ppGpp. Even the small basal level amounts of the nucleotide present in relaxed (relA) strains could be imagined to be essential for cell viability as suggested by the *in vitro* experiments (Stephens, Artz and Ames, 1975; Primakoff and Artz, 1979). Alternatively it could be a dispensable nucleotide, like cAMP, with only regulatory functions (Brickman, Soll and Beckwith, 1973). The properties of such a mutant might provide clues as to regulation by ppGpp as well as regulation of ppGpp synthesis. Relaxed mutants have basal levels of ppGpp that cannot be increased during amino acid starvation but can be derepressed during carbon source starvation (see reviews of Gallant and Lazzarini, 1976; Nierlich, 1978). This behavior could be due to weak residual activity of the relA gene product (stringent factor) operating in a fashion uniquely sensitive to carbon starvation or to a different enzyme pathway for making ppGpp and responding to different signals.

An approach to this question (taken by several laboratories as well as our own) is to isolate mutants that are completely lacking in a functional relA gene product so that if any ppGpp is synthesized during carbon starvation it must be due to another biosynthetic enzyme. For example, we attempted to isolate deletions that would span markers flanking the relA region. Lambda was inserted into the pyrG gene found to be adjacent to the relA gene (Friesen *et al.*, 1976). By inducing these lysogens, we hoped to isolate deletions that would extend through the relA region and include distal flanking genes such as relX. No deletions were found that extended into relA among several hundred cured lysogens. We suspect that there is an essential gene between relA and pyrG. This volume contains a report by Atherly (1979) describing deletions that extend counterclockwise from argA to apparently terminate in relA but not beyond it to include pyrG.

We next proceeded to isolate conditional amber mutations that had a phenotype of impaired ppGpp accumulation when suppression was limited (Cashel, unpublished). Screening for ppGpp accumulation defects at restrictive temperatures in fact yielded mutants with this phenotype mapping in the strep-spc region (Nomura, Morgan and Jaskunas, 1977) as well as relA regions of the chromosome. However, all of these mutants had basal levels of ppGpp that could be depressed during glucose starvation. Similar carbon starvation effects have been reported (Friesen, An and Fiil, 1978) for amber mutants as well as insertion mutants in the relA gene. While these efforts strongly suggest the presence of an enzyme system capable of ppGpp synthesis in addition to the relA encoded protein, there remains the possibility that there is a marginally active amber fragment or deletion fragment of the stringent factor that is responsible for low levels of ppGpp synthesis.

Therefore we were driven to attempt a brute force search for a mutant phenotype consisting of a complete absence of ppGpp, a phenotype we would like to call ppGpp$^{\circ}$. This mutant screen consists of thin layer chromotographic identification of relA2 mutants unable to accumulate ppGpp during glucose starvation. No preselection was imposed but mutants were isolated so as to allow the appearance of a temperature sensitive missense or amber mutation that might be essential for growth. This paper is a preliminary account of the derivation and properties of such a mutant. We know the characterization of this mutant is incomplete, yet it seems sufficient to allow some tentative conclusions about the essential nature of ppGpp and the likelihood of more than one enzyme capable of ppGpp synthesis in E. coli.

MATERIALS AND METHODS

Cells and Growth

The F$^-$, relA2 strain (CF 350) mutagenized is multiply marked to facilitate mapping; it is leu$_{am}$, lac$_{am}$ galK$_{am}$, galE, trp$_{am}$, cdd, nalA, cysC, pyrG, argA, strA and metB. It also bears the supD suppressor locus and mutations rendering suppression temperature sensitive; the strain is a derivative of Dr. Max Oeschger's strain MX403. Nitrosoguanidine mutagenesis was by exposure of tryptone broth glucose grown cells to 100 µg/ml nitrosoguanidine at 30° for 15 min. The cells were then chilled, washed and plated at 30° on the same medium. Single colonies were picked to MOPS glucose (0.4%) media containing 20 µg/ml of each of 18 amino acids, 10 µg/ml cytidine and 4 mM KH$_2$PO$_4$ (Neidhardt, Bloch and Smith, 1974). After overnight growth at 30°, the plates were incubated at 42° for 6 hours before ppGpp assay.

Suppressing revertants were isolated by selecting for simultaneous tryptophan and leucine independence at 42°. Non-suppressing derivatives were isolated among clones resisting galactose killing at 30°. Both suppressor changes were verified by spot tests with known lambda amber mutants, a collection obtained from Dr. Nat Sternberg.

ppGpp Screening assay

Enough cells were transferred by toothpicks from the MOPS plates to make 100 µl of labeling medium visably turbid in the well of a microtiter dish. The MOPS labeling medium contained about 1 µCi/ml of carrier free ^{32}P orthophosphoric acid but no glucose or amino acids. It also contained 300 µg/ml L-valine and 1 mg/ml DL serine hydroxamate (Sigma Chemical Co.) to insure a stringent response is provoked in addition to carbon starvation. Labeling of the cells occurred during a one hour incubation at 42° and was terminated by adding 50 µl of 13 M formic acid. The samples were frozen and thawed once and an aliquot spotted on PEI cellulose thin layers using a plexiglass spotting shield containing holes one cm apart (Bel Art Products). Chromatography was with 1.5 M KH_2PO_4 (pH 3.4) to 5 cm above the origin, which allowed visualization of GTP as well as ppGpp after autoradiography. Several hundred colonies can be screened in this manner by one person in a day. More recently semi-automated devices using microtiter dishes have been adapted to this procedure (Cooke Engineering Co.) making an even higher capacity if necessary. Colonies with no ppGpp activity but noticeable levels of GTP were rechecked by the plate assay using the parental strain as a control.

RESULTS

Mutant Detection

One colony among the first set of 400 screened at random (CF 398) showed the desired ppGpp° phenotype; the biochemical phenotype was confirmed twice on duplicate cultures. Mutant cell growth occurred at both 30° and 42° on MOPS glucose AA Cr plates. The ppGpp° phenotype was next confirmed in liquid cultures grown at 42° and allowed to exhaust the 0.015% dextrose present in the presence of ^{32}P, sampling over the course of three hours after growth stopped. Whereas ppGpp labels almost as intensely as GTP in the CF 350 control culture, no ppGpp was visualized for CF 398. The growth of the mutant culture (55 min doubling time) was slightly slower than the 45 min doubling time for the parental culture before glucose was exhausted.

For this discussion we shall assume (with no direct evidence) that the ppGpp° phenotype is due to a mutation of a single gene we wish to call relS. The pneumonic is rel-less. Figure 1 shows two dimensional chromatograms showing the absence of noticeable ppGpp under conditions which permit easy

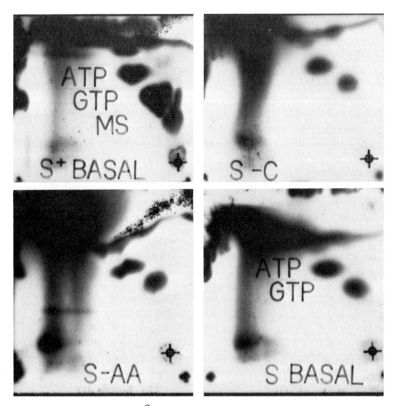

Figure 1. The ppGpp⁰ phenotype. The relA relS mutant nucleotides are subjected to two-dimensional chromatography (4 M sodium formate, 1.5 M potassium phosphate) after glucose starvation (S-C), after amino acid starvation (S-AA) and during balanced growth (S Basal). The expected position of ppGpp on the chromatograph is seen on the upper left panel, labelled S+ Basal.

visualization of even basal levels in relS+ strain (northwest panel). The figure also shows that no ppGpp is seen in the relS strain after amino acid starvation (southwest panel) or after carbon starvation (northeast panel). Basal levels are also absent (southeast panel).

While perhaps promising, these data are not sufficient to convincingly demonstrate a complete absence of ppGpp synthesis. With this goal in mind, we have labeled relS bearing cells with high specific activities of ^{32}P during carbon source starvation (2-5 mCi/ml containing 0.1 mM KH_2PO_4) yielding approximately 35,000 cpm localized to the GTP area of

TABLE I. Hfr Mapping of relS Recipient

Hfr	Mating time	Selection	$relS^+$ Recombinants
KL14	30 min	$metB^+$	no
KL14 rif^R	40 min	$rifR$	no
HfrH	20 min	leu^+	yes
KL16	15 min	$pyrG^+$	no
KL16	30 min	trp^+	no
KL16	60 min	gal^+	yes
HfrC	20 min	leu^+	yes

Localization of relS to the chromosomal region between HfrC and HfrH origin points. The recipient strain was the original CF398, or a relS $pyrG^-$ recombinant derived from the short term KL16 mating. Selections for amino acid markers were with mixtures of a complete set of amino acids lacking the one in question. Selected recombinant colonies, usually 50-100, were purified and screend for the $ppGpp^o$ phenotype.

the chromatogram after two-dimensional chromatography as in Fig. 1. The activity of the ppGpp area of the chromatogram is not significantly different than the adjacent but equivalent sized blank area (150 cpm). Since the ppGpp activity in a $relS^+$ strain undergoing prolonged carbon starvation usually approaches GTP activities, the dramatic extent of the defect in the mutant is evident.

Mapping of relS.

The relS mutation was roughly mapped by Hfr matings to somewhere between the origins of Hfr H and Hfr C donation points (Table I). Preliminary transmission gradient experiments with the KL16 Hfr, selecting for gal^+ recombinants and screening for leu^+ and lac^+ revertants have indicated that relS is located in the leuS-purE region of the chromosome (12-14 minutes). This localization eliminates the spoT, relA, relB or relC genes as the source of the relS mutation.

This preliminary localization has been confirmed by F' matings. For example, the $ppGpp^o$ phenotype was converted to that seen in relA $relS^+$ strains ($ppGpp^-$) by mating to select for lac^+ F'254 merodiploids spanning the 5-14 min region on the E. coli map. Analysis of the lac^- colonies cured of the F'254 episome (by acridine orange and growth at high temperature) revealed restoration of the $ppGpp^o$ phenotype. This indicates not only that the episome was responsible for the phenotypic change but also that the relS allele is recessive.

The position of the relS mutation within the area covered by F'254 was further localized by crosses with episomes that contained sequences overlapping with those of F'254. Since no complementation of the ppGpp^o phenotype was observed with F'42, F'113 or F'152, the relS is placed somewhere between the clockwise end of F'254 and the adjacent terminus of F'113 (12-14 min).

We have attempted P1 transduction of purE (12 min) and leuS (14 min), but so far have not obtained cotransduction with relS. We cannot yet explain this failure since both by conjugal transmission gradients and F' mapping this should be the relS region.

relS Dependent Growth Requirements

Although relS is imprecisely mapped, we can ask whether any special growth requirements are associated with the relS mutation. Such requirements might suggest processes especially sensitive to ppGpp regulation or function. Multiple auxotrophic mutations were in fact encountered with the original relS isolate (CF398) as expected with nitrosoguanidine mutagenesis. Segregation of relS from these requirements in mating experiments was used to decide which requirements in fact are associated with relS.

A leucine requirement was observed at 30° that was removed without changing the ppGpp^o phenotype or the leucine mutation. Dissociation occurred amongst recombinants in matings with Hfr KL14, selecting a rifampicin resistance marker present in the Hfr. The parental leucine amber mutation remains as a leucine requirement at 42°, not 30°.

Another requirement was more difficult to separate from relS. Rifampicin resistant recombinants of the sort just described (relS equals CR435; relS+ equals CF436) differ in having a temperature sensitive requirement for serine, as shown in Table II. This requirement is closely associated with relS in mating experiments; greater than 95% of serine independent recombinants at 42° are also recombinants at the relS locus. However, the occurence of both combinations of relS, ser^{ts} and relS, ser$^+$ indicate that the loci are distinct from one another.

Table II also shows that at 30° (permissive temperature for suppression) the relS strain grows very poorly in the presence of the minimal array of amino acids that allows growth of its relS+ sister strain. This deficiency is overcome by adding isoleucine and valine, but made worse by adding glycine and serine. Since the inhibition by glycine and serine is

TABLE II. Serine Sensitivity and Temperature Sensitive Serine Requirement Present in the relS Mutant Strain CF435 But Absent in relS+ Strain CF436.

	$30°$		$42°$	
	relS	relS+	relS	relS+
met trp arg cys	−	−	−	−
met trp arg cys leu	((+))	+	−	+
met trp arg cys leu ile val	+	+	−	+
met trp arg cys leu gly ser	−	+	−	+
met trp arg cys gly ser ile leu val	+	+	−	+
all-ser	+	+	−	+
all	+	+	+	+

overcome by the ilv group of amino acids, we surmise that this behavior of the relS probably represents a heightened serine sensitivity of the sort reported by Uzan and Danchin (1978). They suggest that one carbon metabolites cause an isoleucine starvation in relA strains. We suppose relA relS strains might be even more sensitive. It is to be stressed that we find no correlation between the temperature sensitive serine requirement and the extreme serine sensitivity; both involve serine, but otherwise seem unrelated.

Thus, no absolute requirements can be associated with the relS mutation although growth in minimal media is slower than in amino acid rich media. The relS mutation is also independent of the degree of suppression. As described in Methods and Materials, sup+ and sup0 revertants of the supD temperature sensitive phenotype have been isolated and no effects were noted on the ppGpp^0 phenotype. The relS mutation also has no influence on the degree of suppression of lambda phage amber mutations as judged by spot tests of SupD suppressible amber mutants (data not shown). This was tested because it might be supposed that the absolute deficiency of ppGpp might lead to

TABLE III. Comparison of Effect of relS Allele on Step Up
and Step Down

		24 Hour growth when plated on:		
Inoculum growth on	relS	Glucose minimal	Gluose casamino acids	Luria broth
Glucose minimal	−	+	++	+++
	+	+	++	+++
Glucose casamino acids	−	(+)	++	+++
	+	+	++	+++
Luria broth	−	−	++	+++
	+	+	++	+++

overproduction of suppressor tRNA and enhance the efficiency of suppression, but apparently this is not the case.

relS Affects Growth After Shift Down

Table III shows that the relS mutant (CF435) has great difficulty growing after being transferred from rich media like Luria broth to minimal media. This effect is dependent upon the relS mutation in that it is absent for the relS$^+$ relA parental type strain even though the latter shows difficulties with such downshifts in comparison with a relA$^+$ relS$^+$ strain. This effect is due partly to amino acid downshift because the growth impairment is seen when casamino acid grown cells are plated on minimal media. The effect is not seen with Luria broth grown cells plated on casamino acids so the relevant downshift seems largely due to an aminoacid downshift. Interestingly the relS can transverse an upshift as well as a relS$^+$ strain. This ability is borne out in more detailed liquid culture experiments (data not shown).

relS Partly Limits Galactose Kinase Expression

As judged by color reactions on MacConkey plates, relS and relS$^+$ cells are equally able to ferment maltose, glycerol and lactose as well as dextrose. However, we have evidence that the relS mutation limits galK gene expression to some extent. In the relS (CF435) strain a galE, galK$_{am}$ combination of genes makes the strain sensitive to galactose suicide when suppression is sufficient to allow expression of galK. At 30° on tryptone plates, cross streaks of galactose give inhibition of the relS⁺ strain but not the relS strain; at 42° both strains are resistant to galactose killing. Since relS has no influence on the degree of suppression of lambda amber mutations, we interpret this to mean that under the partial suppression occurring in the temperature sensitive suppressor strain, relS limits the expression of galK. We have determined that there is not an absolute dependence of galactose gene expression on ppGpp. The relS and relS⁺ strains have been lysogenized with lambda gal8 transducing phage. Galactose fermentation tests on both MacConkeys plates and tetrazolium plates indicate both strains show strong galactose utilization.

DISCUSSION

Genetic characterization of the mutant phenotype leading to an apparently complete block in the ability to accumulate ppGpp is only in its preliminary stages. We have so far been unable to map the relS allele precisely enough to allow transfer out of the heavily mutagenized host into a less perturbed genetic environment, let alone engage in the sort of reversion analysis that would be required to demonstrate that the lesion is the result of mutation of a single gene.

Nevertheless the biochemical phenotype of the mutant is fairly convincing in demonstrating a complete deficiency of ppGpp accumulation. The inability to accumulate ppGpp during a carbon source starvation is certainly unique among E. coli strains we have studied. We would like to interpret mutant growth in the absence of ppGpp as indicating that ppGpp is a dispensable metabolite and provides no essential functions under steady-state growth conditions. Nevertheless, relS strains do grow more slowly than relS⁺ strains in minimal media and are less able to adjust to downshifts of amino acid availability. The strength of this conclusion, of course, depends upon a compelling demonstration of a complete absence of ppGpp in the mutant strain. Perhaps the immunoassay for ppGpp can be applied to this demonstration (Silverman et al., this volume). If the decay rate for ppGpp in the mutant is delayed during glucose starvation as it is in normal relaxed strains during glucose starvation (Gallant, Margason and Finch, 1974; Stamminger and Lazzarini, 1974), then the rates of synthesis of ppGpp in the mutant are even more sensitive to detection. Our most enthusiastic estimate is that there are less than ten molecules of ppGpp per cell in the carbon starved mutant, but we must confess that this estimate involves some questionable assumptions that have yet to be specifically tested. The demonstration that the relS allele is recessive in merodiploids does rule out the possibility that relS might encode something like a constitutive ppGpp degrading activity, which would be expected to be dominant.

Considering the deletion analysis presented in this volume by Atherly, the relA amber and insertion mutants of Friesen, An and Fiil (1978) as well as our own experiences with amber mutants and attempts at deletions (Cashel, unpublished), we would interpret mutant behavior as most likely due to a change in another pathway for ppGpp synthesis. By this view the loss of the ability to increase the ribosomal dependent rate of ppGpp synthesis specifically in response to aminoacyl-tRNA deficiency is specific to the relA dependent pathway. The relB and relC mutations are also relevant to this pathway as are a number of other ribosomal structural protein mutants that interfere with ppGpp synthesis that we have isolated in the strep-spc region of the chromosome. The slow rates of ppGpp synthesis in strong relA mutants are probably due to another enzyme system rather than a residual relA gene product activity. This is not to say that relA cannot be mutated so as to be leaky as shown by Fiil and Friesen (1968) and Block and Haseltine (1973). We suspect instead that the relS mutation we have encountered may well affect an enzyme activity of the sort described for Bacillus brevis by Sy (this volume). It would seem reasonable that no matter how ppGpp is synthesized, degradation systems would apply equally to this diffusable product such that spoT and the slowing of ppGpp degradation during carbon starvation should affect the products of both pathways (Laffler and Gallant, 1974; Stamminger and Lazzarini, 1974).

The evidence that ppGpp participates in regulation of gene expression of a sizable portion of the E. coli genetic repertoire is strong (O'Farrell et al., 1978; Reeh, Pederson and Friesen, 1976). Studies with the relS mutant suggest that the regulatory effects are not as absolute as surmised from in vitro experiments (Stephens, Artz and Ames, 1975; Primakoff and Artz, 1979). Instead the absence of ppGpp would seem to lead to partial impairments. For example, it seems that under conditions of partial suppression of galactokinase amber mutation not enough enzyme can be made in the relS mutant to allow for galactose toxicity, although wild type galactose transductants on the relS mutant are able to ferment galactose readily. By lowering the temperature of growth of the relS strain with the temperature sensitive suppressor, it is possible to obtain sufficient suppression of the galactokinase amber mutation to allow some killing by galactose. Such a scheme should allow estimates of the degree of impairment of gal operon expression in the relS strain.

Apparently the regulatory functions mediated by ppGpp are not absolutely essential for growth; they however do serve to facilitate transient adjustments to changes in growth conditions. It is interesting that adjustment to amino acid downshifts is impaired by the absence of ppGpp but not amino acid upshifts.

We hope that mutants absolutely deficient in ppGpp will ultimately be useful in defining other mechanisms by which cells coordinate genetic expression. Before the existence of such mutants, the regulation of ppGpp could always obscure the search for other regulatory phenomena affecting large sets of genes. In a relA, relS background this should not be the case.

ACKNOWLEDGEMENTS

We are grateful to Dr. Max Oeschger for introducing us to his temperature sensitive suppressor strain and to Dr. Gad Glaser for his encouragement and many discussions in the course of studying this mutant. We again thank Terri Broderick for her able preparation of the manuscript.

REFERENCES

Block, R. and Haseltine, W.A. (1973). Thermolability of the stringent factor in rel mutants of Escherichia coli. *J. Mol. Biol.* 77, 625-629.

Brickman, E., Soll, L. and Beckwith, J. (1973). Genetic characterization of mutations which affect catabolite-sensitive operons in Escherichia coli including deletions of the gene for adenyl cyclase. *J. Bact.* 116, 582-587.

Fiil, N. and Freisen, J.P. (1968). Isolation of relaxed mutants of Escherichia coli. *J. Bact.* 95, 729-731.

Friesen, J.D., An, G. and Fiil, N.P. (1978). Nonsense and insertion mutants in the relA gene of E. coli: cloning relA. *Cell* 15, 1187-1197.

Friesen, J.D., Parker, J., Watson, R.J., Fiil, N.P., Pederson, S. and Pedersen, F.S. (1976). Isolation of a lambda transducing bacteriophage carrying the relA gene of Escherichia coli. *J. Bact.* 127, 917-922.

Gallant, J. and Lazzarini, R.A. (1976). The regulation of ribosomal RNA synthesis and degradation in bacteria in protein synthesis: A series of advances (ed. E. McConkey) 2, 309-359.

Gallant, J., Margason, G. and Finch, B. (1972). On the turnover of ppGpp in Escherichia coli. *J. Biol. Chem. 247*, 6055-6059.

Laffler, T. and Gallant, J. (1974). SpoT, a new genetic locus involved in the stringent response in E. coli. *Cell 1*, 27-30.

Neidhardt, F.C., Bloch, P.L and Smith, D.F. (1974). Culture medium for enterobacteria. *J. Bact. 119*, 736-747.

Nierlich, P.P. (1978). Regulation of bacterial growth, RNA and protein. *Ann. rev. Microbiol. 32*, 393-432.

Nomura, M., Morgan, E.A. and Jaskunas, S.R. (1977). Genetics of bacterial ribosomes. *Ann. Rev. Genet. 11*, 297-347.

O'Farrell, P.H. (1978). The suppression of defective translation by ppGpp and its role in the stringent response. *Cell 14*, 545-557.

Primakoff, P. and Artz, S.W. (1979). Positive control of lac operon expression in vitro by guanosine 5'-diphosphate 3'-diphosphate. *Proc. Natl. Acad. Sci. (Wash.) 76*, 1726-1730.

Reeh, S., Pedersen, S. and Friesen, J.D. (1976). Biosynthetic regulation of individual proteins in relA$^+$ and relA strains of Escherichia coli during amino acid starvation. *Mol. Gen. Genet. 149*, 279-289.

Silverman, R.H., Atherly, A.G., Glaser, G. and Cashel, M. (1979). A radioimmunoassay for ppGpp. In *Regulation of Macromolecular Synthesis by Low Molecular Weight Mediators* ed. G. Koch, D. Richter), Academic Press, New York.

Stamminger, G. and Lazzarini, R.A. (1974). Altered metabolism of the guanosine tetraphosphate, ppGpp, in mutants. *Cell 1*, 85-90.

Stephens, J.C., Artz, S.W. and Ames, B.W. (1975). Guanosine 5'-diphosphate 3'-diphosphate (ppGpp). Positive effector for histidine operon transcription and general signal for amino acid deficiency. *Proc. Natl. Acad. Sci. (Wash.) 72*, 4389-4393.

Sy, J. (1979). Biosynthesis of ppGpp in Bacillus brevis. In *Regulation of Macromolecular Synthesis by low Molecular Weight Mediators*. (ed. G. Koch, D. Richter), Academic Press, New York.

Uzan, M. and Danchin, A. (1978). Correlation between the serine sensitivity and the derepressibility of the ilv genes in Escherichia coli relA$^-$ mutants. *Molec. Gen. Genet. 165*, 21-30.

REGULATION OF MACROMOLECULAR SYNTHESIS BY LOW MOLECULAR WEIGHT MEDIATORS

STUDIES ON THE COORDINATION OF tRNA-CHARGING AND POLYPEPTIDE SYNTHESIS ACTIVITY IN ESCHERICHIA COLI[1]

Wolfgang Piepersberg
Dieter Geyl
Peter Buckel[2]
August Böck

Lehrstuhl für Mikrobiologie
Universität Regensburg

The in vivo correlation between the level of charged tRNA, the extent of inhibition of protein synthesis and the synthesis of ppGpp was determined in strains of Escherichia coli containing a temperature-sensitive valyl- or alanyl-tRNA synthetase. At a stoichiometric ratio of 1:1 between charged and uncharged tRNA protein synthesis is still unrestricted and only a slight increase of the ppGpp content is observed. Marked restriction of protein synthesis and a concomitant rise in ppGpp concentration occurs at charging levels below 30 %. The inhibition of protein synthesis follows the predictions of a substrate (aminoacyl-tRNA) limitation model of translation (O'Farrell 1978) at charging levels above 40 % but deviates from it at lower values. This finding is in accordance with the assumption that the accumulation of ppGpp limits translation in rel^+ strains (O'Farrell 1978). On the other hand, we could not detect any tRNA charging differences in rel^+/rel^- pairs of aminoacyl-tRNA synthetase mutants at non-permissive

[1] Supported by a grant from the Deutsche Forschungsgemeinschaft
[2] Present address: Boehringer Mannheim GmbH, Biochemica Werk Tutzing

*temperatures which should be expected on the basis
of this model. - Under the conditions of a drastic
unbalance of tRNA charging activity relative to ribosomal activity growth becomes linear. This growth inhibition is released when the limiting step is shifted from the tRNA charging reaction to the ribosome.
It is proposed on the basis of these results that
the main purpose of the coordinate regulation of the
components of the translational apparatus under different growth rates is to keep charged tRNA at a
level required for effective competition with uncharged tRNA at the ribosomal A-site. - Results are
also presented which indicate that the basal level
of ppGpp under balanced growth conditions is not
correlated with the increased RNA content of ribosomal assembly mutants.*

According to their regulation relative to growth rate the components of the bacterial translational apparatus may be differentiated into two groups. The first exists of macromolecules providing charged tRNA for the ribosomal A-site (aminoacyl-tRNA syn-thetases, EFTu and tRNA itself) and they are kept in close stoichiometric relation to each other, the second group components are the ribosomal ones, which are regulated coordinately but not parallelly to group I components (for rev. see Neidhardt et al., 1977). The physiological necessity for this coordi-nacy of regulation is not yet clear.

We have attempted to obtain information on this problem by investigating the effects of different charging levels of one species of tRNA on the rate of protein synthesis, on ppGpp formation and on growth. Data are provided which indicate how effec-tively uncharged tRNA competes with charged tRNA at the ribosomal A-site in vivo. In addition, results are summarized which provide information on the physiological need for the coordinate regulation of tRNA charging and polypeptide synthesis activity in vivo.

MATERIALS AND METHODS

The following strains were used in the experiments: BM 113: *relA1 alaS3 pheA pyr T2^R tonA22*; BM 1131: *alaS3 T2^R tonA22, strA^R*; BM 112: *valS relA1 T2^R tonA22* (Buckel et al., 1976); *AT2535 valS*: carries the *valS* lesion of strain BM 112 in the genetic background of AT2535, with the exception of being *pyrB$^+$*; AT2535: *pyrB59, argH1, thi-1, purF1, mtl-2, xyl-2, malA1, ara-13, lacY1 or lacZ4, str8 -9, or -14, tsx-23 or -25, λ^R, tonA2 or 14, supE44*? Unless otherwise stated, all strains were grown in MOPS minimal medium (Neidhardt et al., 1974) supplemented with 1 % bacteriological peptone (DIFCO) and 50 µg/ml of guanosine and uracil each. The bacteriological peptone had been freed from inorganic phosphate (Rubin 1973) and was then supplemented with K_2HPO_4 to a final concentration in the growth medium of 0.264 mM.

Measurement of the synthesis of guanosine-5'-diphosphate-3'-diphosphate (ppGpp) was carried out as previously described (Buckel et al., 1976). The values obtained were normalized to an A_{420} (optical density of the culture) of 1.0. Determination of the level of tRNA charging has also been described (Buckel et al., 1976). The rate of protein synthesis was determined by adding 2 ml of the culture to prewarmed 25 ml flasks containing 4 µC of [^{3}H]leucine; after 3 min, incorporation was stopped by addition of 2 ml 10 % trichloroacetic acid. The incorporation values were corrected to an A_{420} of 1.0. RNA/protein ratios were determined as published (Buckel et al., 1972).

RESULTS AND DISCUSSION

1. THE IN VIVO COMPETITION BETWEEN CHARGED AND UNCHARGED tRNA AT THE RIBOSOME

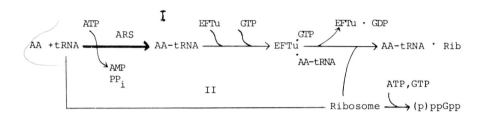

FIGURE 1. The alternative routes for uncharged tRNA in protein synthesis.

Fig. 1 illustrates the alternative routes for uncharged tRNA: path I is followed under conditions of unrestricted supply of amino acids in protein synthesis; path II, as determined by in vitro experiments, leads to the synthesis of the unusual guanosine nucleotides pppGpp and ppGpp (Haseltine et al., 1972). Details on how uncharged tRNA reaches the ribosomal A-site in vivo have not yet been worked out. It is also unknown how effectively uncharged tRNA competes with charged tRNA at the ribosomal A-site, at which level it restricts the rate of protein synthesis and leads to the production of (p)ppGpp.

Information on these questions can be obtained by shifting mutants with temperature-sensitive aminoacyl-tRNA synthetases to different temperatures intermediate between the permissive and the completely restrictive one (Fiil et al., 1972; Buckel et al., 1976). This should lead to different ratios between uncharged and charged tRNA and to different use of routes I and II. As a measurement for path II we have determined the level of ppGpp.

Fig. 2 shows the kinetics of the three parameters determined for strain BM 1131 after a shift from 30°C to 41°C. It shows that a new steady-state is attained at 10 min after the change of temperature. In the following experiments we have therefore

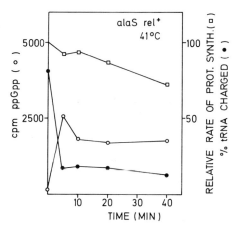

FIGURE 2. Level of charged tRNA^{ala}, of ppGpp and rate of protein synthesis (uncorrected for the temperature effect on the wild-type rate) in strain BM 1131 (alaS3 rel$^+$) after a shift from 30 to 41°C.

chosen to determine these parameters 20 min after the shift. The results are plotted in Fig. 3 as level of ppGpp or rate of protein synthesis versus level of charged tRNA^{val} or tRNA^{ala}.

In case of the temperature-sensitive valyl-tRNA synthetase mutant the level of charged tRNA^{val} at 30°C is already low (as compared to 80-90 % in wild-type strains, not shown) indicating already partial temperature limitation of enzyme activity. Exposing the culture to 33, 35, 37, 39, 41 and 43°C successively lowers the steady-state concentration of charged tRNA^{val}. At 20 % residual charging, there is a drastic rise in the synthesis of ppGpp and also in the inhibition of protein synthesis. The strain with the temperature-sensitive alanyl-tRNA synthetase shows the analogous behaviour with the exception that the level of charging at 30°C is identical to that found in the wild strain and that protein synthesis inhibition and concomitant ppGpp production is first manifested at 39°C.

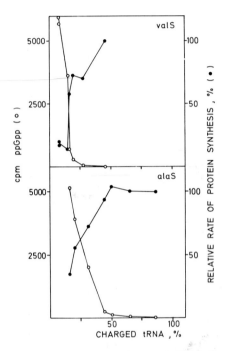

FIGURE 3. Rate of protein synthesis, level of ppGpp and of charged $tRNA^{val}$ (upper graph) and $tRNA^{ala}$ (lower graph) on shifting strains AT2535 valS (top) and BM1131 (bottom) from 30°C to 33°C, 35°C, 37°C, 39°C, 41°C and 43°C. The basal level of ppGpp at 30°C has been subtracted from those obtained at higher temperatures. Relative rate of protein synthesis is corrected for the temperature stimulation of wild-type protein synthesis. A value of 50% e.g. means half the rate of the wild strains at the respective temperature. The righ-most point gives the 30°C value, the next to the left that for 33°C, a.s.o. Strain AT2535 $valS^+$ was shifted for control to 35, 39 and 43°C. It did not accumulate ppGpp as a consequence of the change of the temperature.

Summarizing the results one can state that

(i) upon exposure of mutants with temperature-sensitive aminoacyl-tRNA synthetases it is possible to obtain different and characteristic levels of tRNA charging;

(ii) at a 1:1 ratio between charged and uncharged tRNA protein synthesis is not restricted and only a slightly increased level of ppGpp is observed. An attempt to draw a conclusion on the relative affinities in vivo of charged tRNA (as the EFTu · GTP complex) and of uncharged tRNA (in the free state?) for the ribosome can be made by using ppGpp as an indicator for the saturation of the ribosomal A-site with uncharged tRNA. As half-maximal accumulation of ppGpp occurs at 20 to 30 % charging level an approximately 3- to 5-fold higher affinity of charged relative to uncharged tRNA is indicated.

O'Farrell (1978) has presented the relation between the level of tRNA charging and the inhibition of protein synthesis under the assumption of limitation of translation by the supply of aminoacyl-tRNA. Our experimental curves fit the calculated ones (O'Farrell 1978) in that protein synthesis is first inhibited at a charging well below 50 %, but they deviate from them in that at low aminoacylation levels protein synthesis is inhibited more than it should be for the charging levels measured. This means that within a certain range of aminoacyl-tRNA concentration translation is not limited by the supply of aminoacyl-tRNA but by some other effector, possibly by ppGpp as postulated by O'Farrell (1978).
(It is important to mention in this connection that the determination of the degree of tRNA charging tends to deliver higher values than the actual in vivo ones are. We want to emphasize therefore that every precaution was followed to stop the cells instantaneously as pointed out by Folk and Berg [1970] and Lewis and Ames [1972].)

If ppGpp limits translation in rel$^+$ strains under amino acid restriction but not in rel$^-$ derivatives, rel$^+$ strains should maintain an appreciably higher tRNA charging level for the respective amino acid under starvation conditions (O'Farrell 1978). We have studied this for two rel$^+$/rel$^-$ pairs of aminoacyl-tRNA synthetase mutants (Table I). Within the experimental scatter, tRNA is discharged at the restrictive temperature to the same extent. This means, that either ppGpp is not inhibiting translation in rel^+ strains under amino-acid restriction in vivo or, as already pointed out by O'Farrell (1978) that it is a minor isoaccepting species of tRNA the charging of which limits protein synthesis and which is not resolved by measuring the overall aminoacylation level of tRNA for one amino acid.

Table I. Charging level of tRNA in rel$^+$/rel$^-$ pairs of aminoacyl-tRNA synthetase mutants

Strain	Charging (%) *		
	35°C	39°C	43°C
BM113	43	19	9
BM1131	32	20	11
BM112	19	9	9
AT2535valS	24	16	13

* Charging level at 20 min after a shift from 30°C to the respective temperature.

2. THE PHYSIOLOGICAL NECESSITY FOR COORDINATE REGULATION OF THE COMPONENTS OF THE PROTEIN SYNTHESIS SYSTEM

We have previously demonstrated that growth of temperature-sensitive aminoacyl-tRNA synthetase mutants in rich medium becomes linear upon exposure to temperatures which do not completely block enzyme activity (Buckel et al., 1976). For example, the maximal temperatures permissive for exponential growth of strains AT2535 valS and BM 1131 are 37°C and 39°C, respectively. At higher temperatures growth becomes linear although a considerable proportion of the critical tRNA species is still charged and the inhibition of protein synthesis is far from complete (see Fig. 3). These results indicate that low levels of charged tRNA may not be used for the maintainance of balanced growth.

The following phsyiological and genetic evidence was obtained which indicate the possible nature of growth inhibition.

First, it was shown that low concentrations of bacteriostatic protein synthesis inhibitors phenotypically suppress temperature-sensitive growth (Buckel et al., 1976). Fig. 4 illustrates this for strain BM 113 and chloramphenicol. Measurement of the level of charged tRNA^{ala} indicated almost full charging of tRNA^{ala} when BM 113 was grown at the normally restrictive temperature in the presence of 6 μg chloramphenicol per ml.

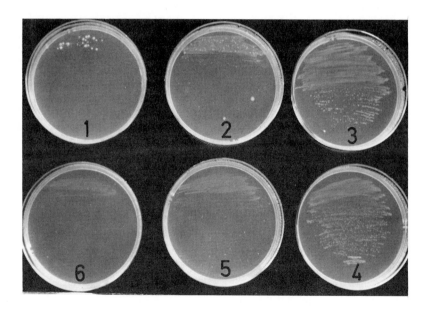

FIGURE 4. Supression of temperature-sensitive growth of strain BM 113 at 40°C by chloramphenicol. 1: no drug; 2: 2 μg/ml; 3: 4 μg/ml; 4: 6 μg/ml; 5: 8 μg/ml; 6: 10 μg/ml.

Secondly, when revertants are isolated which can grow again at the previously restrictive temperature many of them possess a defect in ribosomal assembly (Buckel et al., 1976; Wittmann et al., 1974; Wittmann et al., 1975). Table II gives the characteristics of several of such mutants.

Table II. Characteristics of ribosomal mutations reverting temperature-sensitive growth of synthetase mutants

Strain	Altered ribosomal protein	Map position min	Characteristic
O-1	S5	72	defective assembly
V40-1	S8	72	"
64-2	S20	0.5	"
39-1	S20	0.5	"
V9-4	lacking S20	0.5	"
O-2	?	70	"
4-5	?	70	"

In addition, many strains were isolated which show a normal assembly pattern but exhibit a reduced polypeptide synthesis activity in vitro (Wittmann et al., 1975). The common feature of all these suppressor strains was that the degree of aminoacylation was increased in comparison to the temperature-sensitive aminoacyl-tRNA synthetase mutant.

On the basis of these results we have postulated that it is essential that under all conditions of growth the ribosomal A-site must be saturated to a critical limit with aminoacyl-tRNA (Buckel et al., 1976). It was further proposed that when this is no longer achieved growth becomes linear due to premature termination and the accumulation of faulty peptides in *relA* strains or to the unbalanced synthesis of proteins due to the continuous stringent response in a $relA^+$ background (Buckel et al., 1976). The effect of antibiotic addition or of the presence of ribosomal mutations was seen as shifting the limiting step in protein synthesis from the charging reaction to the ribosome; in case of the ribosomal mutants this occurs by lowering the number of 70S ribosomes per cell via an assembly defect (assembly defective mutants of this kind contain between 15 and 35 % of the normal 70S content, Buckel et al., 1972) or by decreasing ribosomal activity. In each case, charged tRNA accumulates again (Buckel et al., 1976).

In light of these results and the data of O'Farrell (1978) it is conceived that the primary purpose for coordinate regulation of the translational components may be the maintainance of conditions guaranteeing optimal fidelity of protein synthesis.

3. EVIDENCE THAT THE BASAL LEVEL OF ppGpp DOES NOT REGULATE THE SYNTHESIS OF STABLE RNA UNDER BALANCED GROWTH CONDITIONS

The basal level of ppGpp has been implicated as effector for the growth-rate mediated regulation of stable RNA synthesis since it varies inversely relative to growth rate and to the contents of stable RNA (Cashel, 1969; Lazzarini et al., 1971). Ribosomal mutants with an assembly defect provide a means to investigate whether such a correlation exists since they contain an increased amount of rRNA, mostly in form of precursor particles (Buckel et al., 1972). We have determined for this purpose the basal level of ppGpp in three wild strains under steady-state growth in different media and in several ribosomal assembly mutants (Fig. 5). There is a close inverse correlation between ppGpp contents and the RNA/protein ratio in wild strains under balanced growth in different media; the assembly defective mutants, however, deviate greatly in that they accumulate more RNA than appropriate for the respective ppGpp concentration. As these mutants possess a lower amount of mature 70S it is assumed that the previously observed correlation (Cashel 1969, Lazzarini et al., 1971) between ppGpp and the RNA/protein ratio

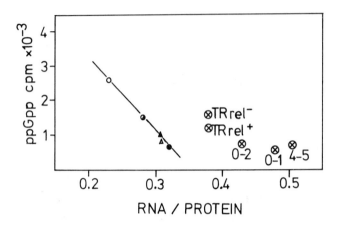

FIGURE 5. Relation between basal level of ppGpp and the RNA/protein ratio in wild strains and in mutants with ribosomal assembly defects during exponential growth at 37°C. (○, ◐, ●): Strain K10 (relA1 T2^R tonA22) in minimal medium with glycerol (○), glucose (◐), glucose + 20 amino acids (40 g/ml each) (●). (△): Strain AB2834 (thi-21 aroE353 mal-352) (▲): Strain AB4511 (aroE24 rel-1 thi-21). AB2834 and AB4511 were grown in glucose minimal medium supplemented with 1 % bacteriological peptone. The latter medium was also used for the assembly mutants O-1, 4-5, O-2 (see Table II and ref. Buckel et al., 1976) and for TRrel$^-$ and TRrel$^+$ which contain the rprotein S5 alteration of strain O-1 in a relA and rel$^+$ genetic background, respectively.

represents a correlation between ppGpp and mature 70S (possibly ribosomal activity) and not an influence of ppGpp on stable RNA synthesis during unrestricted growth.

REFERENCES

Buckel, P., Ruffler, D., Piepersberg, W., and Böck, A. (1972). RNA overproducing revertants of an alanyl-tRNA synthetase mutant of Escherichia coli. *Molec. Gen. Genet. 119*, 323-335.

Buckel, P., Piepersberg, W., and Böck, A. (1976). Suppression of temperature-sensitive aminoacyl-tRNA synthetase mutations by ribosomal mutations: A possible mechanism. *Molec. Gen. Genet. 149*, 51-61.

Cashel, M. (1969).The control of ribonucleic acid synthesis in Escherichia coli. IV. Relevance of unusual phosphorylated compounds from amino acid--starved stringent strains. *J. Biol. Chem. 244*, 3133-3141.

Fiil, N.P., van Meyenburg, K., and Friesen, J.D. (1972). Accumulation and turnover of guanosine tetraphosphate in Escherichia coli. *J. Mol. Biol. 71*, 769-783.

Folk, W., and Berg, P. (1970). Characterization of altered forms of glycyl transfer ribonucleic acid synthetase and the effects of such alterations on aminoacyl transfer ribonucleic acid synthesis in vivo. *J. Bacteriol. 102*, 204-212.

Haseltine, W.A.,Block, R., Gilbert, W., and Weber, K. (1972) MSI and MSII made on ribosome in idling step of protein synthesis. *Nature 238*, 381-384.

Lazzarini, R.A., Cashel, M., and Gallant, J. (1971). On the regulation of guanosine tetraphosphate levels in stringent and relaxed strains of Escherichia coli. *J. Biol. Chem. 246*, 4381-4385.

Lewis, J.A., and Ames, B.N. (1972). Histidine regulation in Salmonella typhimurium. XI. The percentage of transfer RNA^{His} charged in vivo and its relation to the repression of the histidine operon. *J. Mol. Biol. 66*, 131-142.

Neidhardt, F.C., Bloch, P.L., Smith, D.F. (1974).
Culture medium for enterobacteria. *J. Bacteriol. 119*,
736-747.

Neidhardt, F.C., Bloch, P.L., Pedersen, S., Reeh,S.
(1977). Chemical measurement of steady-state levels
of ten aminoacyl-transfer ribonucleic acid synthetases in Escherichia coli. *J. Bacteriol. 129*, 378-387.

O'Farrell, P. (1978). The suppression of defective
translation by ppGpp and its role in the stringent
response. *Cell 14*, 545-557.

Rubin, E.M. (1973). The nucleotide sequence of Saccharomyces cerevisiae 5.8S ribosomal ribonucleic
acid. *J. Biol. Chem. 248*, 3860-3875.

Wittmann, H.G., Stöffler, G., Piepersberg, W., Buckel,
P., Ruffler, D., and Böck, A. (1974). Altered S5 and
S20 ribosomal proteins in revertants of an alanyl-tRNA synthetase mutant of Escherichia coli. *Molec.
Gen. Genet. 134*, 225-236.

Wittmann, H.G., Stöffler, G., Geyl, D., and Böck, A.
(1975). Alteration of ribosomal proteins in revertants of a valyl-tRNA synthetase mutant of Escherichia coli. *Molec. Gen. Genet. 141*, 317-329.

REGULATION OF MACROMOLECULAR SYNTHESIS BY LOW MOLECULAR WEIGHT MEDIATORS

DELETION OF *relA* AND *relX* IN *ESCHERICHIA COLI* HAS NO EFFECT ON BASAL OR CARBON-DOWNSHIFT ppGpp SYNTHESIS

Alan G. Atherly[1]

Department of Genetics
Iowa State University
Ames, Iowa

A mutant of *Escherichia coli* was fortuitously isolated that contains a large deletion encompassing argA, fuc, relX and relA. This mutant (AA-787) is also cold-sensitive for growth and protein synthesis. Unusually high cotransduction of flanking markers (cysC and thyA) indicate loss of about one minute of the E. coli genetic map in strain AA-787. Of the metabolic parameters examined in strain AA-787, protein but not DNA or RNA synthesis was greatly inhibited at 18°C. Ribosome assembly, all ribosome proteins as well as in vitro protein synthesis were normal in strain AA-787 at 18°C. Most interesting, however, guanosine 3'-diphosphate-5'-diphosphate (ppGpp) was synthesized in mutant strain AA-787 at normal basal levels and ppGpp accumulation was stimulated by carbon-source downshift. No ppGpp was obtained using ribosomes isolated from strain AA-787. Although the molecular defect causing cold-sensitivity in strain AA-787 could not be identified, it is clear that deletion of relA and relX has no effect on ppGpp synthesis distinguishable from all other relA point mutants. It must be concluded that another enzyme system exists for ppGpp synthesis.

---

[1] Supported by National Science Foundation Grant No. PCM 76-11012 and the Iowa Agriculture Experiment Station, Ames, IA. Project No. 2299.

I. INTRODUCTION

Guanosine 3'-diphosphate, 5'-diphosphate (ppGpp) has been implicated in the regulation of a number of metabolic events in bacteria: the synthesis of rRNA, tRNA, r-protein, α-subunit of RNA polymerase, elongation factor-Tu, and phospholipids to mention a few (for review, see Cashel, 1975). This compound is synthesized by a ribosome-bound enzyme (stringent factor) from GTP and ATP in the presence of mRNA and a codon-specific uncharged tRNA (Haseltine and Block, 1973; Pederson, et al., 1973). A completely separate pathway for ppGpp synthesis has been implicated. Enzyme extracts from *B. brevis* (Sy, 1976) and *E. coli* are reported to be able to synthesize ppGpp. Evidence that would support this interpretation is that ppGpp accumulation is observed in *relA* strains at significant quantities upon carbon-source downshift. However, all *relA* strains studied are likely point mutants (Block and Haseltine, 1973) and thus may have partly active stringent factor. More recently, insertion and nonsense mutants of *relA* have been described that are still capable of ppGpp synthesis (Friesen, et al., 1978).

An additional locus, *relX*, closely linked to *relA*, is implicated in influencing basal levels of ppGpp and the response of the cell to carbon source downshift (Pao and Gallant, 1978). *relA* normally has a two-fold lowered basal level of ppGpp; *relX* has an additional 4-5 fold decreased level, resulting in about a 10-fold reduced level of ppGpp. *relA-relX* mutants grow normally except they have difficulty adjusting to rapid changes in carbon or energy metabolism. They also show an extreme sensitivity to plating on minimal media containing leucine after growth on rich media. Exactly how *relX* and *relA* are involved in these functions is not known.

Herein, I describe a mutant of *E. coli* deleted in *relA* and *relX* that synthesizes ppGpp at $relA^-$ basal levels and accumulates ppGpp upon carbon-source downshift.

II. MATERIALS AND METHODS

A. Growth Conditions and Bacterial Strains

All strains are derivatives of *Escherichia coli* K12 and are listed in table 1. Chemically defined media was made by supplementing M-9 media (Miller, 1972) with amino acids at 10-50 μg per ml, depending upon their ratio in proteins, thiamine at 1 µg per ml and glucose at 0.2%. For a glucose to succinate downshift 0.03% glucose with 0.3% succinate was used.

B. Isolation of Mutant and Genetic Crosses

No mutagen was present for isolation of the mutant. An overnight culture of strain AA-7852 grown from a single colony was appropriately diluted, 10^6 to 10^7 cells were spread on broth-agar plates, and incubated overnight at 42°C. Several hundred randomly picked colonies were selected and restreaked on duplicate broth-agar plates. One plate was incubated at 42°C, and one at 18°C. Cells that did not grow at 18°C but did grow at 42°C were selected and restreaked. These occurred at a rate of about one per 200 colonies. AA-787 was singled out for further study.

For transfer of cold-sensitive mutation to another strain and to perform a four-factor cross, a lysate of phage P1 was prepared on strain AA-787, ($thyA^+$, and cold-sensitive) by the procedure of Miller (1972). The recipient strain (NF-58) ($argA$ $metB$) was converted to $argA^+$ by mating with Hfr KL-16 and selecting recombinants that did not require arginine for growth and were $relA^+$. A $thyA$ mutant was then selected from this strain (Miller, 1972). The transduction was performed as described by Miller (1972), and $thyA$ recombinants in strain NF58 were obtained by spreading on minimal agar containing methionine and arginine, followed by growth at 30°C for 48 h. These colonies were tested for $argA$ and cold sensitivity by restreaking on similar plates lacking arginine at 30°C and broth-agar plates grown at 18°C. Each colony was identified by a number and recorded. $relA$ was tested in each recombinant by a procedure described by Cashel (personal communication). This procedure involves growing cells overnight on broth-agar plates, picking enough cells with a toothpick to make turbid 200 µl of 40 µM phosphate minimal-glucose (with 50 µg/ml thymidine, 1 mg/ml serine hydroxamate, 300 ug/ml L-valine, and 5 µCi/ml $^{32}PO_4^{3+}$). This mixture is incubated 1 h at 37°C; then 50 ul of 13N formic acid is added, and the mixture is frozen 15 min; 10 ul is spotted 1 cm apart on polyethyleneimine thin-layer sheets and eluted 10 cm with 1.5 M KH_2PO_4, pH 3.4. The spots are visualized by exposure to X-ray film overnight.

C. Polyphenylalanine Synthesis and Preparation of Ribosomes

The procedure for polyuridylic acid-stimulated polyphenylalanine synthesis was exactly as described elsewhere (Menninger et al.1973). Ribosomes (S-30) were prepared by growing cells in broth media at a permissive growth temperature to 5×10^8 cells per ml. In some experiments, cells were transferred to the 18°C for 1 h by adding ice before harvesting. In experiments designed to determine assembly of ribosomes, cells were grown in minimal media containing 20 µg per ml of uridine. After transfer to 18°C the cells were labeled with ^{3}H-uridine for 15 minutes, harvested, washed and ribosomes examined on a sucrose gradient as previously described (Atherly, 1974). Two-dimensional gel electrophoresis of ribosomal proteins was determined using cells grown at 18°C as described previously (Kaltschmidt and Wittmann, 1970).

D. *Enzyme and Gene Activity Assays*

The presence of *lysX* gene activity was measured as described by Jenkins et al (1974). Enolase activity was measured as previously described (Spring and Wold, 1971) and protein determined by the method of Bradford (1976). The presence of *rec* gene activity in strain AA-787 was tested by the procedure of Miller (1972) and peptidyl tRNA hydrolase was assayed by J. Menninger. Synthesis of ppGpp in whole cells was measured as described by Cashel (1969), using 40 µM phosphate media and 0.32 mCi/ml ^{32}P. Cell-free synthesis of ppGpp was determined by the method of Sy (1976).

III. RESULTS

A. *Mutant Selection*

The strain used for mutant selection contains a heat-sensitive mutation in peptidyl-tRNA hydrolase (Atherly and Menninger, 1972). As suggested from the work of Jarvik and Botstein (1975), suppressor mutants were sought that were simultaneously cold-sensitive. Strain AA-7852 *(pth-1)* was spread on plates, incubated at 42°C, colonies picked, restreaked, and incubated at 18°C and 42°C. Nongrowers at 18°C were obtained for further study, and one (AA-787) is the subject of this report. The grandparent strain of the cold-sensitive mutant (AA-787) is strain CP-78. The parent strain, AA-7852, grows poorly at 40°C, but not at all at 42°C. The cold-sensitive mutant, however, grows well at 40°C and 42°C, but not at 18°C in either broth or minimal media.

TABLE I. Strains of Escherichia coli Used in This Study

Strain	Genotype	Reference or source
NF-58	argA, metB, thi, relA$^+$	Fiil, 1969
NF-59	argA, metB, thi, relA	Fiil, 1969
CP-78	argH, thr, leu, his, thi, relA$^+$	Fiil, 1969
AA-7852	derivative of CP-78, pth-1	Atherly and Menninger, 1972
AA-787	derivative of AA-7852, cold-sensitive ΔrelA to argA	This laboratory
AA-38	derivative of NF-58, ΔrelA to argA, metB, thi	Transduction with AA-787
KL-16	Hfr, str$^+$ (clockwise near lysA)	Bachmann et al., 1976
NF-306	F160 lysA to cysC in host NF-305, pyrB43, argG6, metB1, lysA29, his-1, leu-6, recA 1, strA104, specA13, supE44	Fiil, 1969
KLF43/KL259	F143 lysA to tyrA, relA in host KL259, thi-1, tyrA2, pyrD32, his-68, trp-45, thyA33, recA 1	B. J. Bachmann
KLF16/KL110	F116 metC to fuc in host KL110, argG6, metB1, leu6, his1, thyA23, recA-1, supE44	B. J. Bachmann

B. Genetic Studies

The mutation in strain AA-787 was located by interrupted mating with Hfr strain KL-16 by using *thyA* as a reference marker (Atherly, 1979b). The cold-sensitive mutation is slightly distal from *thyA* with reference to the entry point of KL-16 (*lysA* at min 61). This suggested a point near *argA*, *fuc*, or *relA*. Subsequent tests showed all three gene functions to be mutant. The reversion frequency for *fuc*, *argA*, and cold sensitivity were each less than 1×10^{-11}. The nearest functions mapping clockwise *(eno)* and counterclockwise *(recB)* to *relA* and *argA*, respectively, were assayed and found normal (Spring and Wold, 1971; Miller, 1972).

relX was tested in strains CP-78, CP-79, AA-7852 and AA-787 to determine if it was also deleted. These tests were performed by estimating plating efficiency at 30°C on minimal media-agar with 100 ug/ml leucine after growth in L-broth. Pao and Gallant (1978) demonstrated that *relX*, *relA* mutant strains have a greatly reduced plating efficiency under these conditions. From the data presented in table II, it can be concluded that strains CP-79 and AA-787 are *relX*, *relA*: Pao and Gallant (1978) have previously established strain CP-79 to be *relX*, *relA*. Thus, deletion of *relX* and *relA* (strain AA-787) give identical findings as point mutants in *relA* and *relX* (strain CP-79).

TABLE II. *Plating Efficiency on Minimal Agar With Leucine*

Strain and genotype	Plating efficiency
CP-78; $relA^+$, $relX^-$	0.85
CP-79; $relA^-$, $relX^-$	1.25×10^{-2}
AA-7852; $relA^+$, $relX^-$	0.25
AA-787; $\Delta relA$, $\Delta relX$	1.1×10^{-2}

Upon introduction of episome F160 from strain NF-306, cold sensitivity was recessive along with mutant functions in *fuc*, *relA*, and *argA*. On the other hand, the strain was now heat-sensitive due to *pth-1*. Thus, the *pth* gene is recessive to the suppressor mutation in AA-787. Introduction of F143, however, which partly covers the deleted area (counter-clockwise from fuc to metC), did not abolish cold sensitivity. It must be concluded that the gene responsible for cold sensitivity is located between fuc and relA.

The cold-sensitive mutation was more carefully located by a four-factor cross using P1 phage cotransduction. A cotransduction frequency of 42% between thyA, cold sensitivity, and argA indicates a map position near argA (Atherly 1979b) as calculated from the equation of Wu (1966). However, relA, argA, and cold sensitivity all cotransduce at the same frequency, thus, these data suggested a deletion encompassing all three genes.

In separate experiments testing the linkage distance between flanking markers cysC and thyA, a cotransduction frequency of 35% was obtained (Atherly, 1979b). Using the equation of Wu (1966), with these data a distance of 0.56 map units can be calculated between cysC and thyA in strain AA-787. Thus, approximately 1-1.2 map units have been deleted between thyA and cysC (Fig. 1).

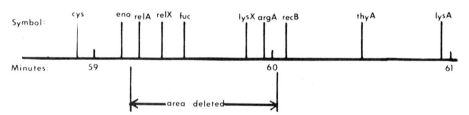

FIGURE 1. A short segment of the E. coli genetic map.

C. Protein and RNA Synthesis

It was of interest to determine if the cold-sensitive pseudorevertant (strain AA-787) from the heat-sensitive hydrolase strain was also defective in protein synthesis at its nonpermissive temperature (18°C). Indeed, protein synthesis is cold-sensitive when compared with its parent strain (Fig. 2A). On the otherhand, RNA synthesis is only slightly inhibited at 18°C (Fig. 2B). In fact, the response of RNA synthesis to reduced protein synthesis is not unlike a "relaxed" response.

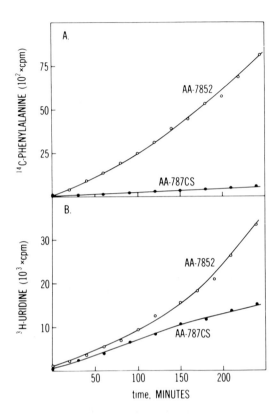

FIGURE 2. *RNA and protein accumulation at 18°C. Cells in exponential growth at 30°C were transferred to 18°C.*

Strain AA-787 is capable of protein synthesis but at a greatly reduced level as seen from β-galactosidase synthesis at 18°C (Table 3). The amount of enzyme synthesized by mutant strain AA-787 is only 2% of the grandparent strain (CP-78) and then only after a lag of 50 min. Yet, the level of β-galactosidase synthesis in strain AA-787 is identical to CP-78 at 35°C.

Polyuridylic acid-stimulated polyphenylalanine synthesis was measured at 15 and 13 mM mg^{+2} as well as at 7 mM Mg^{+2} with addition of N-acetyl phenylalanine tRNA. Rates were linear for at least 30 minutes (data not shown). Polyuridylic acid-stimulated polyphenylanine synthesis is not cold-sensitive when ribosomes from strain AA-787 are used, no matter if the cells are grown at 18°C or 37°C (data not shown). The rate of polyphenylalanine synthesis is about twice that observed using ribosomes from the parent strain (AA-7852). It can be

TABLE III. β-galactosidase Synthesis

Strain	Temp.	Rate (units per ml per min)[a]	Lag before synthesis
CP-78	18	0.0750	15 min.
AA-787	18	0.00210	50
CP-78	35	1.20	2
AA-787	35	1.20	4

Cells of each strain were grown with glycerol as a carbon source to 2×10^8 per ml and induced with 1 mM isopropyl-β-D-thioglactopyranoside. Culture at 18°C also contained 1 mM cyclic adenosine monophosphate. 18°C culture were first grown at 35°C, transferred to 18°C and induced 15 min later.

[a] A unit is defined as a change in optical density at 420mm x 1000 per 10^7 cells/ml. The values given were obtained from the slope of the induction curve.

concluded that in vivo protein synthesis is cold-sensitive in the mutant strain, but the cold-sensitive factor is not involved in polymerization or initiation of polypeptides in a cell-free system.

D. Guanosine 3'-diphosphate 5'-diphosphate (ppGpp) Synthesis

Stringent-factor activity was determined by measuring the cell-free conversion of GTP to ppGpp by using ribosomes from the cold-sensitive mutant (AA-787) and its grandparent strain (CP-78). The data in Table 4 indicate no conversion of GTP to ppGpp for strain AA-787, although considerable ppGpp synthesis is observed by strain CP-78 ribosomes.

Basal levels of ppGpp and the ability to synthesize ppGpp upon down shift were determined by using whole cells. It is known that relA strains can synthesize appreciable quantities of ppGpp upon downshift, but most relA strains used in previous studies are likely point mutants (Block and Haseltine, 1973). If relA alone is responsible for the synthesis of ppGpp, then this strain should not have any ppGpp present. As seen from the data in Fig. 3, the basal

TABLE IV. Cell Free Synthesis of ppGpp

Bacterial strain	Percent conversion of GTP to ppGpp and pppGpp
CP-78	38.6%
AA-787	0.3%

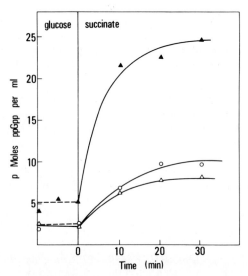

FIGURE 3. Accumulation of ppGpp in cells shifted from glucose to succinate carbon source. Symbols: ▲—▲ NF-58(relA$^+$); 0-0, NF-59(relA); Δ - Δ , AA-38(Δ relA, cold-sensitive).

levels of ppGpp, in a strain possessing this deletion are not different from that found in most relA strains. Furthermore, upon shift from a glucose to succinate carbon source, considerable ppGpp was synthesized in the deletion strain. Results from relA and relA$^+$ strains (NF-58 and NF-59) are given for comparison.

IV. DISCUSSION

Cold-sensitive strain AA-787 was fortuitiously selected from a strain heat-sensitive for peptidyl-tRNA hydrolase (referred to as hydrolase). No mutagen was used. The cold-sensitive mutant was merely picked from a group of revertants occurring at the high temperature. Extensive analysis of strain AA-787 has shown it to be defective in protein synthesis at 18°C, and furthermore, a large segment of the chromosome (one minute or 3.0×10^7 daltons) from relA to argA is deleted. Just exactly why this strain is cold-sensitive is not clear. The only function known mapping in this area of the genome related to protein synthesis is relA (Bachmann et al., 1976). The relA gene product is physically associated with the ribosome (Block and Haseltine, 1975; Cochran and Byrne, 1974) (but only a small percentage of the total population of ribosomes) and is reported to participate in tRNA selection (Richter and Isono, 1976; Richter, 1976). Because of the large segment deleted, one cannot be certain that another ribosome-function gene has not been removed, but two-dimensional polyacrylamide-gel electrophoresis of the ribosomal proteins indicates these to be normal (data not shown). Ribosome assembly at 18°C also is normal (data not included). Cell-free polyuridylic acid-stimulated protein synthesis is not cold-sensitive, even when cells are grown at 18°C before harvesting ribosomes, and, in fact, is stimulated over parent strain rates.

The cold-sensitive phenotype can be retained when transferred to other strains not heat-sensitive for hydrolase function; thus, cold sensitivity is not dependent upon a defective hydrolase. Why strain AA-787 should suppress hydrolase function is the subject of another report (Atherly, 1979a). From results of partial diploids covering the deletion or part of the deletion in strain AA-787, it can be concluded that the cold-sensitive function is located between relA and fuc genes. However, it cannot be distinguished whether cold sensitivity is due to deletion of relA.

Several interesting conclusions can be made from investigation of strain AA-787. The presence of a basal level of ppGpp and the stimulation of ppGpp synthesis upon carbon-source downshift in relA strains may be due to a partly active stringent factor. Strain AA-787, deleted in relA still synthesizes ppGpp similar to archtype relA strains (Borek et al., 1965) even though relX (Pao and Gallant, 1978) is also mutant. These findings strongly argue for the existence of another enzymatic pathway for ppGpp synthesis other than with stringent factor; however, it is possible that the deletion in AA-787 includes a regulatory gene for *relA* and not the structural gene. In support of these findings, Zabos et al. (1976) and Sy (1976) report ribosome-independent synthesis of ppGpp and more recently, Friesen et al. (1978) report nonsense and insertion mutants in *relA* are still capable of ppGpp synthesis.

A few words should be said about *relX*, which is also deleted in strain AA-787. The concentrations of ppGpp, both basal and induced, during downshift were not significantly different between strain NF-58 (*relX$^+$, relA1$^+$*), NF-59 (*relX$^+$ relA1*) and AA-38 (Δ*relX2* and *relA14*) (See Fig. 3). This is in contrast to what was observed by Pao and Gallant (1978), using the same genetic background but a different *relX (relX1)* mutant. Other features of Δ*relX2* and *relX1* are different. *relX1* (in an NF background) is temperature-sensitive (at 42°C) but Δ*relX2* is not; however, strain AA-787 (Δ*relA14, relX2*) has a greatly reduced plating efficiency onto minimal media with 100 µg/ml leucine similar to *relA1 relX1* mutants characterized by Pao and Gallant. The *relX1* mutation was naturally occurring in a strain when found by Pao and Gallant (1978). They ascribe a key role for this gene in restricting ppGpp accumulation during carbon-source downshift. Observations with strain AA-787 and its NF derivative AA-38 do not bear this conclusion out.

Bachman et al. (1976) report the position of the gene *pyrG*, coding for cytidine triphosphate synthetase, as being between *relA* and *argA*. This map position probably is incorrect inasmuch as this enzyme activity was present in strain AA-787. *pyrG* is likely located near min 59, proximal to *relA*. Further, *lysX* is reported in this area of the genome. Strain AA-787 was found not to be mutant for *lysX* (Jenkins et al., 1974). Thus, *lysX* also may be incorrectly located on the *E. coli* genetic map (Bachmann, 1976). It can be concluded that no genes are present between *relA* and *argA* that are necessary for growth on minimal media with glucose, glycerol, or succinate as a carbon source.

ACKNOWLEDGMENTS

I express my appreciation to Mary Mascia for excellent technical assistance and to Robert Silverman and Jack Horowitz for many helpful discussions. My sincere thanks to John Menninger (University of Iowa) for performing the hydrolase assays and to Jack Horowitz and Janice Kolberg for performing two-dimensional gel analysis of ribosomal proteins.

REFERENCES

Atherly, A. G. Ribonucleic acid regulation in amino acid-limited cultures of Escherichia coli grown in a chemostat. J. Bacteriol. 120, 1322 (1974).
Atherly, A. G. Natural premature protein termination can be reduced in Escherichia coli by decreased translation rates. (in press) (1979a).
Atherly, A. G. A mutant of Escherichia coli containing a deletion from relA to argA. J. Bacteriol. 138, 530 (1979b).
Atherly, A. G., Menninger, J. R. Mutant E. coli strain with temperature sensitive peptidyl-transfer RNA hydrolase. Nat. New Biol. 240, 245 (1972).
Bachmann, B. J., Low, K. B., Taylor, A. L. Recalibrated linkage map of Escherichia coli K12. Bacteriol. Rev. 40, 116 (1976).
Block, R., Haseltine, W. A. Thermolability of the stringent factor in rel mutants of Escherichia coli. J. Mol. Biol. 77, 625 (1973).
Block, R., Haseltine, W. A. Purification and properties of stringent factor. J. Biol. Chem. 250, 1212 (1975).
Borek, E., Ryan, A., Rochenbach, J. Nucleic acid metabolism in relation to the lysogenic phenomenon. J. Bacteriol. 69, 460 (1965).
Bradford, M. M. A rapid and sensitive method for the quantitation of microgram quantities of protein utilizing the principles of protein-dye binding. Anal. Biochem. 72, 248 (1976).
Cashel, M. The control of ribonucleic acid synthesis in Escherichia coli. J. Biol. Chem. 244, 3133 (1969).
Cashel, M. Regulation of bacterial ppGpp and pppGpp. Annu. Rev. Microbiol. 29, 301 (1975).
Cochran, J. W., Byrne, R. W. Isolation and properties of a ribosome-bound factor required for ppGpp and pppGpp synthesis in Escherichia coli. J. Biol. Chem. 249, 353 (1974).
Fiil, N. A functional analysis of the rel gene in Escherichia coli. J. Mol. Biol. 45, 195 (1969).
Friesen, J. D., Gynheung, A., Fiil, N. P. Nonsense and insertion mutants in the relA gene of E. coli: Cloning relA. Cell 15, 1187 (1978).
Hansen, M. T., Pato, M. L., Molin, S., Fiil, N. P., von Meyenburg, K. Simple downshift and resulting lack of correlation between ppGpp pool size and ribonucleic acid accumulation. J. Bacteriol. 122, 585 (1975).

Haseltine, W. A., Block, R. Synthesis of guanosine tetra and penta phosphate requires the presence of a codon specific, uncharged transfer ribonucleic acid in the acceptar site of ribosomes. *Proc. Natl. Acad. Sci. USA* 70, 1564 (1973).

Javik, J., Botstein, D. Conditional-lethal mutations that suppress genetic defects in morphogenesis by altering structural proteins. *Proc. Natl. Acad. Sci. USA* 72, 2738 (1975).

Jenkins, C. C., Sparkes, M. C., Jones-Martimer, M. C. A gene involved in lysine excretion in *Escherichia coli*. *Heredity* 32, 409 (1974).

Kaltschmidt, E., Wittmann, H. G. Ribosomal proteins, VII two-dimensional polyacrylamide gel electrophoresis for finger printing of ribosomal proteins. *Anal. Biochem.* 36, 401 (1970).

Menninger, J. R., Walker, C., Tan, P. F., Atherly, A. G. Studies on the metabolic role of peptidyl-tRNA hydrolase. *Molec. Gen. Genet.* 121, 307 (1973).

Pao, C. C., Gallant, J. A gene involved in the metabolic control of ppGpp synthesis. *Molec. Gen. Genet.* 158, 271 (1978).

Miller, J. H. Experiments in Molecular Genetics, Cold Spring Harbor Lab. (1972).

Pederson, R. S., Lund, E., Kjeldgaard, N. O. *Nature New Biol.* 243, 13 (1973).

Richter, D., Isono, K. The mechanism of protein synthesis-initiation, elongation and termination in translation of genetic messages. *Curr. Top. Microbiol. Immunol.*, 76, 81 (1976).

Richter, D. Stringent factor from *Escherichia coli* directs ribosomal binding and release of uncharged tRNA. *Proc. Natl. Acad. Sci. USA* 74, 707 (1976).

Spring, T. and Wold, T. The purification and characterization of *Escherichia coli* enolase. *J. Biol. Chem.* 246, 6797 (1971).

Sy, J. A ribosome-independent, soluble stringent factor-like enzyme from *Bacillus brevis*. *Biochemistry* 15, 606 (1976).

Wu, T. T. A model for three-point analysis of random general transduction. *Genetics* 54, 405 (1966).

Zabos, P., Bauer, P., Schlotthaer, J., Horvath, I. Stringent factor-independent synthesis of pppGpp in *Escherichia coli* strains. *FEBS Lett.* 64, 107 (1976).

**REGULATION OF MACROMOLECULAR SYNTHESIS
BY LOW MOLECULAR WEIGHT MEDIATORS**

MOLECULAR INTERACTIONS IN THE INITIATION OF rRNA SYNTHESIS IN PROKARYOTES

M. Gruber, J. Hamming, B.A. Oostra and G. AB

Biochemisch Laboratorium
The University, Groningen
The Netherlands

We show the presence of (a) positive, specific effector(s) of rRNA synthesis in extracts of fast-growing E. coli. These factors double the initiation frequency of rRNA synthesis in vitro. The factor-induced rRNA initiation is as ppGpp-sensitive as the basal one.

To study molecular events at initiation of rRNA synthesis, we estimate the interaction between RNA polymerase with a restriction fragment containing rRNA promoter(s), using the filter-binding method. Our results have led to a tentative model in which a transition of the primarily formed, labile, RNA polymerase-rRNA promoter complex to a more stable form is the determining step. This step, and thus the binding measured, is salt sensitive. Also, ppGpp acts on this "isomerization".

INTRODUCTION

The synthesis of ribosomal (r) RNA in prokaryotes (Gruber et al., 1978; Nomura et al., 1977) poses an unsolved problem: its rate varies widely with changes in the growth medium. Control of rRNA synthesis in independent of that of (most) other RNA.

This work has been carried out under auspices of The Netherlands Foundation for Chemical Research (S.O.N.) and with financial aid from The Netherlands Organization for the Advancement of Pure Research (Z.W.O.).

Under optimal growth conditions, the seven operons comprising about 1% of the genome account for 70% of total transcription. RNA polymerase, then, is packed as tightly on the rRNA operons as possible: in other words, the rate of rRNA synthesis is then governed by the rate of elongation which apparently can be matched by the initiation frequency. The rRNA promoters under these conditions, behave as very strong ones. We are trying to unravel at least part of the control circuitry and to identify the molecular events.

THE ROLE OF ppGpp

As will be dealt with in other contributions to this volume the stringent response - the suppression of rRNA synthesis in wild-type cells by lack of any amino-acylated tRNA - is concomitant with a negative correlation between the concentration of ppGpp and the rate of rRNA synthesis (Fiil et al., 1972). In 1975, we were able to show (Van Ooyen et al., 1975) that ppGpp specifically depresses rRNA synthesis in vitro in a system containing only E. coli DNA, and RNA polymerase as macromolecular components. The K_i found in vitro was similar to that calculated for the stringent response in vivo. This finding has been confirmed in several laboratories: ppGpp inhibits the transcription of at least four (Jørgensen & Fiil, 1976; Van Ooyen et al., 1976; Travers, 1976; Oostra et al., 1977), but
probably of all seven rRNA operons in vitro. The ppGpp effect is independent of the - bacterial, plasmid or phage - genome containing the rRNA operon. The nucleotide appears to act on initiation (Van Ooyen et al., 1976; Oostra et al., 1977). Clearly, since no other macromolecular components are necessary ppGpp should act by interfering with the formation of the productive RNA polymerase-rRNA promoter complex. With some other promoters, on the other hand, ppGpp stimulates transcription as has been shown for the Lac operon (Primakoff & Artz, 1979).

While these results make a role of ppGpp in vivo even more likely it is probably not the only controlling element involved. The data on rRNA synthesis and ppGpp levels at different growth rates (Friesen et al., 1975) will only fit with an exclusive ppGpp control in vivo if ppGpp-potentiating factors are postulated. Moreover, in the purified system considerably less rRNA is synthesized as fraction of total transcription than in curde extracts (Gruber et al., 1978). The presence of a factor, or factors, positively affecting rRNA synthesis, in these extracts is thus indicated.

OTHER EFFECTORS OF rRNA SYNTHESIS

We have studied the possible existence of other factor(s) in *E. coli*. Cell extracts were treated according to Block (1976) with a Ca^{++}-dependent nuclease which degrades endogenous DNA (and RNA). The treated extracts are added to a transcription system consisting of RNA polymerase and phage DNA containing a rRNA operon. In this way the (possible) effect of the extracts on rRNA and non-rRNA synthesis can be measured (Oostra et al., 1977).

Unavoidably the extracts, and at least some fractions derived from them, contain nuclease activity which may influence the results, but which affects rRNA and non-rRNA to an equal degree as control experiments showed. There is some RNA polymerase present which in all respects tested behaves like added enzyme.

TABLE I. Effect of Nuclease-Treated S-100 on RNA Synthesis

			^3H-UMP inc. (dpm x 10^{-3})		
DNA	Extract	ppGpp	Total RNA	rRNA	rRNA (%)
λ d5ilv	−		62.8	15.7	29.1
λ d5ilv	+		72.8	47.4	69.4
λ rifd18	−	−	41.5	13.0	35.6
λ rifd18	−	+	28.2	4.6	19.4
λ rifd18	+	−	31.4	19.2	66.0
λ rifd18	+	+	21.4	7.5	39.8

For preparing extract and conditions of RNA synthesis see Gruber et al., 1978. RNA polymerase was at 25 µg/ml; KCl at 100 mM; DNA at 10 µg/ml; nuclease-treated S-100, if present, at 10 µl/100 µl; ppGpp, if present, at 0.75 mM.

Table I shows some typical examples of the data obtained. Clearly, the extracts contain factors which selectively stimulate rRNA synthesis. The additional rRNA synthesis is inhibited by ppGpp to the same degree as the original one. Sensitivity to ppGpp is thus unaltered. Moreover, the results are valid for at least two rRNA operons (rrn X and B). Our data strongly suggest the existence of positive control factors. The factors can be enriched to some degree by ammonium sulfate fractionation and glycerol gradient centrifugation, and are most probably proteins since they are inactivated by proteolytic enzymes. The factors can only be considered as regulatory elements if their stimulation of rRNA synthesis is due to an increase in initiation frequency. Table II shows the experiment: we found that the stimulation is reflected in an increased initiation frequency of rRNA. We therefore feel encouraged to continue our studies and try to further purify and characterize the factor(s?).

TABLE II. Frequency of Initiation

Extract	^{3}H-UMP inc. (dpm x 10^{-3})		rRNA(%)	number of initiations/min
	Total RNA	rRNA		
−	25.7	3.2	15.1	1.0
+	26.1	7.2	32.3	2.3

For conditions of RNA synthesis see Table I. KCl was at 125 mM. Two min after adding RNA polymerase heparin was added at 0.5 mg/ml and incubation was continued for 10 min to let RNA chains run to completion.

MOLECULAR EVENTS AT INITIATION

To understand effects of control elements on rRNA synthesis we have analyzed the interaction between RNA polymerase and the rRNA promoter which apparently changes "strength" under different conditions *in vivo*, or *in vitro*. We therefore

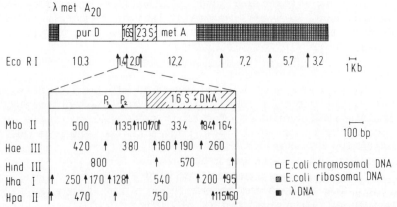

FIGURE 1. *Restriction map of* λ *metA20 DNA (De Boer et al., 1979)*

began a study of the binding of RNA polymerase to the ribosomal RNA promoter(s). The 1.4 kb Eco R1 restriction fragment of phage λmetA20 DNA which contains the ribosomal operon rrn E (Yamamoto & Nomura, 1976) was prepared and labelled. Figure 1 shows the physical map of this phage and the detailed map of this fragment which appears to possess two ribosomal promoters, both active *in vitro* (De Boer et al., 1979), as does rrn B operon. As Figure 1 shows, sub-fragments can be prepared from the 1.4 kb fragment containing only one of both promoters intact. The fragment, or in some experiments, a mixture of fragments, is incubated with RNA polymerase according to Seeburg & Schaller (1977); the binding reaction is stopped by addition of single-stranded DNA trapping all free RNA polymerase. The DNA bound strongly to RNA polymerase is determined by its retention on a membrane filter which retains RNA polymerase, but not free DNA. In experiments with mixtures, the DNA retained is subjected to electrophoresis to determine its composition. In this way the rate and extent of binding and its dependence upon a number of parameters can be measured.

Figure 2 shows "titration" curves with RNA polymerase at two different salt concentrations, and Figure 3 the time course of the reactions at a fixed RNA polymerase concentration. It is apparent from both figures that at 120 mM KCl a "steady state" concentration is rapidly reached which is lower than that at 40 mM; the dependence upon RNA polymerase concentration appears to be similar for both salt concentrations. The rapid and ready attainment of plateau values, and also the shape of the curves, exclude a simple, one-step, equilibrium for the binding reaction. Our results must be due to a more complex mechanism. The relatively most simple one, is a binding reaction with two complexes, I and II, according to the following scheme:

$$DNA + RNA\ polymerase \rightleftharpoons complex\ I \rightleftharpoons complex\ II$$

We assume that complex I forms and decays quickly. It will therefore rapidly disappear when free RNA polymerase is trapped.

A similar scheme for initiation has been postulated earlier by several workers (for a review see Chamberlin, 1976); mostly complex I was considered as the "closed" and II as the "open" complex. Previously (Van Ooyen et al., 1976, Oostra et al., 1977) we assumed the existence of a two-step binding to explain the effect of salt and of ppGpp on the initiation of transcription.

In our model, the effect of salt on the binding would be upon the equilibrium between complex I and II. The higher KCl concentration might affect this equilibrium by increasing the rate of decay of complex II or decrease the rate of its

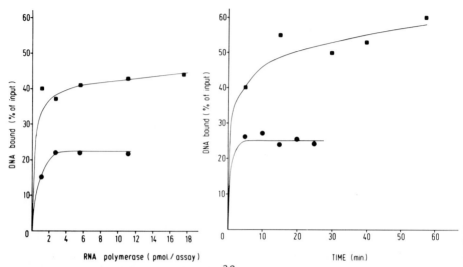

FIGURE 2. RNA polymerase and ^{32}P end-labelled 1.4 kb EcoRI fragment from metA20 DNA (Fig. 1; 0.08 pmol/assay, spec. act. 30 µCi/nmol) were incubated in binding buffer (Seeburg et al., 1977) with 200 µg BSA per ml and 40 mM, and 120 mM KCl respectively in a total volume of 30 µl at 37°C. After 5 min incubation, two volumes of prewarmed binding buffer, containing sonified, denatured calf thymus DNA (final conc. 30 µg/ml) were added. Final KCl concentration became 40 mM in all cases. Incubation was continued for 5 min and the mixture was filtered over nitrocellulose filters, filters were dried and radioactivity was counted (—■— 40 mM KCl, —●— 120 mM KCl). - left

FIGURE 3. RNA polymerase (5.6 pmol/assay) and 1.4 kb fragment were incubated under the same conditions as indicated in the legends of Fig. 2. Complexes were allowed to form for times indicated in the figure, and further treatment was as in the legends of Fig. 2 (—■— 40 mM KCl, —●— 120 mM KCl). - right

formation, or both. Figure 4 shows an experiment in which all four combinations of both salt concentrations were used for measuring the level, as well as the decay of the complex (II) in one experiment. It is clear that 120 mM KCl decreases the extent of complex II formation by accelerating considerably the decay of complex II, independent of the salt concentration at which it was formed. This direct measurement is in agreement with our earlier speculation that the effect of salt on the initiation of rRNA is on the step from the closed to the open complex.

The Initiation of rRNA Synthesis in Prokaryotes

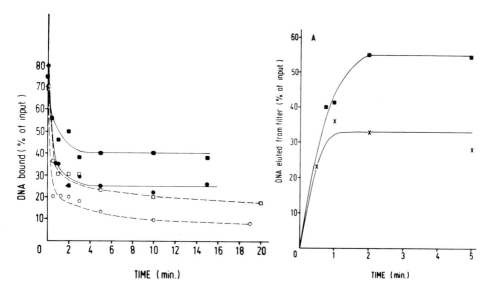

FIGURE 4. Complex formation of RNA polymerase (5.6 pmol/assay) and 1.4 kb fragment under the same conditions as in the legends of Fig. 2. After binding, incubation with calf thymus DNA was performed for the times indicated in the Fig. 4 in presence of different KCl concentrations.
(—■— binding   40 mM KCl, calf thymus DNA 40 mM KCl;
 --□-- binding   40 mM KCl, calf thymus DNA 120 mM KCl;
 —●— binding 120 mM KCl, calf thymus DNA 40 mM KCl;
 --○-- binding 120 mM KCl, calf thymus DNA 120 mM KCl)

FIGURE 5. Complex formation between RNA polymerase (26.4 pmol/assay) and 1.4 kb fragment (0.05 pmol/assay, spec. act. 48 µCi/nmol) in a mixture of 1.4 kb fragment + 1.4 kb x HaeIII + 1.4 kb x Mbo II in presence or absence of ppGpp (0.8 mM) for times indicated in the figure. Conditions were as indicated in the legends of Fig. 2 (—■— 40 mM KCl - ppGpp; —✕— 40 mM KCl + ppGpp).

We are, of course, interested in the effect of ppGpp. Figure 5 shows that it decreases the extent of binding considerably when the 1.4 kb fragment is used. Experiments with the smaller fragments derived from this fragment demonstrated that ppGpp us active with both promoters. Tentative data indicate that it decreases the stability of the complex formed, i.e. increases the decay rate. Whatever the mechanism, we could demonstrate a direct effect of ppGpp on the DNA-RNA polymerase interaction.

CONDLUDING REMARKS

We are at this moment in a phase of our studies were we certainly cannot offer a simple and comprehensive picture of the initiation of rRNA transcription. The proposed model is too simple and is only intended as outline. However, it can explain our results and also our findings that ppGpp doubles the binding of a restriction fragment containing the Lac promoter to RNA polymerase (data not shown). Anyhow, the results obtained appear sufficiently promising to continue these studies, i.e. by measuring the effect of the factor(s) mentioned on the binding reaction which study will join our two approaches.

ACKNOWLEDGMENTS

We are grateful to Drs. H.A. de Boer, S.C. Gilbert, M. Nomura for supplying the λmetA20 lysogen (Yamamoto & Nomura, 1976) and information about the physical map of λmetA20 DNA before publication.

REFERENCES

Block, R. (1976), Synthesis of ribosomal RNA in a partially purified extract from Escherichia coli. *in* "Control of Ribosome Synthesis" (N.O. Kjelgaard & O. Maaløe), p. 226. Munksgaard, Copenhagen.
De Boer, H.A., Gilbert, S.C. & Nomura, M. (1979), DNA sequence of promoter regions for rRNA operon rrnE and rrnA in Escherichia coli. *Cell 17*, 201.
Chamberlin, M.J. (1976), RNA polymerase - An overview. *in* "RNA Polymerase" (R. Losick & M.J. Chamberlin), p. 17. Cold Spring Harbor, New York.
Fiil, N.P., von Meyenburg, K. & Friesen, J.D. (1972), Accumulation and turnover of guanosine tetraphosphate in Escherichia coli. *J. Mol. Biol. 71*, 769.
Friesen, J., Fiil, N.P. & von Meyenburg, K. (1975), Synthesis and turnover of basal level guanosine tetraphosphate in Escherichia coli. *J. Biol. Chem. 250*, 304.
Glaser, G. & Cashel, M. (1979), In vitro transcripts from the rrnB ribosomal RNA cistron originate from two tandem promoters. *Cell 16*, 111.

Gruber, M., Hamming, J., de Lange, F.S.F., Oostra, B.A. & van Ooyen, A.J.J. (1978), Control of E. coli rRNA transcription in vivo and in vitro. *in* "Gene Expression" (B.F.C. Clark *et al.*) *43*, p. 145. Pergamon Press, Oxford and New York.

Jørgensen, P. & Fiil, N.P. (1976), Ribosomal RNA synthesis in vitro. *in* "Control of Ribosome Synthesis" (N.O. Kjeldgaard & O. Maaløe), p. 379. Munksgaard, Copenhagen.

Nomura, M., Morgan, E.A. & Jaskunas, S.R. (1977), Genetics of Bacterial Ribosomes. *Ann. Rev. Genet. 11*, 297.

Oostra, B.A., van Ooyen, A.J.J. & Gruber, M. (1977), In vitro transcription of three different ribosomal RNA cistrons of E. coli; heterogeneity of control regions. *Molec. Gen. Genet. 152*, 1.

Van Ooyen, A.J.J., de Boer, H.A., AB, G. & Gruber, M. (1975) Specific inhibition of ribosomal RNA synthesis in vitro by guanosine 3'-diphosphate, 5'-diphosphate. *Nature 254*, 530.

Van Ooyen, A.J.J., Gruber, M. & Jørgensen, P. (1976), The mechanism of action of ppGpp on rRNA synthesis in vitro. *Cell 8*, 123.

Primakoff, P. & Artz, S.W. (1979), Positive control of lac operon expression in vitro by guanosine 5'-diphosphate-3'-diphosphate. *Proc. Natl. Acad. Sci. USA 76*, 1726.

Seeburg, P.H., Nusslein, C. & Schaller, H. (1977), Interaction of RNA polymerase with promoters from bacteriophage f_d. *Eur. J. Biochem. 74*, 107.

Travers, A. (1977), Modulation of RNA polymerase specificity by ppGpp. *Molec. Gen. Genet. 147*, 225.

Yamamoto, M. & Nomura, M. (1976), Isolation of λ transducing phages carrying rRNA genes at the metA-purD region of the Escherichia coli chromosome. *FEBS Letters 72*, 256.

REGULATION OF MACROMOLECULAR SYNTHESIS
BY LOW MOLECULAR WEIGHT MEDIATORS

THE REGULATION OF RNA POLYMERASE BY GUANOSINE
5'-DIPHOSPHATE 3'-DIPHOSPHATE

P.G. Debenham
R. Buckland
A.A. Travers

MRC Laboratory of Molecular Biology
Hills Road, Cambridge, CB2 2QH
England

E. coli RNA polymerase holoenzyme can exist in a number of interconvertible structural forms each of which has a different promoter preference. These forms differ in sedimentation properties such that more rapidly sedimentating polymerase molecules initiate more efficiently at rRNA promoters in vitro. The nucleotide guanosine 5'-diphosphate 3'-diphosphate (ppGpp), whose accumulation in vivo is correlated with a substantial inhibition of rRNA synthesis, reduces the sedimentation coefficient of RNA polymerase and concomitantly reduces the affinity of the enzyme for rRNA promoters. We suggest that ppGpp alters the promoter selectively of RNA polymerase by converting the enzyme to a form which no longer synthesizes rRNA efficiently.

INTRODUCTION

The nucleotides adenosine 3'-diphosphate 5'-diphosphate (ppApp) and guanosine 3'-diphosphate (ppGpp) have both been implicated as regulators of macromolecular biosynthesis in bacteria. Much evidence suggests that ppGpp is a principal effector of the stringent response in *Escherichia coli* (Cashel, 1969; Fiil et al., 1977) while ppApp is claimed to accumulate immediately prior to sporulation in *Bacillus subtilis* (Rhaese, Grade & Dichtelmüller, 1975). *In vitro* ppApp and ppGpp alter the pattern of promoter selection by RNA polymerase in opposite ways. In particular ppApp increases and ppGpp decreases the optimal salt concentration for the initiation of rRNA synthesis *in vitro* (Van Ooyen, Gruber & Jorgensen, 1976; Travers, 1976; Travers, 1978). One interpretation of these observations is that ppApp and ppGpp increase and decrease respectively the affinity of RNA polymerase for the promoters for rRNA synthesis.

By what mechanism do ppGpp and ppApp alter the promoter preference of RNA polymerase? We show here that the nucleotides alter the sedimentation properties of the enzyme in an opposite way and argue that this change in the pattern of sedimentation is correlated with modulation of promoter preference.

MATERIALS AND METHODS

RNA polymerase was prepared by the method of Burgess & Jendrisak (1975) and contained ∼ 2 moles ω subunit and > 0.75 moles σ subunit/2 moles α subunit.

For analysis by zonal sedimentation polymerase was layered on a 12 ml (SW 40) or 4.6 ml (SW 50.1) linear 15-30% glycerol gradient containing 0.01 M Tris HCl, 0.01 m $MgCl_2$, 0.0001 M dithiothreitol, 0.0001 M EDTA, 0.2 M KCl and nucleoside tetraphosphates as necessary. The gradients were centrifuged at 5° for the indicated times and 3 drop fractions collected. RNA polymerase activity as assayed as described by Burgess & Jendrisak (1975).

rRNA synthesis from λ d_5 *ilv* DNA was assayed as previously described (Travers, 1976).

RESULTS

One target of ppGpp is RNA polymerase holoenzyme (Cashel, 1970; Travers, 1976). To test whether the functional change induced by the nucleotide reflects a detectable structural alteration polymerase holoenzyme was sedimented through a linear 15-30% glycerol gradient containing 10 µM nucleotide. In the absence of ppGpp polymerase activity sedimented as a broad peak with an avarage S value of ∼ 14 S. By constrast, with ppGpp the peak of activity was significantly sharper and shifted towards the trailing edge of the control peak (Fig. 1). This shift in apparant S value was independent of the initial enzyme concentration over the tested range of 0.2-8.0 mg/ml although lower enzyme concentrations resulted in a narrowing

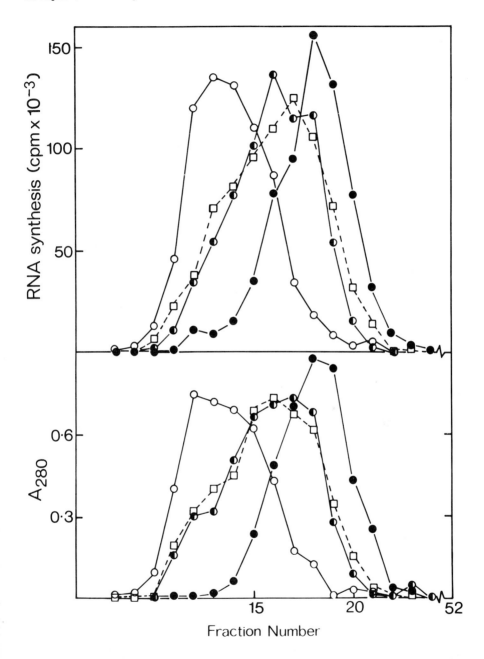

FIGURE 1. Effect of ppApp and ppGpp on sedimentation of RNA polymerase activity and protein. □--□, holoenzyme alone, ●-●, holoenzyme + 10 μM ppGpp o-o, holoenzyme + 10 μM ppApp + 10 μM ppGpp.

of the control peak e.g. Figures 1 and 2. The observed shift in S value, although small is highly reproducible and is apparent when the marker proteins, catalase and ß-galactosidase are sedimented either parallel to or mixed toghether with the RNA polymerase. By contrast in a parallel gradient 10 µM ppApp produced a shift in apparent S value of about the same magnitude in the opposite direction. However, with both nucleotides present at 10 µM the peak of enzyme activity was coincident with that of the control without nucleotide. In all cases the nucleotide(s) did not significantly alter the recovery of enzyme activity. The shifts in activity profile were always accompanied by corresponding shifts in the protein profile (Fig. 1). We conclude that the alterations in sedimentation pattern induced by ppApp and ppGpp are base specific and opposite.

One factor know to affect the sedimentation characteristics of RNA polymerase is the equilibrium between the monomeric and dimeric formes of the enzyme (Richardson, 1966). The sedimentation coefficients ($S_{20,w}$) of these forms are 15 S and 23 S respectively (Berg & Chamberlein, 1970). Dimerisation is strongly favoured by low ionic strength (Richardson, 1966). Accordingly to determine whether the effect of ppGpp is dependent of the monomer-dimer equilibrium the KCl concentration in the gradient was reduced from 0.2 M to 0.025 M, a change which increased the apparent sedimentation coefficient of polymerase to ∼18 S, a value intermediate between the monomeric and dimeric forms. Under these conditions 10 µM ppGpp no longer had any detectable influence of the sedimentation pattern (Fig. 2) even though this nucleotide can strongly influence transcription at these ionic conditions (Travers, 1978). Thus ionic conditions which favour dimerisation diminish or obscure ppGpp induced structural changes in polymerase structure.

Is the failure of ppGpp to induce any detectable structural change in RNA polymerase at low ionic strength a consequence of low ionic strength *per se* or of an increased probability of polymerase-polymerase interactions? To test this point the effect of ppGpp on rRNA synthesis from the *rrnX* cistron contained within λ d_5 *ilv* DNA was assayed over a range of polymerase concentrations. Figure 3 shows that at the highest enzyme concentration tested (∼100 µg/ml) 200 µM ppGpp was without significant effect on the overall extent of rRNA synthesis. However, as the enzyme concentration was lowered to 25 µg/ml the sensitivity of rRNA synthesis to ppGpp was substantially increased. Concomitantly the salt optimum for rRNA synthesis in the absence of the nucleotide was

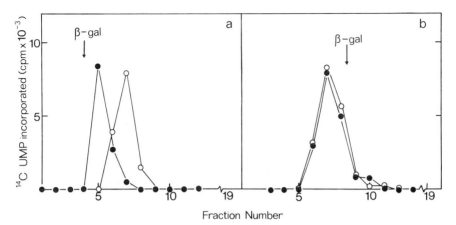

FIGURE 2. Effect of ionic strength on the ppGpp induced alterations in polymerase sedimentation. a, sedimentation at 0.2 M KCl; b, sedimentation at 0.025 M KCl o-o, no added nucleotide; •-•, with 10 µM ppGpp.

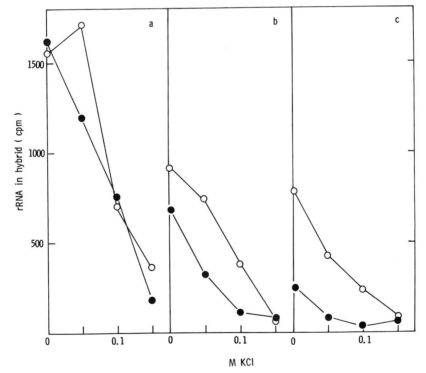

FIGURE 3. Inhibition of rRNA synthesis from λ d_5 ilv DNA by ppGpp at a, 50 µg/ml; b, 15 µg/ml; c, 12.5 µg/ml RNA polymerase. RNA synthesis was for 15' at 30°. o-o, no added nucleotide; •-•, with 100 µM ppGpp.

decreased. We conclude that under the conditions of the reaction increasing polymerase concentrations decrease the sensitivity of the enzyme to ppGpp.

DISCUSSION

We have shown that the opposing functional effects of ppApp and ppGpp on RNA polymerase are paralleled by structural changes apparent as alterations of the sedimentation characteristics of the enzyme. What is the nature of these structural changes? One possibility is an alteration in the subunit composition of the enzyme. A mutation, alt-1, affecting the σ subunit of RNA polymerase alters both the transcriptional specificity and the sedimentation profile of the enzyme in a manner parallel that of 10 µM ppGpp (Travers et al., 1978). This mutant enzyme has the same subunit composition $\alpha_2\beta\beta'\sigma$, as the wild type polymerase thus suggesting that changes in polymerase structure similar to those induced by the nucleotides can occur in the absence of alterations of polypeptide composition. Other possibilities include a perturbation of the monomer-dimer equilibrium or the induction of conformational changes. The data show that when the possibility of enzyme-enzyme interactions is increased by lowering the KCl concentration the effect of ppGpp is no longer apparent. Similarly the effect of ppGpp on polymerase function is negated by increasing enzyme concentration. Such an increase also increases the optimum salt concentration for rRNA synthesis and in this respect mimics the effect of ppApp. The effect of enzyme concentration on the efficacity of the nucleotides as effectors is more apparent when observed in the analytical ultracentrifuges. At $\sim$ 0.5 mg/ml polymerase ppGpp decreases the $S_{20,w}$ of the enzyme by 0.8 $\pm$ 0.2 S while ppApp has no significant effect. Coversely at $\sim$ 1.0 mg/ml polymerase ppGpp has no significant effect and ppApp increases the $S_{20,w}$ by 0.4 $\pm$ 0.1 S (Butler, Buckland & Travers, unpublished observations). Under all conditions the enzyme sediments in these experiments primarily as the monomeric form. Further, other experiments suggest that RNA polymerase holoenzmye can exhibit functional heterogeneity such that the subpopulation of enzyme molecules which sediments more rapidly than avarage initiates efficiently at rRNA promoters while the subpopulation which sediments more slowly transcribes rRNA rather poorly (Travers, Buckland & Debenham, submitted). There is thus a correlation between the sedimentation coefficient of an RNA polymerase molecule and its template preference, a correlation which is maintained in the presence of the nucleoside polyphosphates ppApp and ppGpp.

Our interpretation of these observations is that RNA polymerase holoenzyme can exist in a number of different states, each of which has a different promoter preference. In the absence of added effectors the position of the equilibrium between the different states is determined by the probability of interactions between molecules of polymerase holoenzyme. This probability is increased at low ionic strength or high temperature. On this model ppGpp acts by converting the enzyme to a form with a relatively lower self-association constant. However, the data suggest that the nucleotide can only do so when the equilibrium already favours this form. Similarly, ppApp converts the enzyme to a structural form with a higher self-association constant. Since even in the presence of ppApp the polymerase sediments predominately as the monomeric form we propose that the observed changes in sedimentation position reflect, at least in part, conformational differences.

ACKNOWLEDGEMENTS

P.G.D. thanks the Medical Research Council for a scholarship.

REFERENCES

Berg, D. & Chamberlin, M. (1970), Physical studies on ribonucleic acid polymerase from *Escherichia coli* B. *Biochemistry* 9, 5055.
Burgess, R.R. & Jendrisak, J.J. (1975), A procedure for the rapid large-scale purification of *Escherichia coli* DNA dependent RNA polymerase involving Polymin P precipitation and DNA cellulose chromatography. *Biochemistry 14*, 4634.
Cashel, M. (1969), The control of ribonucleic acid synthesis in *Escherichia coli*. IV. Relevance of unusual phosphorylated compounds from amino acid starved strains. *J. Biol. Chem.* 244, 3133.
Cashel, M. (1970), Inhibition of RNA polymerase of ppGpp, a nucleotide accumulated during the stringent response to amino acid starvation in *E. coli*. *Cold Spring Harbor Symp. Quant. Biol. 35*, 407.
Fiil, N., Willumsen, B.M., Friesen, J.D. & von Meyenburg, K. (1977), Interaction of alleles of the *relA, relC* and *spoT* genes in *Escherichia coli*: analysis of the interconversion of

GTP, ppGpp and pppGpp. *Mol. Gen. Genet. 150*, 87.
Rhaese, H.J., Grade, R. & Dichtelmüller, H. (1976), Studies on the control of development. Correlation of initiation of differentiation with the synthesis of highly phosphorylated nucleotides in *Bacillus subtilis*. *Eur. J. Biochem. 64*, 205.
Richardon, J.P. (1966), Some physical properties of RNA polymerase. *Proc. Nat. Acad. Sci. USA 55*, 1616.
Travers, A. (1976), Modulation of RNA polymerase specificity by ppGpp. *Mol. Gen. Genet. 147*, 225.
Travers, A. (1978), ppApp alters transcription selectively of *Escherichia coli* RNA polymerase. *FEBS Lett. 94*, 345.
Travers, A.A., Buckland, R., Goman, M., Le Grice, S.F.J. & Scaife, J.G. (1978), A mutation affecting the σ subunit of RNA polymerase changes transcriptional selectively. *Nature 273*, 354.
Van Ooyen, A.J.J., Gruber, M. & Jorgensen, P. (1976), The mechanism of action of ppGpp on rRNA synthesis *in vitro*. *Cell 8*, 123.

REGULATION OF MACROMOLECULAR SYNTHESIS BY LOW MOLECULAR WEIGHT MEDIATORS

SYNTHESIS AND DEGRADATION OF THE PLEIOTROPIC EFFECTOR, GUANOSINE 3',5'-BIS(DIPHOSPHATE) IN BACTERIA

Dietmar Richter

Institut für Physiologische Chemie
Abteilung Zellbiochemie
Universität Hamburg
Hamburg, W. Germany

Metabolism of guanosine 3',5'-guanosine(diphosphate) [ppGpp] has been studied in Escherichia coli, Bacillus stearothermophilus, Bacillus subtilis and Bacillus brevis, and is best summarized as follows: 1. While the mechanism of pyrophosphate transfer from ATP to GDP was identical in the strains assayed, there were differences in the activation of the stringent factor (SF) catalyzing this reaction. In E. coli and B. subtilis, SF was associated with ribosomes and activated by a ribosome-mRNA-uncharged tRNA complex. In B. stearothermophilus and B. brevis SF was preferentially present in the high-speed supernatant fraction and functioned independently of ribosomes. Formation of ppGpp was specific in that neither ATP as pyrophosphate donor nor GDP (or GTP) as acceptor could be replaced by other nucleotides. 2. Degradation of ppGpp was catalyzed by the spoT gene product, a guanosine 3',5'-bis(diphosphate) 3'-pyrophosphohydrolase (ppGppase). Double labeling experiments revealed that ppGppase specifically degraded (p)ppGpp from the 3'-position yielding (p)ppG and P~Pi. Like SF the ppGppase was associated with ribosomes, released by 1 M K-acetate extraction and insoluble in buffers of low ionic strength. 3. Studies on ppGpp metabolism in relA and spoT mutant strains indicated that ppGpp was degraded to ppGp, pGpp, and pGp. In Escherichia coli strain AB259 (relA, spoT) ppGp and pGpp were re-utilized for synthesis of ppGpp.

INTRODUCTION

In auxotrophic strains of *Escherichia coli*, deprivation of an essential amino acid and consequently lack of the appropriate amino acylated tRNA initiates a regulatory process termed stringent response that is governed by the *relA* gene and defined by the curtailment of synthesis of rRNA, tRNA and some mRNA species as well as other intermediary metabolic reactions (Gallant & Lazzarini, 1976). Haseltine et al. (1973) identified the product of the *relA* gene as a ribosome-bound enzyme termed stringent factor that transfers pyrophosphate from ATP to GDP or GTP yielding the unusual nucleotides guanosine 3',5'-bis(diphosphate) [ppGpp] and guanosine 3'-diphosphate, 5'-triphosphate [pppGpp], respectively. Both, ppGpp and pppGpp accumulated when $relA^+$ strains were starved for an essential amino acid. This and other experiments implicated that (p)ppGpp functions as pleiotropic effector controlling both gene expression and certain metabolic reactions. Experiments with *spoT* mutant strains of *E. coli* (which are characterized by a reduced rate of ppGpp decay and a lack of pppGpp formation) suggested that ppGpp rather than pppGpp functions as a signal in this regulation process (Gallant & Lazzarini, 1976; Nomura et al., 1977).

In order to permit rapid fluctuation of the intracellular concentration of ppGpp in response to environmental changes one should expect that the signal is metabolically labile (Tomkins, 1975). The present communication summarizes the enzymes involved in the metabolism of ppGpp and compares the reactions which lead to the synthesis or decay of ppGpp in various bacterial species.

MATERIALS AND METHODS

Conditions for studying synthesis and degradation of ppGpp were according to published procedures (Heinemeyer & Richter, 1978b; Richter & Geis, 1978; Fehr et al., 1979).

RESULTS

Synthesis of ppGpp

While the mechanism of synthesis of ppGpp - donation of pyrophosphate from ATP to GTP (or GDP) - proceeds identically in all bacterial systems assayed so far, the enzymes involved differ in their physico-chemical properties as well as in their dependence on co-factors. In *E. coli* and *Bacillus subtilis* stringent factors were activated by a ribosome complex (Richter et al., 1979), whereas in *Bacillus stearothermophilus* this complex was not required (Fehr et al., 1979).

Ribosome-Dependent Synthesis of ppGpp

In general, stringent factor extracted from ribosomes of *E. coli* or *B. subtilis* catalyzed the synthesis of ppGpp only when activated by codon-specific binding of uncharged tRNA to ribosomes (Table I). Detailed studies with the *E. coli* system showed that tRNA cycles on and off the ribosome with each ATP hydrolyzed, while the stringent factor remained ribosome-bound. Release of tRNA was dependent upon hydrolysis of ATP suggesting that binding and release of tRNA was driven by a pulsating ribosomal contraction mechanism (Richter, 1976).

TABLE I. Ribosome-Dependent and Independent Synthesis of (p)ppGpp

Stringent factor isolated from	Components added to the assay mixture		% of (p)ppGpp synthesized
	70S complex	tRNA^{Phe}	
E. coli	+	+	70.1
	+	−	0.5
	−	+	0.7
	−	−	0.9
B. subtilis	−	−	0.7
	+	+	49.3
	+	−	2.0
	−	+	0.9
B. stearothermo-philus	+	−	65.1
	+	+	60.5
	−	+	59.3
	−	−	61.4
B. brevis	−	−	41.5
	+	+	40.0
	+	−	43.2
	−	+	41.8

Ribosome-Independent Synthesis of ppGpp

In *Bacillus stearothermophilus* a different stringent factor has been found which functions independently from the ribosome-mRNA-tRNA complex (Fehr et al., 1979). In contrast to *E. coli* or *B. subtilis* this enzyme is not associated with ribosomes but present in the high-speed supernatant fraction. So far the *in vivo* mechanism for triggering the accumulation of ppGpp in this strain has not yet been elucidated.

Degradation of ppGpp

The mechanism of degradation of ppGpp proceeds at least by two different pathways. A specific ppGpp degradation is catalyzed by the product of the *spoT* gene, a guanosine 3',5'-bis(diphosphate) 3'-pyrophosphohydrolase (ppGppase), which yields ppG and pyrophosphate (Heinemeyer & Richter, 1978 a, b). Another ppGpp degrading enzyme has been found in $spoT^+$ and *spoT* mutant strains which hydrolyzes ppGpp from the 3' and 5' end yielding ppGp, pGpp and pGp as products (Heinemeyer et al., 1978). This enzyme may be regarded as phosphatase and identical with the known acid phosphatase of *E. coli*.

Hydrolysis of ppGpp by a Specific ppGppase

In vivo studies have shown that ppGpp is rapidly degraded when amino acid deprived *E. coli* cells are supplemented with the lacking amino acid suggesting an effective degradation mechanism (for rev. see Gallant & Lazzarini, 1976; Nomura et al., 1977). Furthermore, genetic studies revealed that decay of ppGpp is controlled by the *spoT* gene. *spoT* mutant strains do not only degrade ppGpp at a lower rate than wild-type strains, they also lack pppGpp suggesting that degradation of ppGpp is preceded by its phosphorylation to pppGpp (Gallant & Lazzarini, 1976). Using an *in vitro* system from *E. coli* the product of the *spoT* gene has been identified as a ribosome-associated ppGppase that releases pyrophosphate from the 3'-position of ppGpp (or pppGpp) yielding ppG (or pppG)(Heinemeyer & Richter, 1977; Heinemeyer & Richter, 1978 a, b; Heinemeyer et al., 1978). Interestingly, ppGppase was inhibited by uncharged or charged tRNA regardless of whether or not ribosomes were present (Table II). The inhibitory effect observed with charged tRNA could be due to uncharged tRNA still present in the Phe-tRNA preparation. The physiological significance of the tRNA inhibition is not yet clearly understood, however it is intriguing to speculate

Table II. Effect of Ribosomes and tRNA on the Decay of ppGpp by ppGppase

Components added	Experiment						
	1	2	3	4	5	6	7
ppGppase$_{E.coli}$	+	+	+	+	+	−	−
ppGppase$_{B.stear.}$	−	−	−	−	−	+	+
70S-mRNA complex	−	+	+	+	−	−	−
tRNA^{Phe}	−	−	+	−	+	−	+
Phe-tRNA^{Phe}	−	−	−	+	−	−	−
ppGpp degraded in %	49.0	48.1	23.2	34.2	29.1	30.5	18.9

The assay composition was the same as reported (Richter et al., 1979).

that the same "co-factor" which triggers the production of ppGpp suppresses the decay of ppGpp.

The strictly manganese-dependent ppGppase has been found in the ribosomal fraction of isogenic pairs of $spoT^+$ but not of $spoT$ strains (Heinemeyer & Richter, 1978b). A ppGppase with identical specificities as the E. coli enzyme has been isolated from Bacillus brevis, Bacillus subtilis and Bacillus stearothermophilus (Table II). In all of these strains, the ppGppase has been found in the ribosomal fraction where it can be released by K-acetate extraction. The tentative molecular weight of the enzyme isolated from the strains mentioned above has been estimated as 65,000-70,000 (Richter & Geis, 1978; Richter et al., 1979). Dissociation of ribosomes at low ionic conditions and subsequent sucrose density gradient centrifugation showed that some ppGppase activity was released into the supernatant fractions; significant activity remained with the 50S and to a lesser extent with the 30S ribosomal subunit (E. coli, B. stearothermophilus). Similar experiments with B. subtilis revealed a different distribution of ppGppase activity co-migrating with the 30S subunit. As was recently shown ppGppase tends to form aggregates in buffer of low ionic strength which may reach a size sufficiently large enough to co-migrate with either of the two ribosomal subunits (Richter & Geis, 1978).

Hydrolysis of ppGpp by an Unspecific Phosphatase

Although *spoT* mutant strains show little or no ppGppase activity they utilize ppGpp by use of a phosphatase which splits off phosphates from the 3' and 5'end yielding ppGp, pGpp and pGp (Heinemeyer et al., 1978). This enzyme, however, is not restricted to *spoT*[+] strains. Whether this phosphatase has any physiological significance in particular in *spoT* strains has not yet been resolved.

Fate of ppGpp in a relA, spoT Mutant Strain

In general the mechanism for synthesis and degradation of ppGpp seems to be identical in all bacteria which possess this signal device. Whether both types of stringent factor, ribosome-dependent and independent are present in *E. coli* remains to be elucidated. A ribosome-independent stringent factor could be responsible for ppGpp production in *relA* mutant strains known to accumulate ppGpp upon nitrogen and carbon source shift-down but not upon amino acid starvation (Gallant & Lazzarini, 1976). However, attempts to identify a stringent factor in *relA* strains have been unsuccessful so far. Data shown below indicate that, in *relA* strains, synthesis of ppGpp may proceed by quite a different mechanism which is closely coupled to the ppGpp degrading reaction. Using an *relA*, *spoT* mutant strain AB259 we found that ppGpp is rapidly converted into ppGp and pGpp and to a lesser extent into pGp. In some experiments greater than 90% of the labeled ppGpp was degraded within one minute. Surprisingly upon longer incubation ppGpp reappeared and after 10 minutes of incubation approximately 60% of the original ppGpp could be regained. At the same time ppGp, pGpp and pGp decreased (Fig.1). The decay of ppGpp was also observed at pH 2.5 (Fig. 1) indicating the presence of an acid phosphatase. Whether ppGp and pGpp serve as substrates in the "back-reaction" is similarly unsolved as to where the high-energy bond to form ppGpp is deprived from. Recently, in *E. coli* (Pao & Gallant, 1979) and *B. subtilis* (Nishino et al., 1979) a "third magic spot" has been reported that accumulates during various down-shift conditions. Magic spot 3 has been identified as a guanosine nucleotide with the isomeric structure ppGp or pGpp. It is intriguing to speculate that these two compounds are intermediary metabolic products of a ppGpp turnover reaction as observed in our *in vitro* assay system. Degradation and resynthesis of ppGpp has been followed for up to 60 minutes of incubation. During this period of time about every 10 minutes there is an optimum of ppGpp synthesis followed by

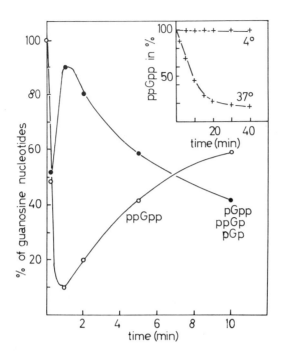

FIGURE 1. Decay of ppGpp in spoT, relA strain AB259. S-100 fraction (Heinemeyer & Richter, 1978b) was assayed; at the time indicated 8 µl aliquots were withdrawn, stopped with 2.5 µl of 8.8% formic acid and chromatographed in 1.5 M KH_2PO_4, pH 3.4. The inserted figure shows the decay of ppGpp by a phosphatase present in the S-100 fraction. The pH of the assay mixture was adjusted to 2.5 with formic acid. Incubation was carried out at 4°C or 37°C.

an optimum of ppGpp decay. Apparently this synthesis-degradation process oscillates although the amplitude of the waves became reduced with longer incubation time. Interestingly in the presence of ATP this oscillation process could be suppressed. ATP did not block decay of ppGpp per se, however, as indicated in Figure 2, new, yet unidentified degradation products appeared which may no longer be suitable for reutilization in ppGpp formation.

FIGURE 2. Two-dimensional polyethyleneimine thin layer chromatography of the degradation products of ppGpp (Heinemeyer & Richter, 1978b). Decay of ppGpp was studied in the presence (right panel) or absence (left panel) of 4 mM ATP. Spots 1 and 2 were identified as ppGp and pGpp, respectively; spots 3 and 4 have not been identified yet.

DISCUSSION

The metabolic pathway of the ppGpp cycle can be summarized as follows:

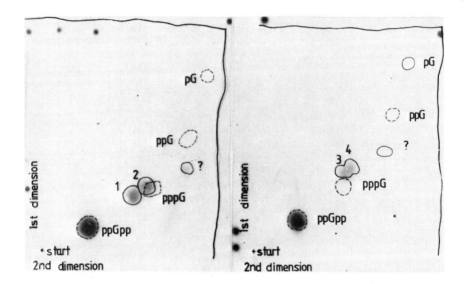

The net reaction requires two energy-rich bonds, pyrophosphate may either be cleaved to inorganic phosphate or exchanged with ATP. As indicated with more purified ppGppase preparations the ATP $\rightleftharpoons$ P$\sim$P$_i$ (Heinemeyer & Richter, 1978a; Heinemeyer & Richter, 1978b) exchange reaction is not an essential step in the ppGpp cycle. Stringent factor is activated when uncharged tRNA is bound to ribosomes. In contrast to the *E. coli* and *B. subtilis* system stringent factor from *B. stearothermophilus* or *B. brevis* does not depend on the ribosomal activation process. A similar ribosome-independent mechanism of ppGpp synthesis may exist in *relA* mutant strains known to response to nitrogen and to carbon shift-down but not to amino acid starvation. So far attempts to isolate either type of stringent factor from *relA* strains have been unsuccessful.

In general, the mechanism for specific decay of ppGpp by the *spoT* gene product seems to be identical in all bacteria assayed. In addition ppGpp can be degraded by unspecific phosphatases that hydrolize ppGpp to pGpp, ppGp, and pGp (Heinemeyer et al., 1978; Tetu et al., 1979). In *relA*, *spoT*

$$ppGpp \xrightarrow[\text{3'and/or 5'specific ?}]{\text{phosphatase}} \begin{matrix} \nearrow^{P_i} ppGp \\ pGpp \end{matrix} \xrightarrow{P_i} pGp$$

mutant strains ppGp and pGpp seemed to be reutilized for the formation of ppGpp. Whether this reaction reflects the physiological significance of ppGpp metabolism in these mutant strains remains to be studied.

ACKNOWLEDGMENTS

S. Fehr, M. Geis, F. Godt, E.A. Heinemeyer, and R. Harder were involved in the various projects reported and discussed here. The author thanks Dr. J.A. Bilello for help in editing the manuscript and Deutsche Forschungsgemeinschaft for financial support.

REFERENCES

Fehr, S., Godt, F., Isono, K. & Richter, D. (1979), A ribosome-independent stringent factor form *Bacillus stearothermophilus* and a low molecular weight substance inhibitory to its activity. *FEBS Lett.* 97, 91.
Gallant, J. & Lazzarini, R.A. (1976), The regulation of ribosomal RNA synthesis and degradation in bacteria. *In*

Protein Synthesis: A Series of Advances (ed. E. McConkey), 2, 309.

Haseltine, W.A., Block, R., Gilbert, W. & Weber, K. (1972), MS I and MS II made on ribosomes in idling step of protein synthesis. *Nature 238*, 381.

Heinemeyer, E.A. & Richter, D. (1977), *In vitro* degradation of guanosine tetraphosphate (ppGpp) by an enzyme associated with the ribosomal fraction from *Escherichia coli. FEBS Lett. 84*, 357.

Heinemeyer, E.A. & Richter, D. (1978a), Mechanism of the *in vitro* breakdown of guanosine 5'-diphosphate 3'-diphosphate in *Escherichia coli. Proc. Natl. Acad. Sci. USA 75*, 4180.

Heinemeyer, E.A. & Richter, D. (1978b), Characterization of the guanosine 5'-triphosphate 3'-diphosphate and guanosine 5'-diphosphate 3'-diphosphate degradation reaction catalyzed by a specific pyrophosphorylase from *Escherichia coli. Biochemistry 17*, 5368.

Heinemeyer, E.A., Geis, M. & Richter, D. (1978), Degradation of guanosine 3'-diphosphate 5'-diphosphate *in vitro* by the *spoT* gene product of *Escherichia coli. Eur. J. Biochem. 89*, 125.

Nishino, T., Gallant, J., Shalit, P., Palmer, L. & Wehr, T. (1979), Regulatory nucleotides involved in the rel function of *Bacillus subtilis. J. Bacteriol.* submitted.

Nomura, M., Morgan, E.A. & Jaskunas, S.R. (1977), Genetics of bacterial ribosomes. *Ann. Rev. Genet. 11*, 297.

Pao, C.C. & Gallant, J. (1979), A new nucleotide involved in the stringent response in *Escherichia coli. J. Biol. Chem. 254*, 688.

Richter, D. (1976), Stringent factor from *Escherichia coli* directs ribosomal binding and release of uncharged tRNA. *Proc. Natl. Acad. Sci. USA 73*, 707.

Richter, D. & Geis, M. (1978), Isolation of a pyrophosphorylase from *Bacillus subtilis* and *Bacillus stearothermophilus* that specifically degrades guanosine 3'-diphosphate 5'-diphosphate. *FEBS Lett. 96*, 247.

Richter, D., Fehr, S. & Harder, R. (1979), The guanosine 3,5'-bis(diphosphate) [ppGpp] cycle. Comparison of synthesis and degradation of guanosine 3',5'-bis(diphosphate) in various bacterial systems. *Eur. J. Biochem.* in press.

Tetu, C., Dassa, E. & Boquet, P.-L. (1979), Unusual pattern of nucleoside polyphosphate hydrolysis by the acid phosphatase (optimum pH = 2.5) of *Escherichia coli. Biochem. Biophys. Res. Commun. 87*, 314.

Tomkins, G.M. (1975), The metabolic code. *Science 189*, 760.

**REGULATION OF MACROMOLECULAR SYNTHESIS
BY LOW MOLECULAR WEIGHT MEDIATORS**

BIOSYNTHESIS OF GUANOSINE TETRAPHOSPHATE
IN *BACILLUS BREVIS*

Jose Sy[1]

The Rockefeller University
New York, New York

Two distinct ppGpp synthetic enzymes, a soluble and a ribosome-bound synthetase, have been identified in Bacillus brevis. Both catalyze the transfer of the pyrophosphoryl group from ATP to the 3'OH of GTP or GDP. The ribosome-bound enzyme (mol. wt. 76,000) requires a ribosome-mRNA-uncharged-tRNA complex for activity. It is inhibited by thiostrepton and tetracycline and is therefore similar to the relA gene product, the stringent factor, of Escherichia coli. The B. brevis stringent factor, in contrast to the E. coli factor, is not stimulated by 20% methanol. The soluble enzyme (mol. wt. 55,000) is neither activated nor inhibited by the ribosomal complex. Both enzymes are present in about equal amounts in log phase cells. However, the ribosome-bound stringent factor activity is drastically reduced in stationary cultures where cells are sporulating.

INTRODUCTION

Guanosine 5'-diphosphate,3'-diphosphate (ppGpp), the putative effector of stringent response, is accumulated in *E. coli* under two general conditions: starvation for a required amino acid and starvation for a carbon sourse (Gallant & Lazzarini, 1976; Nierlich, 1978). Accumulation during amino acid starv-

---

[1] This work was supported by grants from the National Science Foundation (PCM 77-17683 to J.S.) and the National Institutes of Health (GM-13971 to F.Lipmann).

ation is due mainly to increased synthesis, i.e., the activation of ATP:GTP(GDP) pyrophosphotransferase (stringent factor), the product of the *relA* gene (Pedersen & Kjeldgaard, 1977). Accumulation during carbon source deprivation is principally due to reduced degradation, i.e. the inhibition of ppGpp 3'-pyrophosphohydrolase, the product of the *spoT* gene (An et al., 1979). The *E. coli* stringent factor[2] requires a complex of ribosome-mRNA and a message-coded, uncharged tRNA for activity. In the absence of the ribosomal complex, stringent factor activity may be stimulated by the addition of 20% methanol (Sy et al., 1973). Starvation for glucose in *relA*$^-$ cells leads to the accumulation of ppGpp, although these cells do not contain any detectable stringent factor activity nor do they accumulate ppGpp in response to amino acid starvation. A recent study on nonsense and insertion mutations in the *relA* gene further indicates the necessity for an alternate *relA* independent, ppGpp synthetic pathway in *E.coli* (Friesen et al., 1978). However, in vitro studies have so far failed to detect any ppGpp synthesis in *relA*$^-$ cell-free extracts.

We have previously reported on the findings of a ribosome-independent ppGpp synthetic enzyme in *B. brevis* (Sy, 1976; Sy & Akers, 1976). However, no stringent factor activity was found in the *B. brevis* ribosomal fraction (Sy et al., 1974). I wish to report here the finding of a *B. brevis* stringent factor which indicates that this organism contains two separate enzymes that synthesize ppGpp.

MATERIALS AND METHODS

Beef extract and peptone were obtained from Difco, and [α-^{32}P]GTP from Amersham.

Assay for ppGpp Synthesis

A high Mg^{2+} concentration was used for the assay of ribosome-dependent synthesis of ppGpp. Reaction mixtures were incubated at 30°C and contained in 50 μl: 50 mM Tris-OAc, pH 8.1; 20 mM Mg(OAc)$_2$; 4 mM dithiothreitol; 0.2 mM [α-^{32}P]GTP (7-35 Ci/mol); 4 mM ATP; 100 μg/ml of poly(A,U,G); 100 μg/ml of tRNA *(E. coli)*; and the appropriate ribosomal fractions.

---

[2] Stringent factor will represent the ribosome-dependent ATP:GTP(GDP) pyrophosphotransferase, and ppGpp-synthetase the ribosome-independent enzyme.

After 15, 30, and 45 min of incubation, 10 μl samples were removed and the reactions stopped by the addition of 10 μl of 1 M HCOOH. Following centrifugation, the supernates were chromatographed on polyethyleneimine-cellulose thin layer sheets, and ppGpp was determined as previously described (Sy et al., 1973). For the assay of the ribosome independent synthesis of ppGpp, the reaction mixtures contained 10 mM Mg^{2+} instead of 20 mM and no poly(A,U,G) or tRNA. Other procedures were as described above.

Preparation of B. brevis Extracts

Ten-liter fermentors containing 10 g of beef extract and 10 g of peptone per liter were inoculated with 25 ml of an overnight culture of B. brevis (ATCC 8185). The cultures were incubated at 37°C, stirred at 450 rpm, and aerated at 6 liters/min. Absorbances at 650 mμ were determined at intervals. At the various time points, 1-5 liters of culture were withdrawn, chilled rapidly by the addition of ice, and centrifuged at 10,000 rpm for 30 min. The cells were washed with buffer A (40 mM Tris-OAc, pH 8.1, 14 mM $Mg(OAc)_2$, 60 mM KOAc, 1 mM dithiothreitol) and resuspended with an equal volume of buffer A containing 2.5 μg/ml DNAse (RNAse-free, Worthington). The cells were then disrupted by a French press at 15,000 psi, and the extracts were centrifuged at 30,000 x g for 30 min. The S-30 supernates were further centrifuged at 40,000 rpm for 4 hr in a spinco 40 rotor. The pellets (crude ribosomes) were re-suspended in a minimal volume of buffer A and assayed for stringent factor activity. Ammonium sulfate was added to the S-100 supernates until 45% saturation. The precipitates containing all the ppGpp-synthetase activity were redissolved in buffer A and dialyzed against 1 liter of Buffer A.

Preparation of Stringent Factors, ppGpp-Synthetase, and Washed Ribosomes

E. coli stringent factor and ammonium chloride washed ribosomes were obtained as previously described (Sy et al., 1973). Highly purified ppGpp-synthetase from B. brevis was also obtained as described (Sy & Akers, 1976). B. brevis stringent factor and washed ribosomes were produced by the following procedure. Crude ribosomes from late log-phase cells were extracted twice with 1.0 M NH_4Cl in buffer A. The combined supernates were saturated with ammonium sulfate to 45%, and the protein precipitate was dissolved in a minimal amount of buffer A. The protein solution was dialyzed against 1 liter of buffer A for 5 hr. The precipitate that formed, which had 80-90% stringent factor activity, was centrifuged out and redissolved in buffer A containing 0.5 M NH_4Cl. The twice-washed ribosomes were resuspended in buffer A.

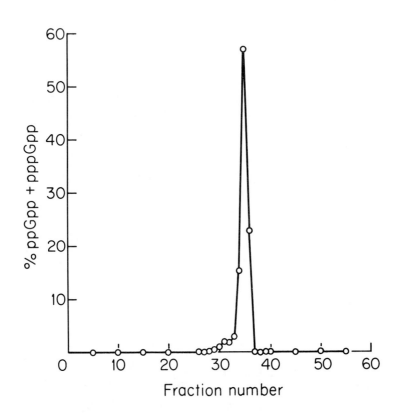

FIGURE 1. Gel electrophoresis of ppGpp-synthetase. The 0-45% ammonium sulfate fraction of B. brevis S-100 (70 µg) was electrophoresed in 10 cm of 7.5% polyacrylamide gel as described by Davis (1964). After electrophoresis, the gel was frozen and sliced into 1-mm segments. Each slice was then incubated at 30°C for 3 hr with 50 µl of 50 mM Tris-OAc, pH 8, 4 mM dithiothreitol, 2 mM ATP, 0.2 mM [α-^{32}P]GTP (7.5 Ci/mol), 2.2 mM $Mg(OAc)_2$, 0.4 mg/ml of bovine serum albumin. After incubation, 5 µl were withdrawn and applied to polyethyleneimine-cellulose sheets and the ppGpp + pppGpp formed was assayed as described (Sy et al., 1973).

RESULTS

The ribosome-independent ppGpp-synthetase can be readily demonstrated in crude extracts as well as in polyacrylamide electrophoresis of the enzyme preparations (Fig. 1). The enzyme (mol. wt. 55,000) catalyzes a pyrophosphoryl transfer from ATP to the 3'OH of GDP or GTP and is not inhibited by thiostrepton or tetracycline (Sy, 1976; Sy & Akers, 1976). During these studies we noticed that *B. brevis* ribosomal fractions carried little or no stringent factor-like activity. Since the brevis cells used in these studies were all from the stationary phase where cells are sporulating (Lee et al., 1975) we have re-examined the question of the presence of stringent factor using log phase cells instead. When crude ribosomal fractions were isolated from log phase cells they were found to be highly active in ppGpp synthesis. Figure 2

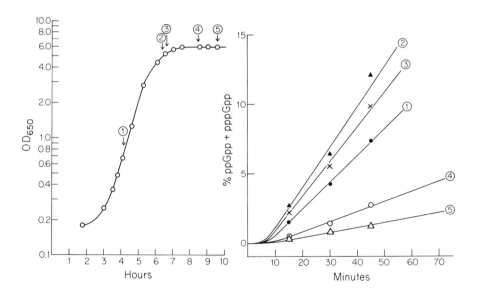

FIGURE 2. Specific activity of stringent factor in various growth phases. B. brevis cell extracts were obtained from the various time points of the growth curve (left panel). The ribosomal fractions were prepared by centrifugation for 4 hr at 40,000 rpm in a Spinco 40 rotor and assayed for stringent factor activity (right panel) as described in Materials and Methods. Numbers in the right panel correspond to the time points at which cells were harvested. The amount of ribosomes used in the assays were: .1, 35 µg; 2, 39 µg; 3, 33 µg; 4, 39 µg; and 5, 40 µg.

shows that the specific activity of ppGpp synthesis in the ribosomal fractions depended on when the cells were harvested. With stationary phase cultures, very little synthetic activ-

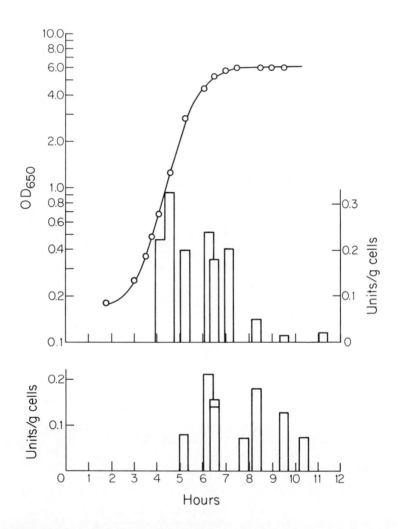

FIGURE 3. Total activity of stringent factor and ppGpp-synthetase at various growth phases. The specific activity of stringent factor (upper panel) at various time points of the growth curve was determined as described in Fig.2. The 0-45% ammonium sulfate fractions from S-100 were prepared and the ppGpp-synthetase activity (lower panel) was assayed as described in Materials and Methods. One unit of activity is defined as the synthesis of 1 μm of ppGpp per min.

ity was detected. This low activity was not due to inhibition of stringent factor activity since addition of E. coli stringent factor to the ribosomal fractions led to a rapid synthesis of ppGpp. In contrast to the stationary phase cells, the ribosomal fractions from log phase cells contained 3- to 10-fold more ppGpp synthetic activity.

Figure 3 shows that the total stringent factor activity per gram of cell dropped 6- to 10-fold in going from log to stationary growth. In contrast, the total ppGpp-synthetase activity was relatively constant although the data points were more scattered due to the presence of interfering activity in the ammonium sulfate fractions. There were approximately similar amounts of stringent factor and ppGpp-synthetase activity in the log phase cells, while ppGpp-synthetase predominated in the stationary cells.

To ascertain that the ppGpp synthetic activity found in the ribosome fraction was like the stringent factor, the enzyme was partially purified as described in Materials and Methods. Table I shows that the B. brevis stringent factor requires ribosome, mRNA, and uncharged tRNA for maximal activity. These requirements are identical to those of the E. coli stringent factor. Furthermore, both enzymes are inhibited by the same ribosomal inhibitors (Table II). Thiostrepton and tetracycline, both inhibitors of tRNA binding, were potent inhibitors, while micrococcin, an inhibitor of translocation, was not.

TABLE I. Requirements of B. brevis Stringent Factor

Addition	% ppGpp + pppGpp
Ribosomes	0
Stringent factor	4.7
Ribosomes + stringent factor	13.0
Ribosomes + stringent factor + poly(A,U,G) + tRNA	49.9

Reactions in 25 µl contained 50 mM Tris-OAc, pH 8.1, 4 mM dithiothreitol, 20 mM Mg(OAc)$_2$, 8 mM phosphoenolpyruvate, 3 µg of pyruvate kinase, 4 mM ATP, 0.2 mM [α-^{32}P]GTP (23 Ci/mol), and, if added, 10 µg of poly(A,U,G), 10 µg of tRNA, 89 µg of ammonium chloride-washed B. brevis ribosomes, and 4 µg of B. brevis stringent factor. Incubation was at 30°C for 40 min, and the ppGpp + pppGpp formed was determined as described in Materials and Methods.

TABLE II. Effects of Antibiotics on Stringent Factor

System	Additions	% GTP converted
B. brevis	None	62.5
	Thiostrepton (4×10^{-5} M)	1.6
	Tetracycline (1 mM)	1.9
	Micrococcin (8×10^{-6} M)	59.1
E. coli	None	35.2
	Thiostrepton (4×10^{-5} M)	0
	Tetracycline (1 mM)	0
	Micrococcin (8×10^{-6} M)	36.1

Reaction mixtures in 25 µl containing 50 mM Tris-OAc, pH 8.1, 4 mM dithiothreitol, 20 mM Mg(OAc)$_2$, 8 mM phosphoenolpyruvate, 3 µg of pyruvate kinase, 4 mM ATP, 0.2 mM [α-^{32}P]GTP (23 Ci/mol), 6.6 µg of poly(A,U,G), 6.6 µg of tRNA, and the indicated antibiotics, were incubated with either 89 µg of B. brevis-washed ribosomes + 4 µg of B. brevis stringent factor, or with 170 µg of E. coli-washed ribosomes + 3 µg of E. coli stringent factor. Incubations were at 30°c for 40 min, and the reactions were stopped by the addition of HCOOH. The ppGpp + pppGpp formed was assayed as described in Materials and Methods.

TABLE III. Ribosome Specificity of Stringent Factors

	Additions	% GTP converted
Ribosomes	Stringent factor	
B. brevis	–	0
–	B. brevis	0
B. brevis	B. brevis	62.5
E. coli	B. brevis	12.8
E. coli	–	0
–	E. coli	0
E. coli	E. coli	35.2
B. brevis	E. coli	41.9

Reaction conditions were similar to those in Table II.

The *B. brevis* stringent factor differs from the *E. coli* enzyme in its ribosome preference, in that it prefers the homologous ribosome over the coli ribosome. The *E. coli* stringent factor, on the other hand, utilizes both kinds of ribosome equally well (Table III). As we have previously reported, the *E. coli* stringent factor, in the absence of the ribosomal complex, has a very low synthetic activity which can be greatly stimulated by incubation in the presence of 20% methanol (Sy et al., 1973). Table IV indicates that, unlike the *E. coli* enzyme, the basal rate of the *B. brevis* stringent factor is not stimulated by methanol. Under similar conditions the ppGpp-synthetase is slightly inhibited by methanol.

The molecular weight of the *B. brevis* stringent factor was determined by sucrose density gradient centrifugation using bovine serum albumin and ppGpp-synthetase as markers (Fig. 4). It has an apparent molecular weight of 76,000, a value very close to that for the *E. coli* enzyme.

DISCUSSION

Two ppGpp synthetic activities in *B. brevis* were identified in these studies: the ribosome-bound stringent factor which requires an uncharged tRNA/mRNA/ribosome complex for

TABLE IV. Methanol Effect on B. brevis Stringent Factor

Time (hr)	% GTP converted	
	- methanol	+ methanol
1	5.1	5.8
2	10.5	11.2
4.25	24.8	24.3
17	73.0	69.2

Reaction mixture in 50 µl containing 50 mM Tris-OAc, pH 8.1, 4 mM dithiothreitol, 10 mM Mg(OAc)$_2$, 4 mM ATP, o.2 mM [α-^{32}P]GTP (35 Ci/mol), 50 µg of bovine serum albumin, 8 mM phosphoenolpyruvate, 3 µg of pyruvate kinase, 4 µg of B. brevis stringent factor, and with or without 20% methanol, were incubated at 30°C. At the indicated time points, 5 µl was withdrawn and the reaction stopped by the addition of 5 µl HCOOH (2 M). The ppGpp + pppGpp formed was assayed as described in Materials and Methods.

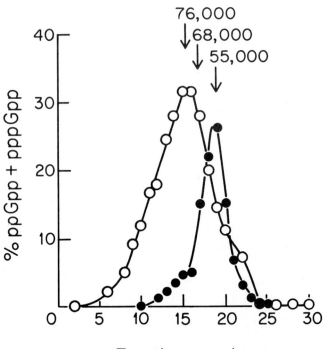

FIGURE 4. *Sucrose gradient centrifugation of ppGpp-synthetase and stringent factor. B. brevis ppGpp-synthetase (6 μg) and stringent factor (50 μg) were mixed and centrifuged for 16 hr at 50,000 rpm in a Spinco SW 56 rotor in a 5-20% sucrose gradient containing 40 mM Tris-OAc, pH 8.1, 5 mM dithiothreitol, 250 mM KCl, 0.1 mM EDTA; 140-μl fractions were obtained, and each fraction was assayed for the two enzymatic activities as described in Materials and Methods.*

activity and should be the enzyme responsible for the increase in ppGpp synthetic activity during amino acid starvation; and the supernatant enzyme, i.e., ppGpp-synthetase, which is probably responsible for the basal level synthesis of ppGpp. It is not currently known whether the ppGpp-synthetase is under any metabolic control.

Maximal activity for the stringent factor requires the presence of high Mg^{2+} concentration (20 mM), while for the ppGpp-synthetase, the optimal Mg^{2+} concentration is that which titrates the nucleoside polyphosphate as a mono Mg^{2+} salt (Sy & Akers, 1976). Very little stringent factor activity can be demonstrated at Mg^{2+} concentrations of 10 mM or lower, and the ppGpp-synthetase activity is greatly inhibited at 20 mM Mg^{2+}. Therefore, the two enzymes can be readily distinguished using high and low Mg^{2+} assay conditions.

There is a possibility that the ppGpp-synthetase (mol.wt. 55,000) may be a proteolytic product of the stringent factor (mol.wt. 76,000) with the ribosome binding site cleaved. However, the relatively constant amount of ppGpp-synthetase throughout the growth phase does not support this notion since one would expect the total synthetase activity to increase as the stringent factor activity drops.

Rhaese and Groscurth (1974) have reported that in *Bacillus subtilis* the ppGpp synthetic activity of the ribosomal fraction in sporulating cells was replaced by a ppApp synthetic activity. No ppApp synthesis was detected with *B. brevis* ribosomes prepared from all phases of the growth curve, including sporulating cells (data not shown), although the ppGpp synthetic activity of the ribosomal fraction did drop in sporulating stationary cells (Fig. 1 and 2).

ACKNOWLEDGEMENT

The author is grateful to Dr. Fritz Lipmann for his continuous interest and encouragement.

REFERENCES

An, G., Justesen, J., Watson, R.J. & Friesen, J.D. (1979), Cloning the *spoT* gene of *E. coli*: Identification of the *spoT* gene product. *J. Bacteriol. 137,* 1100.
Davis, B. (1964), Disc electrophoresis-II-method and application to human serum proteins. *Ann. N.Y. Acad. Sci. 121,* 404.
Friesen, J.D., An, D. & Fiil, N.P. (1978), Nonsense and insertion mutants in the *relA* gene of *E. coli*: cloning *relA*. *Cell 15,* 1187.
Gallant, J. & Lazzarini, R.A. (1976), The regulation of ribosomal RNA synthesis and degradation in bacteria. *in* Protein synthesis (E.H. McConkey, ed.), vol. 2, p. 309, Marcel Dekker, Inc., New York.
Lee, S.G., Littau, V. & Lipmann, F. (1975), The relation between sporulation and the induction of antibiotic synthesis and of amino acid uptake in *B. brevis*. *J. Cell Biol. 66,* 233.
Nierlich, D.P. (1978), Regulation of bacterial growth, RNA and protein synthesis. *Annu. Rev. Microbiol. 32,* 393.

Pedersen, F.S. & Kjeldgaard, N.O. (1977), Analysis of the *relA* product of *E. coli*. *Eur. J. Biochem. 76*, 91.

Rhaese,H.J. & Groscurth, R. (1974), Studies on the control of development. In vitro synthesis of HPN and MS nucleotides by ribosomes from either sporulating or vegetative cells by *B. subtilis*. *FEBS Lett. 44*, 87.

Sy, J., Ogawa, Y. & Lipmann, F. (1973), Nonribosomal synthesis of guanosine 5',3'-polyphosphates by the ribosomal wash of stringent *E. coli*. *Proc. Natl. Acad. Sci. USA 70*, 2145.

Sy, J., Chua, N.H., Ogawa, Y. & Lipmann, F. (1974), Ribosome specificity for the formation of guanosine polyphosphates. *Biochem. Biophys. Res. Commun. 56*, 611.

Sy, J. (1976), A ribosome-independent, soluble stringent factor-like enzyme isolated from a *B. brevis*. *Biochemistry 15*, 606.

Sy, J. & Akers, H. (1976), Purification and properties of guanosine 5',3'-polyphosphate synthesis from *B.brevis*. *Biochemistry 15*, 4399.

**REGULATION OF MACROMOLECULAR SYNTHESIS
BY LOW MOLECULAR WEIGHT MEDIATORS**

A RADIOIMMUNOASSAY FOR ppGpp

Robert H. Silverman[1]
Alan G. Atherly

Department of Genetics
Iowa State University
Ames, Iowa

Gad Glaser[2]
Michael Cashel

Laboratory of Molecular Genetics
National Institute of Child Health and Development
National Institutes of Health
Bethesda, Maryland

A sensitive, specific radioimmunoassay for ppGpp has been developed. Rabbits were injected with ppGpp chemically coupled to rabbit gamma globulin. Serum from these rabbits contained antibody directed against ppGpp. Competition binding experiments were performed between the antibody, high specific activity [^{32}P]-labeled ppGpp, and various nucleotides. As little as 0.1 picomole unlabeled ppGpp measurably displaced the [^{32}P]-ppGpp from the antibody. The limit of detection for ppGpp using this assay is about 1,000-fold lower than previous methods.

ppGpp resulted in about a 100-fold greater displacement of [^{32}P]-ppGpp than ppGp demonstrating that the β-3'-phosphate of ppGpp is important for the specificity of the nucleotide binding. The 5'-phosphates of ppGpp were less important.

---

[1] present address: National Institute for Medical Research, The Ridgeway, Mill Hill, London NW7 1AA, England.
[2] present address: Hadassah-Hebrew University Medical School, Jerusalem, Israel.

Various other nucleotides tested including ppG, pppG, pppm⁷G, pppA, ppApp and ppI were required in about 10,000-fold higher concentrations than ppGpp to displace a similar amount of [³²-P]-ppGpp.

I. INTRODUCTION

In bacteria, ppGpp is believed to coordinate growth-related processes including ribosome production [see Cashel (1975) for review]. To determine whether a similar control mechanism exists in higher organisms, investigators have search for ppGpp in eucaryotic cell types. Despite some conflicting reports, the weight of evidence favors an absence of concentrations greater than 0.1-1.0 micromolar ppGpp in higher eucaryotes [see Silverman and Atherly (1979) for review].

In $vivo$ studies on the presence of ppGpp have employed [$^{32}PO_4$]-labeling followed by acid extraction and one- or two-dimensional thin-layer chromatography which is the method successfully used to detect ppGpp in microorganisms. However, the lower limit of detection for ppGpp by this assay is only in the micromolar range. We have developed a radioimmunoassay for ppGpp which is about 1,000 fold more sensitive than the [$^{32}PO_4$]-labeling technique which may allow investigations of possible ppGpp in the subnanomolar range.

II. MATERIALS AND METHODS

A. *Preparation of Antibody to ppGpp*

Antibody to ppGpp was prepared as described for making antibody to 3-0-succinyl digitoxigenin by Oliver *et al.* (1968) with only minor modifications. Rabbit gamma globulin (50 mg) was dissolved in water (25 ml), filtered and mixed with 25 mg of EDC [1-ethyl-3-(3-dimethylaminopropyl) carbodiimide hydrochloride] from Sigma Chemical Co. and 20 mg of the lithium salt of ppGpp. The pH was adjusted to 5.5 and the solution incubated at 25°C for one day in the dark, then dialysed against physiological saline for two days. Calculations from measurements of absorbance and protein yielded about 70 μmoles of ppGpp bound per μmole protein. The injection schedule was 0.5 ml of this solution injected into footpads (0.25 ml/footpad) four times over the course of a month. Ten days after the last injection, the seven rabbits were bled and the serum removed.

B. *Preparation of* [*32P*]*-ppGpp*

   1. Method 1. High specific activity [^{32}P]-ppGpp was prepared from *E. coli* strain 10B6 (*rel* A$^+$, *val*s^{ts}, *thy* A). Cells were grown at 30°C in Tris-glucose minimal medium (Irr and Gallant, 1969) supplemented with casamino acids and containing 50 µM potassium phosphate and 2 mM KCl. At a density of 1.7 x 10^8 cells/ml, 20 mCi [^{32}PO$_4$] (New England Nuclear) was added to 5.0 ml of cells. After one doubling, the culture was shifted to 42°C to inactivate the heat-sensitive valyl-tRNA synthetase. After 10 min, the cells were centrifuged into a pellet and extracted twice with 100 µl 2N formic acid at 0°C. The extract was prepared by adding formic acid to the cell pellet, followed by centrifugation at low speeds. The acid-extract was streaked along a 20 cm line on a polyethyleneimine (PEI)-cellulose sheet (Brinkman). Elution was with 1.5 M KH$_2$PO$_4$, pH 3.4. The position of the [^{32}P]-ppGpp was determined by exposing the chromatogram to medical X-ray film (Kodak) for 15 min. The region on the chromatogram where the [^{32}P]-ppGpp migrated was cut out and eluted 11 h at 2-4°C in 4 ml 1M triethylammonium-bicarbonate, pH 9.0. This was evaporated to dryness under vacuum at 30°C. The residue was resuspended in 1 ml H$_2$O and re-evaporated 5 times. The [^{32}P]-ppGpp was then dissolved in 0.5 ml 10 mM KPO$_4$, pH 7.0. The specific acitivty of the [^{32}PO$_4$] in the [^{32}P]-ppGpp was taken to be the same as the [^{32}PO$_4$] in the growth medium, which was 80 Ci/mmole PO$_4$. Gallant and Harada (1969) have reported that exogenous [^{32}PO$_4$] is equilibrated with the nucleoside triphosphate pools in about a half of a generation. The purity of the [^{32}P]-ppGpp was determined to be 82% by PEI-cellulose chromatography.

   2. Method 2. The [^{32}P]-ppGpp was prepared as in Method 1 except for the following: 21 mCi [^{32}PO$_4$] was added to 2.5 ml of cells, the [^{32}P]-ppGpp was eluted from the PEI-cellulose in 6 ml 2M triethylammonium-bicarbonate, pH 8.1 for 16 h at 4°C, the [^{32}P]-ppGpp was then lyophilized, suspended in H$_2$O, lyophilized again and finally resuspended in 2 ml H$_2$O. The specific activity of the [^{32}PO$_4$] in the media was 166 Ci/mmole and the purity of the [^{32}P]-ppGpp was 65%.

C. *Nucleotides*

   Guanosine 5'-diphosphate 3'-diphosphate (ppGpp), guanosine 5'-triphosphate 3'-diphosphate (pppGpp), guanosine 5'-phosphate 3'-diphosphate (pGpp), guanosine 5'-diphosphate 3'-phosphate (ppGp), adenosine 5'-diphosphate 3'-diphosphate (ppApp) and adenosine 5'-phosphate 3'-diphosphate (pApp) were purchased from ICN. Adenosine 5'-triphosphate (ATP), Inosine 5'-diphosphate (IDP), guanosine 5'-diphosphate (GDP), guanosine 5'-triphosphate (GTP) and 5-phosphorylribose-1-pyrophosphate (PRPP) were purchased from Sigma Co. 7-methylguanosine 5'-triphosphate (pppm^{7}G) and 2'-O-methylguanosine-5'-phosphate were purchased from P.L. Biochemicals, Milwaukee, WI.

D. *Radioimmunoassay for ppGpp*

1. *Method 1.* The following were contained in a final volume of 250 µl: 50 mM KPO$_4$, pH 8.0, 0.1 pmole [^{32}P]-ppGpp (about 10,000 cpm); 6 µl anti-ppGpp serum; and 10 to 100 µl of the nucleotide to be tested. The antibody was added last. The mixture was incubated at 2-4°C for 20 hr. Twenty-four µl of goat anti-rabbit gamma globulin (Research Products International) was added to each tube. After 1 h on ice, 1 ml 50 mM KPO$_4$, pH 8.0 was added and the mixture was centrifuged at 30,000 x g for 10 min. The supernatant was discarded and 5 ml Bray's solution was added to the pellet. The pellet was disintegrated with a pasteur pipette and transferred to a vial for radioactivity determination on a Packard Model 3320 scintillation spectrometer.

2. *Method 2.* The assay was performed as in Method 1, except the following components were added in the amounts indicated: 0.01 pmole [^{32}P]-ppGpp (about 5,000 cpm), 10 mM KPO$_4$, pH 8.0, 1 µl anti-ppGpp serum, 9 µl goat anti-rabbit gamma globulin, and 1 ml 5 mM KPO$_4$, pH 8.0 was added just before centrifugation.

III. RESULTS

Antibody to ppGpp was prepared by immunizing rabbits with ppGpp chemically coupled rabbit gamma globulin as described in Materials and Methods. The results of a competition binding assay between the antibody, [^{32}P]-ppGpp, and various guanine nucleotides are shown in Figure 1. The antibody is specific for ppGpp, although the phosphoryl groups at the 5' position appear less important than those at the 3' position (pGpp and pppGpp are fairly effective in displacing [^{32}P]-ppGpp as compared to ppGp). pppG was, by comparison, inefficient in displacing [^{32}P]-ppGpp from the antibody. A variety of other nucleotides were tested for competition binding with [^{32}P]-ppGpp and antibody as shown in Figure 2. These nucleotides

A Radioimmunoassay for ppGpp 111

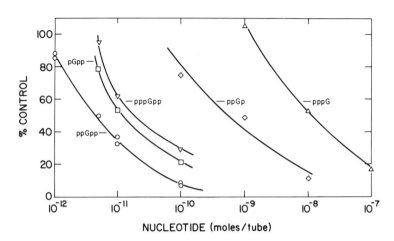

FIGURE 1. Competition binding with various guanine nucleotides. The assays were performed by Method 1 (see Materials and Methods). % control refers to the amount of radioactivity precipitated in the presence of added unlabeled nucleotide/radioactivity precipitated in the absence of unlabeled nucleotide x 100; (100% = 4,700 cpm).

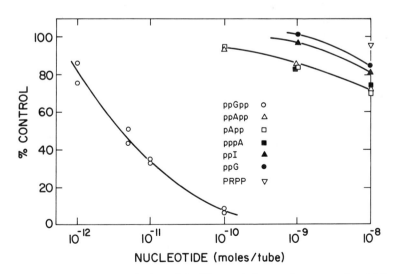

FIGURE 2. Competition binding with various nucleotides. The assays were performed by Method 1 (see Materials and Methods). (100% = 4,350 cpm).

were required in about 10,000-fold greater amounts than unlabeled ppGpp to compete a comparable amount of [^{32}P]-ppGpp. Because ppApp was among these poor displacers of [^{32}P]-ppGpp, it would appear that the antibody is highly specific for the guanine moiety of ppGpp. Other nucleotides and compounds tested included 2'-O-methyl GMP, pm^{7}G, pG and polyphosphate (data not shown), all of which were required in at least 10,000-fold excess over unlabeled ppGpp to displace a similar amount of [^{32}P]-ppGpp.

By increasing the specific activity of the [^{32}P]-ppGpp, and by reducing the amount of [^{32}P]-ppGpp, antibody and KPO$_4$ in the assay mixture, we were able to increase the sensitivity of the assay by about 10-fold as shown in Figure 3. The lower limit of detection for ppGpp by this method was 10^{-13} moles or less.

IV. DISCUSSION

The sensitivity of the radioimmunoassay for ppGpp allows detection of subnanomolar quantities without the necessity of prelabeling with a radioisotope. 10^{-13} moles of ppGpp was accurately measured, as this was present in 250 µl of incubation mixture the concentration of ppGpp measured was 10^{-13} moles/250 x 10^{-6} l = 4 x 10^{-10}M. The sensitivity can be increased if the extracted ppGpp is concentrated from a larger volume of extract. Thus, previously undetectable levels of ppGpp could be accurately estimated, much below 0.1-1.0 micromolar, which is the limit of detection using prelabeling techniques (Silverman and Atherly, 1979). In some instances it is very inconvenient or difficult to prelabel with radioisotopes (e.g., during embryogenesis and in tissue extracts).

Both ATP and GTP are present in most cells at sufficiently high concentrations to interfere with the radioimmunoassay. If ATP or GTP were present at about 10^{-4}M, a concentration of ppGpp below 10^{-8}M could not be distinguished from the ATP or GTP. It may be necessary to chromatographically separate the nucleotide pools from cells and then perform the radioimmunoassay on the fractions. Alternately, it may be possible to prelabel cells and then use the antibody to precipitate the ppGpp present. This could be followed by chromatography of nucleotides precipitated.

The development of radioimmunoassays for other unusual nucleotides, e.g. Ap$_4$A, and 2'-5' adenosines, may also be possible. Because these nucleotides are likely to be present in

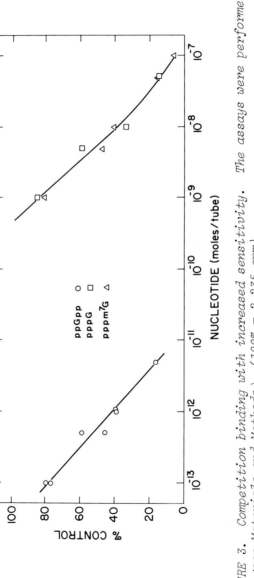

FIGURE 3. Competition binding with increased sensitivity. The assays were performed by Method 2 (see Materials and Methods). (100% = 2,235 cpm).

very low levels, radioimmunoassay may be used in future studies on the role of unusual nucleotides in the regulation of cellular events.

ACKNOWLEDGMENTS

Robert H. Silverman thanks Dr. Alice Wertheimer for valuable suggestions made in the preparation of the manuscript.

REFERENCES

Cashel, M. Regulation of bacterial ppGpp and pppGpp. *Ann. Rev. Microbiol.* <u>29</u>, 301 (1975).
Gallant, J., Harada, B. The control of ribonucleic acid synthesis in *Escherichia coli*. III. The functional relationship between purine ribonucleotide triphosphate pool sizes and the rate of ribonucleic acid accumulation. *J. Biol. Chem.* <u>244</u>, 3125 (1969).
Irr, J., Gallant, J. The control of ribonucleic acid synthesis in *Escherichia coli*. II. Stringent control of energy metabolism. *J. Biol. Chem.* <u>244</u>, 2233 (1969).
Oliver, G.C., Parker, B.M., Brasfield, D.L., Parker, C.W. The measurement of digitoxin in human serum by radioimmunoassay. *J. Clin. Invest.* <u>47</u>, 1035 (1968).
Silverman, R.H., Atherly, A.G. The search for guanosine tetraphosphate (ppGpp) and other unusual nucleotides in eucaryotes. *Microbiol. Reviews* <u>43</u>, 27 (1979).

REGULATION OF MACROMOLECULAR SYNTHESIS
BY LOW MOLECULAR WEIGHT MEDIATORS

GUANOSINE 3', 5'-BIS (DIPHOSPHATE) SEARCH
IN EUKARYOTES

Robert Silverman [1]
Alan Atherly

Department of Genetics
Iowa State University
Ames, Iowa

Dietmar Richter

Institut für Physiologische Chemie
Abteilung Zellbiochemie
Universität Hamburg
Hamburg, W.Germany

In almost every instance where ppGpp was found in eucaryotes there is evidence to the contrary. Nevertheless, the following conclusions can be drawn from the available data. (i) Lower eucaryotes may synthesize ppGpp in response to various environmental stresses (starvation, heat shock, etc.). (ii) Mammalian cells in culture probably do not synthesize ppGpp at levels higher than about 0.1 to 1.0 μM. (iii) Mouse embryos don't synthesize detecable quantities of ppGpp. (iv) Isolated eucaryotic ribosomes do not synthesize ppGpp. (v) Mitochondria and chloroplast synthesis of ppGpp and pppGpp have not been ruled out; if a ppGpp-synthesizing enzyme is present, however, it probably is either loosely associated with the ribosomes or not associated with the ribosomes at all.

Most of the research reviewed herein placed the limits of detection of ppGpp in the micromolar range. Nanomolar levels, if present, would have gone undetected. More sensitive methods

---

[1] Present address: National Institute for Medical Research, Mill Hill, London NW7 1AA, U.K.

(e.g., radioimmunoassay) will be required to determine if ppGpp is present in higher eucaryotes.

INTRODUCTION

Bacteria possess the ability to adjust a number of diverse cellular processes rapidly in response to the availability of amino acids and other nutrients. Thus, starving an amino acid auxotroph for its required amino acid results in the so-called stringent response, consisting of the restriction of transcription of ribosomal ribonucleic acid (rRNA) and transfer RNA (tRNA), carbohydrate synthesis, phospholipid synthesis, lipid synthesis, phosphorylation of glycolytic intermediates, de novo nucleotide synthesis, transport of nucleobases and glycosides, and polyamine synthesis in addition to an increased rate of protein decay (see Cashel, 1975 for review). The bacterial cell emerges as an extremely efficient and complex organism capable of coordinating a network of interconnecting, yet biochemically distinct, biosynthetic pathways.

In 1969, Cashel and Gallant discovered two spots on autoradiograms from chromatograms of $^{32}PO_4$-labeled *Escherichia coli* extracts, which they called magic spots I and II (Cashel & Gallant, 1969). These substances were synthesized by stringent (rel^+) strains of *E. coli* in response to amino acid starvation. Isogenic relaxed (rel) strains upon amino acid starvation failed to induce the synthesis of these compounds and failed to give a stringent response. Magic spots I and II have since been identified as guanosine 5'-diphosphate 3'-diphosphate (ppGpp) and guanosine 5'-triphosphate 3'-diphosphate (pppGpp), respectively (see Lipmann & Sy, 1976 for review).

The stringent response is a remarkable device that bacteria have evolved to coordinate growth-related processes with the protein synthetic capacity of the cell. The relaxed bacterium may be analogous to a cancer cell in that neither can regulate its growth when common sense dictates that it should do so. Thus, the stringent response may be viewed as a model for the control of growth, which may or may not tell us something about how cellular growth in higher organisms is regulated. To determine if such a mechanism exists in eucaryotes, many investigators have attempted to ascertain whether guanosine polyphosphates or other unusual highly phosphorylated nucleotides are present in organisms other than procaryotes.

Before analyzing the literature, a few points concerning nucleotide hunting expeditions should be considered. Gallant

& Margason (1972) pointed out the hazards of identifying nucleotides solely on the basis of one-dimensional chromatography, a warning sometimes ignored by researchers in the field. They found an adenine-containing nucleotide in *Bacillus subtilis* that comigrated with ppGpp in one dimension, using KH_2PO_4, but was resolved from ppGpp in other chromatographic systems, as others have also found (Loewen, 1976). In addition, condensed inorganic phosphate or polyphosphate is ubiquitous in nature (Harold, 1966) and can easily be mistaken for nucleotides in one-dimensional chromatograms of $^{32}PO_4$-labeled cells extracts. Also, contamination of eucaryotic cells with bacteria or mycoplasmas must be rigorously avoided. The stability of nucleoside polyphosphates on polyethyleneimine, a frequently used thin-layer ion exchanger, is poor. It is known that 20 to 30% degradation of ppGpp is routinely found. Loewen (1976) found an adenine derivative that migrates near ppGpp on polyethyleneimine that is completely degraded.

In vivo assays for ppGpp usually consist of culturing the organisms or cells in $^{32}PO_4$, extraction with acid (usually formic acid), and, finally, thin-layer chromatography of the extract. Various techniques have been used to attempt to induce ppGpp synthesis in vivo in eucaryotes. By analogy to bacteria, amino acid or nutrient starvation is frequently done.

THE SEARCH FOR ppGpp IN EUCARYOTES

Lower Eucaryotes

Buckel and Böck (1973) have looked for ppGpp in amino acid auxotrophic strains of the mold *Neurospora crassa (leu)* and the chlorococcal green alga *Ankistrodesmus braunii (arg)*. In both strains, RNA synthesis is stringently coupled to protein synthesis. The organisms were labeled with $^{32}PO_4$ for various durations of amino acid starvation. Chromatograms of the cell extracts revealed no label in the region where ppGpp migrates. These findings were not altered by increasing the specific activity of the $^{32}PO_4$, increasing the duration of amino acid starvation, or using two-dimensional chromatography. In addition, Alberghina et al. (1973) failed to detect ppGpp in *N. crassa* grown on a poor carbon source, glycerol. In a similar study McMahon & Langstroth (1972), *Chlamydomonas reinhardi* failed to synthesize ppGpp in response to amino acid starvation of an auxotrophic strain or by addition of the arginine analog canavanine to a wild-type strain.

However, Heizmann & Howell (1978) examined the same auxotrophic strain of *C. reinhardi* but used slightly different detection techniques and found significant amounts of ppGpp. Heizmann and Howell found that addition of carrier pppA and pppG to samples (as in McMahon & Langstroth, 1972) obscured the presence of ppGpp. It is not known why this effect was observed. They convincingly demonstrated the presence of ppGpp, using several criteria: two-dimensional chromatographic comigration with *E. coli* ppGpp, resistance to periodate oxidation, presence of guanine, acid and base lability, and alkaline phosphatase (EC 3.1.3.1) sensitivity. Heizmann and Howell (1974) did not demonstrate the source of ppGpp synthesis, but effects of cycloheximide on ppGpp synthesis suggest a ribosomal site. Whether it is cytoplasmic or chloroplast, or both, was not established.

ppGpp synthesis also has been reported in *Euglena*, by Heizmann (1974). The time-dependent induction of two spots, which migrate slower than pppG on polyethyleneimine-cellulose and which could be labeled with $^{32}PO_4$ or ^{3}H-guanosine, was demonstrated. However, no further characterization of these compounds was reported.

Klein (1974) has examined cultures of the cellular slime mold *Dictyostelium discoideum* for the presence of ppGpp. This organism, when in the amoeba phase on its life cycle is induced to differentiate by starvation. The differentiation is characterized by profound morphological and biochemical changes. Amoebae prelabeled with $^{32}PO_4$ and growing exponentially in a medium containing 100 μg of penicillin and streptomycin per ml were centrifuged and suspended in a buffer of 2-N-morpholinoethanesulfonate in an attempt to induce ppGpp synthesis. One- and two-dimensional chromatograms of extracts from these cells revealed a labeled compound that comigrated with authentic ppGpp. The lability of the compound to phosphomonoesterase digestion and nitrous acid oxidation resembled that of bacterial ppGpp. Only starved cells synthesized the compound. No evidence was presented, however, indicating that the compound contained a guanine moiety. The compound may be an isomer of ppGpp or some other nucleoside polyphosphate. The addition of antibiotics reduces the possibility of bacterial contamination. Others, however, have been unable to repeat these findings (W.A. Haseltine & A. Jacobson, unpublished data, as cited by Jacobson & Lodish, 1975).

Kudrna & Edlin (1975) searched for ppGpp in growing and amino acid-starved *S. cerevisae*, but none was found. The authors suggested that the mechanism of coupled RNA and protein synthesis in yeast is different from that in bacteria. In contrast to the previous report, Pao et al. (1977) claim to have found ppGpp in *S. cerevisae* subjected to heat shock. Subjecting cells to 38° C for 6 min induced synthesis of the compound. None was found at room temperature. That heat shock was required to elicit synthesis of the compound may explain why Kudrna & Edlin (1975) failed to detect it in yeast. Pao et al. (1977) reported an absence of bacterial contamination in their yeast cultures. The yeast compound was identical as ppGpp by the following criteria: (i) comigration with authentic ppGpp in two separate two-dimensional chromatographic systems; (ii) adsorption to charcoal; (iii) insensitivity to periodate oxidation, indicating the esterification of the 2' or 3' position of the ribose moiety; (iv) alkaline hydrolysis at the same rate as that of ppGpp and producing the same breakdown products as those of ppGpp; (v) zinc-activated inorganic pyrophosphatase (EC 3.6.1.1) yielding the same breakdown products as those of ppGpp; (vi) oxytetracycline was used (2 mg/ml); and (vii) 3-phosphoglycerate phosphokinase (EC 2.7.2.3), which phosphorylated ppGpp to pppGpp *in vitro*, also phosphorylated the yeast compound to a substance that comigrates with pppGpp. This evidence is the most complete characterization of ppGpp in any of the studies involving eucaryotes. Provided that the bacterial contamination tests were adequate (these were: plating in MacConkey or rich nutrient agar plates), the synthesis of ppGpp in yeast subjected to heat shock is well substantiated. Pao et al. suggest that the ppGpp present is synthesized in the mitochondria, although they give no direct evidence for this. In *E. coli*, oxytetracycline inhibits the nonribosomal synthesis of ppGpp in addition to the ribosome-dependent reaction (Silverman & Atherly, 1978). Therefore, nothing can be concluded about the site of ppGpp synthesis in yeast. Synthesis of ppGpp in yeast is not stimulated by deacylated tRNA, since heat shock in a strain of yeast with a temperature-sensitive isoleucyl-tRNA synthetase (EC 6.1.1.5) produced no more ppGpp than did the parent wild-type strain (Pao et al., 1977).

HIGHER EUCARYOTES

Smulson (1970) was the first to publish an attempt to detect ppGpp in eucaryotic cells. $^{32}PO_4$-labeled HeLa cells were amino acid starved by incubation in the absence of amino acids or with the isoleucine analog O-methylthreonine. Neither treatment induced detectable levels of ppGpp.
Tomkins and co-workers examined 3T3 cells subjected to various nutritional downshifts and were unable to detect any ppGpp synthesis (Mamont et al., 1972). Cells were grown in $^{32}PO_4$ for 15 h before starvation for serum alone or for amino acids and serum. Formic acid extracts of these cells revealed no label in the region where ppGpp migrates on two-dimensional chromatograms. Tomkins and co-workers reported a limit of detection

for ppGpp at 0.5% of the ppGpp level. Also, Thammana et al. (1976) induced isoleucine starvation in synchronized 3T3 cells by using a medium lacking the amino acid or a medium containing O-methylthreonine, an inhibitor of isoleucyl-tRNA synthetase. Extracts of cells from the beginning of the S phase into G2 were analyzed. No ppGpp was detected, regardless of the concentration of O-methylthreonine or the length of $^{32}PO_4$ labeling. Primary cultures of mouse embryonic fibroblasts gave the same negative results. Their limit of detection for ppGpp was placed at 0.5 to 1% of the counts of pppG.

Fan et al. (1973) examined a proline-requiring line of CHO cells and were unable to detect unusual nucleotides during normal growth or during amino acid or serum starvation. Similarly, Stanners & Thompson (1974) in collaboration with J.D. Friesen failed to detect ppGpp synthesis in CHO cells temperature sensitive for leucyl-tRNA synthetase at the nonpermissive temperature. Further evidence for an absence of ppGpp in mammalian cells was provided by Jolicœur et al. (1974). They reported an absence of ppGpp in Landschutz tumor cells incubated with $^{32}PO_4$ in either complete or amino acid-deficient medium.

Rapaport & Bucher have looked for ppGpp in normal and regenerating rat livers by injection of rats with $^{32}PO_4$, ^3H-hypoxanthine, or ^3H-guanosine (Rapaport & Bucher, 1976). Frozen livers were acid extracted, and the nucleotides were separated. Despite a high level of labeling of the nucleotide pool, no ppGpp or pppGpp was detected.

Embryonic material also has been examined for ppGpp synthesis. Irr et al. (1974), using the ribosome-dependent ppGpp assay worked out for $E.\ coli$ ribosomes, claimed to have found that ribosomes from 10- and 11-day mouse embryos are capable of synthesizing ppGpp, whereas ribosomes isolated from older embryos or from adult mouse livers were inactive. The compound synthesized by mouse embryonic ribosomes comigrated with authentic ppGpp in three different chromatographic systems and was labeled with either (α-^{32}P)pppG or ^3H-pppG. The amounts of ppGpp synthesis reported were significant, 37 to 69% of the rate found with $E.\ coli$ ribosomes. However, other groups have repeated these experiments with the same mouse strain without success (Silverman & Atherly, 1977; Martini, Irr & Richter, 1977; Polland & Parker, 1977). Ribosomes isolated from 10- to 13-day embryos were tested for ppGpp synthesis, and none was found. $E.\ coli$ ribosomal high-salt wash, a source of stringent factor, was then added to the embryonic ribosomes in an attempt to stimulate the stringent factor. The results varied from minimal stimulation of ppGpp synthesis to no stimulation. It was concluded that mouse embryonic ribosomes alone are incapable of synthesizing ppGpp and that the lack of significant stimulation of *E. coli* stringent factor by the embryonic ribosomes indicated that it was unlikely that a ppGpp-synthesizing factor dissociated from the ribosomes during isolation. The latter conclusion gains support from the fact that the supernatant from the 100,000 x g centrifugation of 11-day-embryo extracts failed to stimulate the embryonic ribosomes to synthesize ppGpp. In addition, Silverman and Atherly demonstrated the absence of detectable ppGpp in acid-soluble pools isolated from mouse embryos of various stages cultured in the presence of $^{32}PO_4$.

No ppGpp synthesis was detected in vivo in sea urchin eggs or embryos in two studies (Brandhorst & Fromson, 1976; Perrone et al., 1976). In the former study, ribosomes isolated from sea urchin embryos failed to synthesize ppGpp in vitro.

Stimulation of ppGpp synthesis by using eucaryotic ribosomes and *E. coli* stringent factor has been attempted to establish whether the potential for ppGpp synthesis exists on the eucaryotic ribosomes. Failures to stimulate *E. coli* stringent factor with eucaryotic ribosomes have been reported for Ehrlich ascites (Beres & Lucas-Lenard, 1975), yeast cytoplasm, calf brains, reticulocytes (Richter, 1973; Martini & Richter, 1978), and mouse embryos (Martini et al., 1977; Martini & Richter, 1978). Slight stimulation of stringent factor, however, has been reported for ribosomes isolated from wheat germ (Beres & Lucas-Lenard, 1975) and yeast mitochondria (Richter, 1973), and a large stimulation has been reported for *C. reinhardi* chloroplast ribosomes, but not for cytoplasmic ribosomes (Sy et al., 1974). The latter two reports are discussed in greater detail in the following section. In contrast Pollard & Parker (1977) found that addition of *E. coli* stringent factor to mouse embryonic ribosomes or to eucaryotic ribosomes from different organisms and tissues resulted in ppGpp synthesis. The reasons for the apparent conflict with the other reports may be due to a non-specific reaction of the *E. coli* stringent factor not unlikely the stimulation of stringent factor by methanol (Sy et al., 1973). Martini and Richter (1978) reported that although mouse embryo or rabbit reticulocyte ribosomes fail to stimulate *E. coli* stringent factor to synthesize ppGpp, the ribosomal proteins prepared from these ribosomes did stimulate ppGpp synthesis by stringent factor.

In contrast to the preceding negative evidence, Rhaese (1975) claimed to have found ppGpp and other nucleoside polyphosphates in various mammalian cell lines. The only autoradiograms presented, however, were of one-dimensional chromatograms of extracts from $^{32}PO_4$-labeled cells. No further characterization of these compounds was presented. It is well known that mammalian cells synthesize condensed inorganic phosphates (Griffin et al., 1965) and other slow-migrative compounds (Goh & LeJohn, 1977), which could account for Rhaese's spots. Others have examined the same cell lines used by Rhaese and have not found ppGpp (Fan et al., 1973; Stanners & Thomson, 1974; Thammana et al., 1976; R.A. Lazzarini, unpublished data, as cited by Pao et al., 1977; R.S. Esworthy, unpublished data). Clearly, the weight of the evidence is not in Rhaese's favour. This is especially evident in view of the fact that Rhaese discovered his compounds by merely labeling growing cells with $^{32}PO_4$; i.e., no special treatments were required to induce the spots.

Although the evidence favours an absence of greater than about 0.1 to 1.0 M ppGpp in mammalian cells, the possibility of very low levels, perhaps synthesized by mitochondria, has not been ruled out.

Mitochondria and Chloroplasts

Because there are some similarities between bacteria and mitochondria and chloroplasts (e.g., the susceptibility of protein synthesis to certain antibiotics), these organelles have been investigated as possible sources of ppGpp synthesis.

Richter (1973) demonstrated that yeast mitochondrial ribosomes could slightly stimulate ppGpp synthesis by *E. coli* stringent factor. As mentioned, ribosomes isolated from yeast cytoplasm, reticulocytes, or calf brains failed to stimulate stringent factor.

Horvath et al. (1975) reported ppGpp and pppGpp synthesis in isolated rat liver mitochondria and pppGpp synthesis in isolated spinach chloroplasts, both incubated in the presence of $^{32}PO_4$, however, later retracted these data (personal communication to D.R.).

Similarly, Sy et al. (1974), using ribosomes isolated from *C. reinhardi*, demonstrated that neither the cytoplasmic nor the chloroplast ribosomes were capable of ppGpp synthesis. However, addition of *E. coli* stringent factor to the chloroplast ribosomes resulted in abundant ppGpp synthesis. No ppGpp synthesis was found when the cytoplasmic ribosomes plus stringent factor were used. Thiostrepton, which inhibits *E. coli* ribosome-dependent ppGpp synthesis, also inhibited the chloroplast ribosome plus stringent factor assay.

The possibility exists that mitochondria and chloroplasts may contain stringent-factor-like enzymes that are loosely associated with the ribosomes and thus are lost during ribosome isolation. Nonribosomal ppGpp synthesis in mitochondria also is a possibility. B. brevis and B. stearothermophilus contain a ribosome-independent guanosine 5',3'-polyphosphate synthetase (Sy & Akers, 1976; Fehr et al., 1979).

REFERENCES

Alberghina, F.A.M., Schiaffonati, L., Zardi, L. & Stuarani, E. (1973), Lack of guanosine tetraphosphate accumulation during inhibition of RNA synthesis in Neurospora crassa. Biochim. Biophys. Acta 312, 435.
Beres, L. & Lucas-Lenard, J. (1975), Studies in the in vitro synthesis of ppGpp and pppGpp. Biochim. Biophys. Acta 395, 80.
Brandhorst, B. & Fromson, D. (1976), Lack of accumulation of ppGpp in sea urchin embryos. Dev. Biol. 48, 458.
Buckel, P., Böck, A. (1973), Lack of accumulation of unusual guanosine nucleotides upon amino acid starvation of two eukaryotic organisms. Biochim. Biophys. Acta 324, 184.
Cashel, M. (1975), Regulation of bacterial ppGpp and pppGpp. Annu. Rev. Microbiol. 29, 301.
Cashel, M. & Gallant, J. (1969), Two compounds implicated in the function of the R.C. gene in E. coli. Nature (London) 221, 838.
Fan, K., Fisher, K.M. & Edlin, G. (1973), Effect of amino acid and serum deprivation on the regulation of RNA synthesis in cultured Chinese hamster ovary cells. Exp. Cell Res. 82, 111.
Fehr, S., Godt, F., Isono, K. & Richter, D. (1979), A ribosome independent stringent factor from Bacillus stearothermophilus and a low molecular weight substance inhibitory to its activity. FEBS Lett. 97, 91.
Gallant, J. & Margason, G. (1972), Amino acid control of messenger ribonucleic acid synthesis in Bacillus subtilis. J. Biol. Chem. 247, 2289.
Goh, S.H. & LeJohn, H.B. (1977), Genetical and biochemical evidence that a novel dinucleoside polyphosphate coordinates salvage and de novo nucleotide biosynthetic pathways in mammalian cells. Biochem. Biophys. Res. Commun. 74, 256.
Griffin, J.B., Davidian, N.M. & Penniall, R. (1965), Studies of phosphorus metabolism by isolated nuclei. J. Biol. Chem. 240, 4427.
Harold, F.M. (1966), Inorganic polyphosphates in biology: structure, metabolism, and function. Bacteriol. Rev. 30, 772.
Heizmann, P. (1974), Rôle des synthèses protéiques dans la formation du ribosome chloroplastique chez l'Euglene. Biochimie 56, 1357.

Heizmann, P. & Howell, S. (1978), Synthesis of ppGpp and chloroplast ribosomal RNA in *Chlamydomonas reinhardi*. *Biochim. Biophys. Acta* 517, 115.

Horvath, I., Zobas, P.I., Szabados, G.Y. & Bauer, P. (1975), In vitro synthesis of guanosine polyphsophates in rat liver mitochondrial preparations. *FEBS Lett.* 56, 179.

Irr, J.D., Kaulenas, M.S. & Unsworth, B.R. (1974), Synthesis of ppGpp by mouse embryonic ribosomes. *Cell* 3, 249.

Jacobson, A. & Lodish, H.F. (1975), Genetic control of development of the cellular slime mold *Dictyostelium discoideum*. *Annu. Rev. Genet.* 9, 145.

Jolicoeur, P., Lemay, A., Labrie, F. & Steiner, A.L. (1974), Phosphorylation of chromosomal and ribosomal proteins and intracellular levels of cyclic 3',5'-adenosine monophosphate and cyclic 3',5'-guanosine monophosphate during amino acid starvation in Landschutz tumour cells. *Exptl. Cell Res.* 89, 231.

Klein, C. (1974), Presence of magic spot in *Dictyostelium discoideum*. *FEBS Lett.* 38, 149.

Kudrna, R. & Edlin, G. (1975), Nucleotide pools and regulation of ribonucleic acid synthesis in yeast. *J. Bacteriol.* 121, 740.

Lipmann, F. & Sy, J. (1976), The enzymatic mechanism of guanosine 5'3' polyphosphate synthesis. *Prog. Nucleic Acid Res. Mol. Biol.* 17, 1.

Loewen, P.C. (1976), Novel nucleotides from *E. coli* isolated and partially characterized. *Biochem. Biophys. Res. Commun.* 70, 1210.

McMahon, D. & Langstroth, P. (1972), The effects of canavanine and of arginine starvation on macromolecular synthesis of *Chlamydomonas reinhardi*. *J. Gen. Microbiol.* 73, 239.

Mamont, P., Hershko, A., Kram, R., Schachter, L., Lust, J. & Tomkins, G.M. (1972), The pleiotypic response in mammalian cells: search for an intracellular mediator. *Biochem. Biophys. Res. Commun.* 48, 1378.

Martini, O. & Richter, D. (1978), Eukaryotic ribosomal proteins stimulate *Escherichia coli* stringent factor to synthesize guanosine 5'-diphosphate,3'-diphosphate (ppGpp) and guanosine 5'triphosphate,3'-diphosphate (pppGpp). *Molec. Gen. Genet.* 166, 291.

Martini, O., Irr, J. & Richter, D. (1977), Questioning of reported evidence for guanosine tetraphosphate synthesis in a ribosome system from mouse embryos. *Cell* 12, 1127.

Pao, C.C., Paiella, J. & Gallant, J.A. (1977), Synthesis of guanosine tetraphosphate (magic spot I) in *Saccharomyces cerevisea*. *Biochem. Biophys. Res. Commun.* 74, 314.

Pirrone, A.M., Roccheri, M.C., Bellanca, V., Acierno, P. & Guidice, G. (1976), Studies on the regulation of ribosomal RNA synthesis in sea urchin development. *Dev. Biol.* 49, 311.

Pollard, J.W. & Parker, J. (1977), Guanosine tetra- and pentaphosphate synthesis by bacterial stringent factor and eukaryotic ribosomes. *Nature (London) 267*, 371.
Rapaport, E. & Bucher, N.L.R. (1976), Two new adenine nucleotides in normal and regenerating rat liver. in W.H. Fishman & S. Sell (eds.), Onco-developmental gene expression, p. 13. Academic Press, New York.
Rhaese, H.J. (1975), Studies on the control of development synthesis of regulatory nucleotides, HPN and MS, in mammalian cells in tissue cultures. *FEBS Lett. 53*, 113.
Richter, D. (1973), Formation of guanosine tetraphosphate (magic spot I) in homologous and heterologous systems. *FEBS Lett. 34*, 291.
Silverman, R. & Atherly, A.G. (1977), Mouse embryos fail to synthesize detectable quantities of guanosine 5'-diphosphate 3'diphosphate. *Dev. Biol. 56*, 200.
Silverman, R. & Atherly, A.G. (1978), Unusual effects of 5a,6' anhydrotetracycline and other tetracyclines: inhibition of guanosine 5'-diphosphate 3'-diphosphate metabolism, RNA accumulation and other growth-related processes in *Escherichia coli*. *Biochim. Biophys. Acta. 518*, 267.
Smulson, M. (1970), Amino acid deprivation of human cells: effects on RNA synthesis, RNA polymerase, and ribonucleoside phosphorylation. *Biochim. Biophys. Acta 199*, 537.
Stanners, C.P. & Thompson, L.H. (1974), Studies on a mammalian cell mutant with a temperature-sensitive leucyl-tRNA synthetase. in B. Clarkson & R. Baserga (eds.), Control of proliferation in mammalian cells, p. 191. Cold Spring Harbor Laboratory, Cold Spring Harbor, New York.
Sy, J. & Akers, H. (1976), Purification and properties of guanosine 5',3'-polyphosphate synthetase from *Bacillus brevis*. *Biochemistry 15*, 4399.
Sy, J., Chua, N.H., Ogawa, P. & Lipmann, F. (1974), Ribosome specificity for the formation of guanosine polyphosphates. *Biochem. Biophys. Res. Commun. 56*, 611.
Thammana, P., Buerk, R.R. & Gordon, J. (1976), Absence of ppGpp production in synchronized BALB/C mouse 3T3 cells on isoleucine starvation. *FEBS Lett. 68*, 187.

Part II
PURINE NUCLEOTIDES AND SPORULATION

REGULATION OF MACROMOLECULAR SYNTHESIS BY LOW MOLECULAR WEIGHT MEDIATORS

INITIATION OF BACTERIAL AND YEAST SPORULATION BY PARTIAL DEPRIVATION OF GUANINE NUCLEOTIDES

Ernst Freese, Juan M. Lopez and Elisabeth B. Freese

Laboratory of Molecular Biology
NINCDS/NIH, Bethesda, Maryland

Microbial differentiation usually starts when rapidly metabolizable carbon or nitrogen sources or phosphate are exhausted. Subsequently, the cells use either internal reserves or slowly metabolize external carbon or nitrogen sources to synthesize specific developmental macromolecules. The principle that nutritional deprivation is needed for certain types of differentiation applies also to plants, as for the case of pollen formation within a tapetum, and maybe to early embryonic and germ cell development in animals. We will demonstrate in this paper (A) that the sporulation of Bacillus subtilis can be initiated by a partial deficiency of purine and in particular guanine nucleotides and that under all conditions under which sporulation has so far been observed as a result of nutritional deprivation, the concentrations of GDP and GTP decrease during the initiation process. At the end of the paper (B) we will show that in an appropriate growth medium the sporulation of Saccharomyces cerevisae can also be induced by inhibitors of purine synthesis.

RESULTS WITH *BACILLUS SUBTILIS*

Sporulation Caused by General Nutritional Deprivation

The first morphologically observable change in a sporulating *Bacillus* cell is the formation of an asymmetric septum; the double membrane subsequently engulfs the small cell compartment which eventually produces a spore that is resistant to heat and organic solvents.

TABLE I. Sporulation under Different Conditions

Condition	S/V in %
Standard strain 60015:	
No addition to A[1]	$2.2*10^{-2}$
No addition to A + aa[2]	$1.3*10^{-2}$
Nutrient sporulation medium	60
20 mM glutamate as sole carbon source	34
4 mM nitrate as sole nitrogen source	6.5
2.1 mM hadacidin in A	4.8
2.1 mM hadacidin in A + aa	$7.6*10^{-3}$
150 µM mycophenolic acid in A	5.0
1.8 mM decoyinine in A	30
2 mM psicofuranine in B	4.0
Relaxed strain 61852:	
No addition to A	$4.3*10^{-4}$
1.8 mM decoyinine in A	17

The compounds were added at $A_{600}=0.5$. S and V were measured 10 hr later. V was always between 10^8 and $2*10^9$/ml.
1) A = S6 medium (6) + 1% D-glucose
2) aa = addition of 19 amino acids (12)
3) B = S7 medium containing 20 mM glutamate (7) + 1% D-glucose

  The sporulation development usually starts at the end of growth in a nutrient sporulation medium or as a result of cell transfer to a medium containing glutamate or lactate as sole carbon source, nitrate (optimal 5 mM) as sole nitrogen source or a medium containing no phosphate (Table I). Recent reviews concerning sporulation in *Bacillus* can be found in (Piggot & Coote, 1976; Freese, 1972; Freese, 1976; Sonenshein & Campbell, 1978).
  Whereas these general nutritional deficiencies could cause the decrease of any one of many metabolites, we and other colleagues in our laboratory wanted to determine which specific compound(s) can control the initiation of *B. subtilis* sporulation. For this purpose, we have used media containing excess of glucose, malate or other easily metabolizable carbon sources, ammonium ions and phosphate, and we attempted to initiate sporulation by analogs and inhibitors of many biochemical reactions or by starvation of auxotrophic mutants. Most of the used compounds, such as inhibitors of DNA, RNA or protein synthesis or amino acid analogs, were ineffective at

any concentration. But most purine analogs or inhibitors of purine synthesis (including plant kinetins and chemotherapeutic drugs) induced sporulation of *B. subtilis*, when they were used at concentrations that caused partial inhibition of purine synthesis and therefore of growth (Freese et al., 1978; Heinze et al., 1978).

Sporulation Caused by a Purine Deficiency Resulting from a Block in the Common Part of the Branched Purine Pathway

Figure 1 shows the components of the branched path of *de novo* purine nucleotide synthesis essential for this paper. The inhibitors used are indicated by numbers. General purine synthesis can be inhibited by a number of compounds; one of them is amethopterin (methotrexate) whose effect on sporulation is shown in Figure 2 in order to allow comparison with results in yeast shown below (Fig. 8). Amethoperin is a folic acid analog which inhibits folic acid reductase and thereby all methylation and hydroxymethylation reactions of the cell. Tetrahydrofolate is needed for two steps in the purine pathway but also for the synthesis of thymine, glycine and certain vitamins. If all these compounds except purines are added to the medium a specific purine deficiency can be generated whose extent depends on the amount of amethopterin added to the culture. Figure 2 shows how the doubling times of *B. subtilis* increase with the amethopterin concentration, and it demonstrates that a culture containing an intermediate amethopterin concentration of about 10 μM produces optimal sporulation (measured 10 hr after amethoptherin addition). If a specific thymine deficiency is created by leaving out thymine but adding hypoxanthine, thymineless death occurs and no sporulation is observed (Heinze et al., 1978).

A block in the common purine pathway can also be obtained by a mutation. If such a purine requiring mutant (auxotroph) is grown in the presence of a purine, e.g. hypoxanthine, and (at $A_{600}=0.5$) the cells are washed on Millipore filters and resuspended in purine-free medium, one can measure residual growth and observe sporulation 10 hr later. By the use of many purine auxotrophs, such experiments have shown that leaky purine auxotrophs sporulate better than stringent ones (Freese et al., 1978; Freese et al., 1979). One can also generate different rates of purine synthesis and thus of growth in a stringent purine auxotroph by supplying the medium with a suboptimal amount of a precursor of purine nucleotides. Adenine, hypoxanthine, or their nucleosides cannot be used for this purpose because they are actively transported and metabolized with such a low K_m that only very low (μM) concentrations are

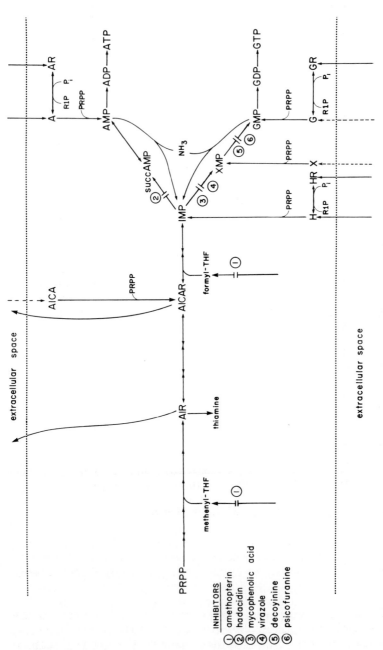

FIGURE 1. Purine pathway of B. subtilis. The arrows from the "extracellular space" indicate active transport with a low K_m (full line) and transport with a high K_m (dashed line). Abbreviations: AIR=5'PR-5-aminoimidazole, AICAR=5'PR-5-amino-4-imidazole carboxamide, R1P= ribose-1-P, H-hypoxanthine, HR=inosine, X-xanthine, THF=tetrahydrofolic acid.

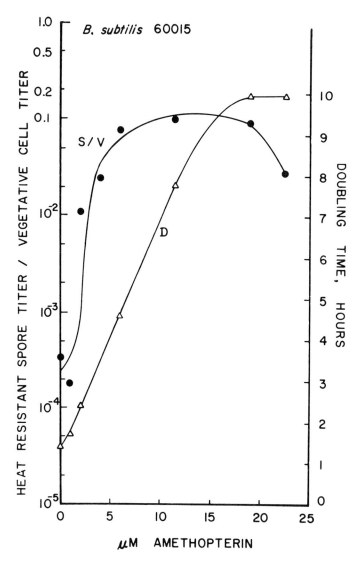

FIGURE 2. Sporulation induction after amethopterin addition at $A_{600}=0.5$; for details see Heinze et al., 1978. S/V = frequency of spores per cell at t_{10}. D = doubling time.

not growth rate limiting; but, at these low concentrations the compound is used up within minutes. In contrast, 5-amino — 4-imidazolecarboxamide (AICA), which enters the general purine pathway (Fig. 1), is so inefficiently transported or metabolized by the cell that the growth rate of a purine mutant depends on the millimolar concentrations of this compound in the medium. Figure 3 shows that the frequency of sporulation increases until the concentration of AICA reaches about 1.8 mM and decreases at higher concentrations.

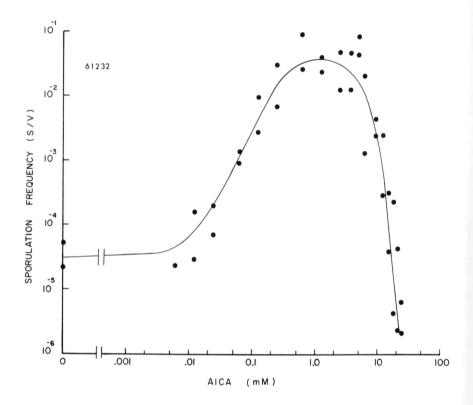

FIGURE 3. Sporulation of the purine auxotroph 61232 at different concentrations of AICA. The highest AICA concentration allowed growth at the normal rate. For details see Freese at al., 1979.

The Effect of a Block in One of the Branched Portions of the Purine Pathway

Now we would like to know whether inhibition or mutational deficiency of either the GMP ot the AMP branch of the purine path alone can initiate sporulation. The GMP branch can be inhibited by several compounds. Mycophenolic acid inhibits IMP dehydrogenase, the first enzyme in GMP synthesis, and induces sporulation (Fig. 4). Sporulation can also be induced

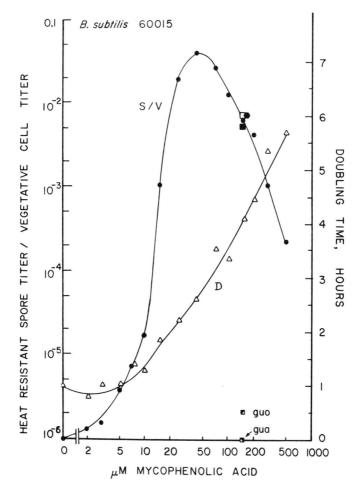

FIGURE 4. Sporulation induction by mycophenolic acid. Strain 60015 was grown in S7 medium (Freese et al., 1979) + 1% glucose + 0.1% vitamin-free casein hydrolysate. At A_{600} = 0.5, the culture was distributed into 125 ml flasks containing different amounts of mycophenolic acid. At optimal sporulation, S was $5*10^7$/ml and V $1.3*10^8$. In other experiments (at 50 μM mycophenolic acid), S was similar but V was sometimes up to 10 times higher (different degrees of lysis?). Addition of 500 μM guanine (■) or guanosine (◪) together with 150 μM mycophenolic acid abolished sporulation induction, whereas addition of adenine (■), adenosine (□) or hypoxanthine (⬢) did not.

by decoyinine (Mitani et al., 1977) or less effectively, by psicofuranine (Zain-ul Abedin et al., 1979), both inhibitors of GMP synthetase (Table I). Decoyinine is the most effective inducer of B. subtilis sporulation: it even enables the sporulation of citric acid cycle and certain other mutants which ordinarily cannot sporulate (Freese et al., 1979) and of manganese deficient cells (Vasantha & Freese, 1979).

GMP synthesis can also be prevented by the use of guanine auxotrophs. These mutants sporulate well in the absence of guanine and surprisingly well also in its presence because guanine is very badly transported into cells (Freese et al., 1979). However, in the presence of excess (1 mM) guanosine, which is transported and metabolized efficiently, all sporulation is prevented. When such cells are transferred to flasks containing medium with different amounts of guanosine, the results shown in Figure 5 are obtained (Freese et al., 1979). The frequency of sporulation observed without any guanosine depends on the leakiness of the gua mutation but is generally quite high. The added guanosine is readily converted (by guanosine phosphorylase) to guanine, so that low guanosine concentrations are rapidly used up, subsequently allowing excellent sporulation. But at sufficiently high concentrations (> 500 µM), enough guanosine remains in the medium throughout growth that the intracellular GMP concentration remains high, allowing rapid growth and preventing sporulation.

The AMP branch can be inhibited by hadacidin, an aspartate analog. In minimal medium, hadacidin optimally induces 5-10% sporulation if it is used at the partially inhibitory concentration of 2.1 mM (Table I) (Mitani et al., 1977). Its effect on growth and sporulation is prevented by adenine or aspartate. Interestingly, a combination of other amino acids also prevents the sporulation induction by hadacidin without preventing the growth inhibition (Table I). All specific adenine auxotrophs available so far show very little sporulation after adenine removal (or in its presence), and their sporulation has not been significantly increased by the addition of adenine at any concentration (Freese et al. 1979). [Mutants blocked at different steps of the pyrimidine pathway also do not sporulate upon pyrimidine removal from the medium; a mutant blocked before orotic acid can grow at different rates when this compound is added at different concentrations, but it produces only a low spore titer at any orotic acid concentration.] Apparently, mutants deficient in GMP synthesis are special in their ability to allow excellent sporulation in the absence of guanosine.

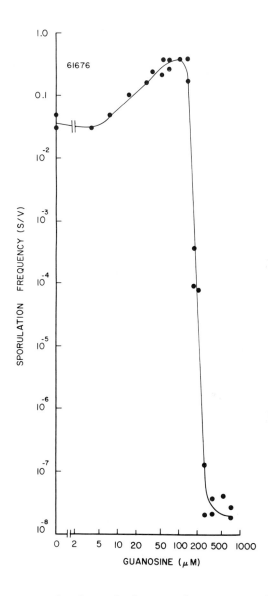

FIGURE 5. Sporulation of the guanine auxotroph 61676 at different concentrations of guanosine. Cell were grown as in Fig. 5 but with 1 mM guanosine. At $A_{600}=0.5$ they were washed with guanosine-free medium and transferred to flasks containing medium and different concentrations of guanosine. S/V determined at t_{10}.

Change of Nucleotide Pools During Initiation of Sporulation

To determine which nucleotides always decrease during the initiation of sporulation, nucleotide pools (expressed as pmole/A_{600}) were measured in ^{32}P labelled cells under many different conditions leading to sporulation. Some of the results are displayed in Figure 6. More detailed data will appear elsewhere (Lopez et al., 1979). The biosynthetic equilibria are such that the concentrations of the nucleoside triphosphates generally are 10 times higher than of the diphosphates and those in turn are 2 times higher than of the monophosphates. Therefore, only the results of nucleoside triphosphate determinations are here reported. The relative concentrations of ATP:UTP:GTP:CTP normally are approximately 7:1.6:1.4:1.

Partial inhibition of general purine synthesis (in the presence of AICA) causes a decrease in the cellular concentration of the purine nucleoside triphosphates; and more slowly of the pyrimidine nucleoside triphosphates (Fig. 6a). GTP decreases by a larger percentage than the other triphosphates. Hadacidin, the inhibitor of AMP synthesis causes a significant decrease not only of ATP but also of all other nucleoside triphosphates (Fig. 6b). However, in the presence of an amino acid mixture, the decrease of GTP is avoided although ATP still decreases by 90%; no sporulation is induced under these conditions (Table I). Apparently, the remaining ATP concentration is high enough that guanosine nucleotides can still be synthesized, which indicates that their synthesis is only indirectly controlled by ATP. Both decoyinine addition and guanosine removal from a guanine auxotroph decrease GTP and UTP but not ATP or CTP (Fig. 6c, d). Thus neither the increase nor the decrease of ATP or CTP are required for the initiation of sporulation. The decrease of UTP observed under most conditions is also not required for sporulation as follows from the facts that the concentration of this compound increases when sporulation is initiated by shift down to glutamate as sole carbon source (Fig. 6e) and that a uracil mutant cannot efficiently sporulate when the concentration of uracil nucleotides is decreased to any value (by changing the orotic acid supply); GTP increases under these conditions (Lopez et al., 1979). 6-azauracil can also induce sporulation and produces a decrease of both UTP and GTP (Lopez et al., 1979). Partial nitrogen deficiency causes a decrease in GTP and UTP (Fig. 6f) and phosphate deficiency results in a decrease of all four nucleoside triphosphates (Lopez et al., 1979).

Our results show that the intracellular concentration of GTP (and GDP) decreases under all conditions under which

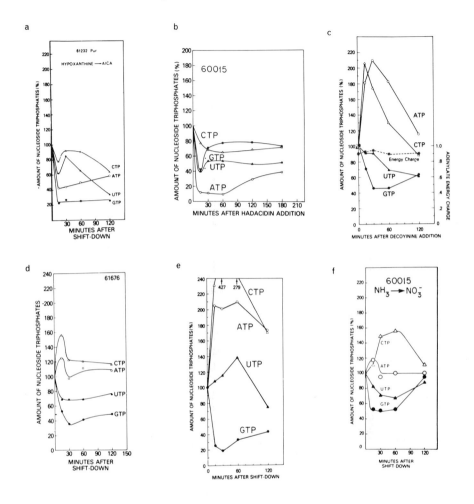

FIGURE 6. Alterations in the intracellular concentrations of nucleoside triphosphates after growth of B. subtilis for more than two generations in 20-50 μmole $^{32}P_i$ to $A_{600}=0.5$ and subsequent exposure to different conditions producing optimal sporulation (S/V in Table I). For details see Lopez et al., 1979. a) Purine auxotroph (61232 in 1.85 mM AICA). b) Strain 60015 after 2.1 mM hadacidin addition to S6+1% glucose. c) Same after 1.8 mM decoyinine addition. d) Guanine auxotroph (61676) grown as in Fig. 6 and transferred to medium without guanosine. e) 60015 shifted from a medium with 1% casein hydrolysate to one with 20 mM L-glutamate as sole carbon source. f) 60015 shifted to 3 mM nitrate as sole nitrogen source.

sporulation is observed. The decrease of ATP or UTP, if not accompanied by a decrease of GTP, does not initiate sporulation. While the results obtained with inhibitors or mutants of the GMP path prove that a partial deficiency in the synthesis of guanine nucleotides is sufficient to initiate sporulation, the pool measurements suggest the possibilty that the decrease of GTP (or GDP) is necessary for the initiation of sporulation, at least under all sporulation conditions presently known.

A few speculations about the possible role of GTP may be instructive. One possibility is that GTP is converted into a compound whose increase or decrease controls sporulation. Two compounds which normally increase as a result of amino acid starvation are ppGpp and pppGpp (Gallant & Lazzarini, 1976). But these compounds or their change are apparently not needed for sporulation. They have not been found to increase at the end of growth in nutrient sporulation medium (Fortnagel & Bergmann, 1974); although they showed a transient increase if this medium also contained glucose, they
were not observed in a relaxed (rel) mutant even after certain step-down conditions (Rhaese et al.,1976). Under these conditions, the concentration of GTP decreases and sporulation occurs. Addition of decoyinine causes a decrease of these compounds in rel$^+$ strains, and they are not observed in rel$^-$ strains (Lopez et al., 1979). We have observed another ^{32}P labeled compound which differs from any of the commercially avaibable highly phosphorylated nucleotides (Fig. 7). Its intracellular concentration decreases after addition of 6-azauracil, decoyinine or mycophenolic acid, or after shift-down from glucose to lactate, all conditions that initiate sporulation. Conceivably, the decrease of GTP could also cause the change of some other compound of high or low molecular weight (such as an increase of pppAppp (Rhaese et al.,1976)) which in turn initiates sporulation. However, we want to emphasize that there is no *a priori* reason to assume that differentiation must depend on the production or disappearance of any unusual compound (Freese, 1976). Since differentiation must have evolved by mutations of replicating cells, it could have been controlled early in evolution by one of the continuously present compounds such as GTP. In some organisms this control of differentuation may have been retained as such while in more evolved organisms it may now be generated by a superposition of more complex and therefore better controllable mechanisms.

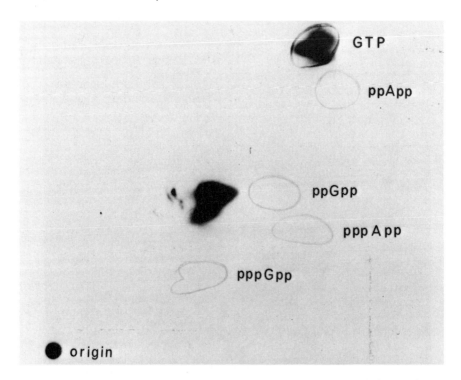

FIGURE 7. Two-dimensional separation of highly phosphorylated nucleotides in formic acid extracts of B. subtilis. 60015 cells, grown in S6+1% glucose and $^{32}P_i$, were extracted and the extract was chromatographed on polyethyleneimine thin layer plates   (Lopez et al., 1979). First dimension: 0.75 M KH_2PO_4 + 1.25 M guanidine-HCl, pH 3.4 by H_3PO_4. Second dimension: 60 min electrophoresis in 0.1 M potassium phosphate buffer + 7 M urea, pH 7.0, at 60 mA. Before electrophoresis the plates were washed for 10 min in methanol. The pencil lined standards (2.5 µg each) were visualized under UV.

RESULTS WITH *SACCHAROMYCES CEREVISIAE*

The yeast *Saccharomyces cerevisae* is usually grown in a rich medium and sporulation is measured after transfer of the cells to potassium acetate, i.e. in a medium containing a relatively poor carbon source and no nitrogen or phosphate (Fowell, 1967; Roth & Halvorson, 1969). Addition of NH_4Cl to this transfer medium prevents sporulation. To measure sporulation induction, we have used strain Y55 in a medium (MNCP) containing 100 mM potassium MES buffer, pH 5.5, yeast nitrogen base (6.7 mg/ml) including all vitamins, 1% vitamin-free casein hydrolysate, and 100 mM sodium pyruvate. This medium is essentially free of purines. Sporulation was detected under the phase contrast microscope by the visual appearance of asci containing 2 or more spores. The normal product of meiosis consists of an ascus with four spores. Such a tetrad can often not be seen as such because three spores cover the visibility of the fourth. We will for simplicity's sake call an ascus containing three or four spores a "tetrad". If the above MNCP medium contains all methylated and hydroxy-

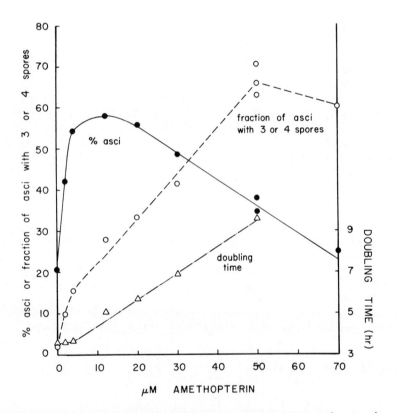

FIGURE 8. Sporulation of yeast caused by amethopterin. Saccharomyces cerevisae strain Y55 was grown in MNCP medium + 100 μg/ml thymine + 100 μg/ml glycine + 50 μg/ml L-tryptophan. At $A_{600}=1$, different amounts of amethopterin (1mM dissolved in 0.02 N HCl) were added. At t_{55} the frequency of asci per (normal plus empty) cells and the fraction of asci with 3 or 4 spores were determined under a phase contrast microscope (1000-fold magnification).

methylated compounds except purines, amethopterin inhibits growth and causes sporulation (Fig. 8). The frequency of asci first increases and then decreases with increasing amethopterin concentration. Interestingly, the fraction of asci with tetrads, which is only 0-5% without induction, increases with the amethopterin concentration even when the frequency of asci/cell already decreases again (Fig. 8). In contrast to *B. subtilis* where induction apparently starts very soon after addition of the inhibitor, yeast shows a significant increase in the spore titer only 30 or more hr after exposure to the inhibitor. Since yeast can sporulate well already 18 hr after transfer to potassium acetate, some compound, present in the MNCP medium (or accumulated in the cells before transfer), apparently has to run out before sporulation can take place.

Of the specific inhibitors of GMP synthesis, decoyinine and psicofuranine are ineffective in yeast. Virazole causes the production of about 60% asci if it is used at high concentrations (15 mM), but only 20% of these asci contain tetrads. Mycophenolic acid (5 mM) causes the production of 60-70% asci after 40-50 h and 80-100% of these asci contain tetrads (Fig. 9). At higher concentrations, mycophenolic acid precipitates in the medium. In the presence of the inhibitor, asci appear a few hr later than in the control. Conceivably this delay may be necessary to allow the four products of meiosis to separate before they are encased into spore walls.

Sporulation of yeast is more complex than that of *B. subtilis*. Firstly, yeast undergoes meiosis which requires different and more biochemical reactions than the asymmetric septation of *B. subtilis*. Secondly, all sporulation is prevented, even in the presence of mycophenolic acid, by the addition of glucose, fructose or mannose (until these more rapidly metabolizable carbon sources are completely used up). Presumably, these carbohydrates produce an excess of some other compound(s) which controls the onset of differentiation.

REFERENCES

Piggot, P.J. & Coote, J.G. (1976), Genetic aspects of bacterial endospore formation. *Bacteriol. Rev. 40*, 908.
Freese, E. (1972), Sporulation of *Bacilli*, a model of cellular differentiation. in "Current topics in developmental biology", (A.A. Moscona & A. Monroy), p. 85. Academic Press, New York.
Freese, E. (1976), Metabolic control of sporulation. in "Spore Research 1976", (A.N. Barker, G.W. Gould & J. Wolf), p. 1. Academic Press, London.

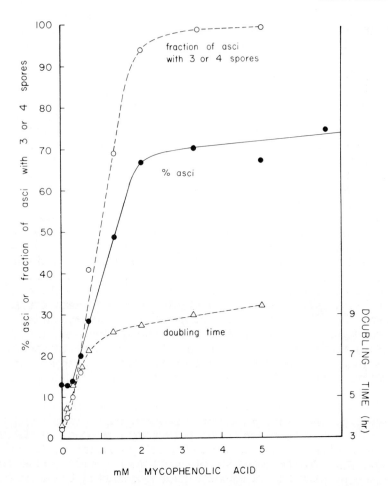

FIGURE 9. Sporulation of yeast caused by mycophenolic acid. Y55 was grown in MNCP. At $A_{600}=1$, the cells were centrifuged, resuspended in 2 times concentrated medium and added in 5 ml portions to 125 ml flasks containing 5 ml of different amounts of mycophenolic acid (dissolved at 15 mM in KOH at pH 12 and diluted with H_2O). The frequencies of asci with 2, 3 or 4 spores were determined at t_{44}. 500 μM guanine prevented sporulation whereas adenine or hypoxanthine did not.

Sonenshein, A.L. & Campbell, L.M. (1978), Control of gene expression during sporulation. in "Spores VII" (G. Chambliss & J.C. Vary), p. 179. American Society for Microbiology, Washington, D.C.
Freese, E., Heinze, J., Mitani, T. & Freese, E.B. (1978),

Limitation of nucleotides induces sporulation. *in* "Spores VII" (G. Chambliss & J.C. Vary), p. 277. American Society for Microbiology, Washington, D.C.

Heinze, J.E., Mitani, T., Rich, K.E. & Freese, E. (1978), Induction of sporulation by inhibitory purines and related compounds. *Biochim. Biophys. Acta 521*, 16.

Freese, E., Heinze, J.E. & Galliers, E.M. (1979), Partial purine deprivation causes sporulation of *B. subtilis* in the presence of excess ammonia, glucose and phosphate. *J. Gen. Microbiol.* - in press.

Zail-ul Abedin, M., Lopez, J.M. & Freese, E. (1979), Induction of bacterial sporulation by glycosyl adenines and zeatin. - in preparation.

Mitani, T., Heinze, J.E. & Freese, E. (1977), Induction of sporulation in *Bacillus subtilis* by decoyinine or hadacidin. *Biochem. Biophys. Res. Comm. 77(3)*, 118.

Freese, E.B., Vasantha, N. & Freese, E. (1979), Induction of sporulation in developmental mutants of *Bacillus subtilis*. *Molec. Gen. Genet. 170*, 67.

Vasantha, N. & Freese, E. (1979), The role of manganese in growth and sporulation of *B. subtilis*. *J. Gen. Microbiol.* - in press.

Lopez, J.M., Marks, C.L. & Freese, E. (1979), The decrease of guanosine nucleotides initiates sporulation of *Bacillus subtilis*. *Biochim. Biophys. Acta* - in press.

Gallant, J. & Lazzarini, R.A. (1976), The regulation of rRNA synthesis in microorganisms. *in* "Protein Synthesis" (E. Mc Conkey), p. 309. Marcel Dekker Inc., New York.

Fortnagel, P. & Bergmann, R. (1974), The snythesis of MS1 and MS2 by *Bacillus subtilis*. *Biochem. Res. Comm. 56*, 264.

Rhaese, H.J., Grade, R. & Dichtelmüller, H. (1976), Studies on the control of development. Correlation of initiation of differentiation with synthesis of highly phosphorylated nucleotides in *B. subtilis*. *Eur. J. Biochem. 64*, 205.

Fowell, R.R. (1967), Factors controlling the sporulation of yeast. II. The sporulation phase. *J. App. Bacteriol. 30*, 450.

Roth, R. & Halvorson, H.O. (1969), Sporulation of yeast harvested during logarithmic growth. *J. Bacteriol. 98*, 831.

REGULATION OF SPORULATION BY HIGHLY PHOSPHORYLATED NUCLEOTIDES IN BACILLUS SUBTILIS

Hans J. Rhaese, Reinhard Groscurth,
Roman Vetter, Hannelore Gilbert

Institut für Mikrobiologie
Universität Frankfurt
Frankfurt am Main

A highly phosphorylated nucleotide is synthesized by a membrane bound enzyme of Bacillus subtilis upon removal of carbon sources from various growth media. This substance has been tentatively identified as adenosine-5',3'(2')-bis-triphosphate, p_3Ap_3. The enzyme synthesizing this substance is named adenosine-bis-triphosphate (abt) synthetase. Isolation of a temperature sensitive abt synthetase from a temperature sensitive sporulation mutant with a lesion in the spoOF gene seems to indicate that a mutation in the structural gene coding for this enzyme affects sporulation. A DNA fragment of 1.3 million daltons was isolated from EcoR1 cleaved B. subtilis DNA which carries essential parts of the abt gene as shown by transformation of an abt negative mutant. This fragment was cloned in the B. subtilis plasmid pBS 161-1.

INTRODUCTION

The simple procaryote *Bacillus subtilis* can be induced to sporulate by replacing the growth medium with a medium lacking easily fermentable carbon sources. In order to elucidate the molecular mechanism of initiation of sporulation, we searched for low molecular weight effectors, which may be synthesized in response to this shift down and which may be responsible for the observed changes in transcription needed to synthesize molecules involved in sporulation.
   Our observation published in several papers (Rhaese and Groscurth, 1974; Rhaese *et al.*, 1976;

Rhaese and Groscurth, 1976), that synthesis of unusual highly phosphorylated nucleotides, abbreviated HPN, is always correlated with sporulation seems to indicate that these substances play an important role in initiation of sporulation in *B.subtilis*. The discovery of a cytoplasmic membrane bound enzyme, which is able to synthesize a highly phosphorylated nucleotide, whose apparent structure is adenosine-5',3'(2')bis-triphosphate (p_3Ap_3), and the observation, that this enzyme can be inhibited by phosphorylated metabolites of carbon sources, are the basis of our sensor model of initiation of differentiation described previously (Rhaese, 1978).

The observation that the membrane bound enzyme abt synthetase, synthesizing adenosine-bis-triphosphate, p_3Ap_3, is always present in *B.subtilis* independent of the life cycle of this bacterium (Rhaese and Groscurth, 1979), indicates that its activity is regulated by external carbon sources. The close association of p_3Ap_3 synthesis and sporulation indicates that the highly phosphorylated nucleotide mentioned above is needed for sporulation.

In this communication we further confirm our previous findings by showing that synthesis of HPN depends entirely on the absence of carbon sources and is independent of other nutrients in the growth medium. In addition, we try to demonstrate that sporulation depends on the prior synthesis of p_3Ap_3 by showing that mutations in the structural gene for abt synthetase affect sporulation.

MATERIALS AND METHODS

Bacteria

The sporulating *B.subtilis* strain 60015 (*try, met*) and early blocked asporogenous mutants JH 649 (*try, phe, spoO*) and JH 756b (*try, phe, spoO^{ts}*) were used in this study.

Media and Growth Conditions

Growth conditions in semisynthetic yeast extract containing medium (SYM) and in sporulation medium (SSM) will be described in detail elsewhere (Groscurth and Rhaese, in preparation). Nutrient broth (NB) contains per liter 8 g DIFCO nutrient broth and 4 g sodium chloride. Mandelstam's sporulation and replacement media were used as described (Sterlini and Mandelstam, 1969).

Biochemical Methods

All methods concerning labelling, analysis, and quantitative measurements of highly phosphorylated nucleotides have been described in detail (Rhaese et al., 1975). DNA was isolated according to published procedures (Saito and Miura, 1963). Isolation of plasmid pBS 161-1 and treatment of DNA and plasmid with EcoR1 and Pst restriction endonucleases was performed as described (Kraft et al., 1978). Agarose gel electrophoresis in a Hoefer Scientific gel apparatus as well as visualization of UV absorbing material will be described elsewhere (Rhaese, Vetter, and Gilbert, in preparation).

Transformation

An efficient transformation procedure, especially suitable for early blocked sporulation mutants has been developed by Chang and Cohen (Chang, S. and Cohen, S.N., personal communication). We are grateful to Drs. Chang and Cohen for allowing us to use their method prior to publication.

RESULTS

Synthesis of a Highly Phosphorylated Nucleotide upon Shift Down from Rich to Poor Growth Media

In order to show that synthesis of the highly phosphorylated nucleotide p_3Ap_3 depends entirely on carbon source starvation and is independent of the growth medium itself, we grew the sporulating *B.subtilis* strain 60015 in various growth media and resuspended the cells after removal of the medium by centrifugation in a medium lacking carbon sources.

As can be seen in figure 1, the highly phosphorylated nucleotide p_3Ap_3 (the spot near the origin) is immediately synthesized when cells grown in SYM in the presence of 0.5 mCi/ml of $H_3^{32}PO_4$ are resuspended in SSM containing again 0.5 mCi/ml of $H_3^{32}PO_4$. Samples were removed at 15 min intervals, extracted with formic acid and chromatographed on polyethyleneimine cellulose thin layer plates using 1.5M KPO_4-buffer, pH 3.4.

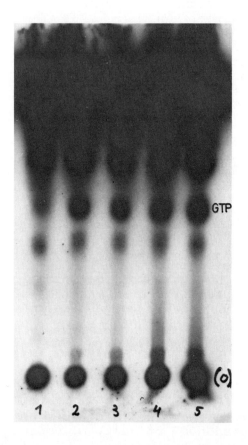

Figure 1. Autoradiogram of a PEI thin layer chromatogram of ^{32}P-labelled cells extracted with 1 M potassium formate, pH 3.4 as described in Materials and Methods. Column 1 contains an extract of cells removed one minute after shift-down. Column 2, 3, 4, and 5 show samples taken at 15 minutes intervalls after shift-down. The highly phosphorylated nucleotide is the spot near the origin (o).

The same procedure was followed using other growth and sporulation media (Figure 2). Regardless of whether cells were grown in SYM, NB or Mandelstam's medium, synthesis of p_3Ap_3 begins almost immediately after replacement of the growth medium by sporulation medium. For example, cells grown in SYM and resuspended in SSM produce p_3Ap_3 at a remarkable rate (—●—). Cells grown in NB (—■—) synthesize somewhat less p_3Ap_3. There is some difference in synthesis of p_3Ap_3 between cells grown in SYM and shifted to SSM (—●—) and those, grown in Mandelstam's medium and shifted to Mandelstam's replacement medium (—▲—). But the addition of glucose to Mandelstam's medium followed by a shift to Mandelstam's replacement medium (—△—), resulted in a somewhat higher rate of synthesis (Figure 2).

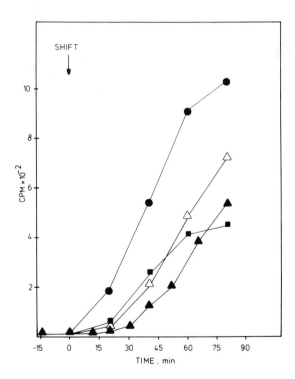

Figure 2. Increase in synthesis of the highly phosphorylated nucleotides p_3Ap_3 after a shift of ^{32}P labelled cells from SYM to SSM (—●—), NB to SSM (—■—), Mandelstam growth medium to replacement sporulation medium (—▲—), and Mandelstam's medium containing 10 mM glucose to replacement sporulation medium (—△—).

Chromatographic comparison of the radioactive material migrating near the origin in the chromatogram shown in figure 1 with UV absorbing material isolated from unlabelled 10 l cultures shows that these substances seem to be identical. The apparent identity of these substances with a chemically synthesized adenosine-bis-triphosphate, p_3Ap_3, was also established by chromatographic means.

All data obtained so far seem to indicate, even though further physicochemical analysis are required, that the substance mentioned in figure 1, which is produced in response to a shift-down, is p_3Ap_3.

Isolation of Early Blocked Temperature Sensitive Sporulation Mutants of B.subtilis

To show that p_3Ap_3 is synthesized by the enzyme abt synthetase after deprivation of carbon sources to initiate sporulation and not needed for some other unknown physiological functions we searched for asporogenous mutants unable to synthesize p_3Ap_3.

The early blocked asporogenous mutant JH649, isolated by Dr.Hoch, is such a mutant (Rhaese et al. 1978). This mutant has been mapped and the mutation found to be in the *spoOF* gene, which can be cotransformed with the *ctrA* and *furC* markers, respectively.

Since spontaneous revertants of this mutant regained the ability to synthesize p_3Ap_3 simultaneously with the ability to sporulate and since this mutant was shown by Dr.Hoch (personal communication) to have a single site mutation in the *spoOF* gene but that it is otherwise isogenic with the sporogenous, p_3Ap_3 synthesizing wild-type strain, we suspected that the mutation is in the structural gene for abt synthetase.

When we mutagenized wild-type DNA with hydroxylamine (1M), treated competent cells of the sporogenous *ctrA* type as described previously (Rhaese and Groscurth, 1979), and selected for temperature sensitive sporulation mutants, we found among cytidine-independent transformants one which was able to sporulate at 30° but not at 42°. We then tested the ability of this mutant to synthesize p_3Ap_3 *in vivo* at 30° the same way as described above (Figure 2). We found that there is no difference between wild-type and mutant (data not shown). However, when we raised the temperature to 42°, the mutant was no longer able to synthesize p_3Ap_3 (Figure 3, —▲—), whereas the wild-type shows no difference in p_3Ap_3 synthesizing capacity at those conditions (—△—).

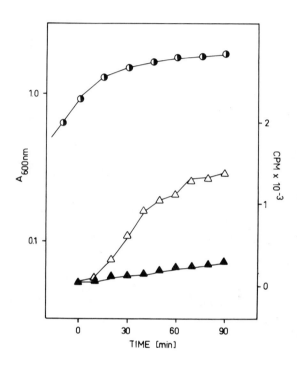

Figure 3. Growth of B.subtilis strains 60015 and mutant JH756b (—◐—) before and after shift-down into SSM at time 0. Synthesis of the highly phosphorylated nucleotide was measured at 42° both in 60015 (—△—) and the temperature sensitive mutant (—▲—).

To show that the temperature sensitivity of the *in vivo* synthesis of p_3Ap_3 and of sporulation is due to a mutation in the structural gene of abt synthetase, we isolated and tested the ability of membrane vesicles to synthesize p_3Ap_3 at 30° and 42°. As is shown in figure 4, p_3Ap_3 is synthesized at 42° to some extent (Figure 4, —●—). However, when the temperature is lowered to 30° a significant increase is observed (Figure 4, —○—).

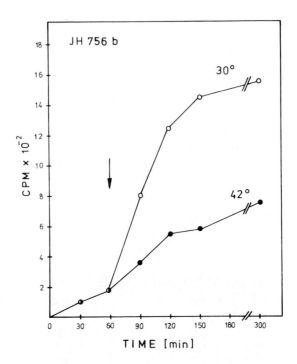

Figure 4. Synthesis of the highly phosphorylated nucleotide by membrane vesicles of the temperature sensitive sporulation mutant in vitro at 42° (—●—) and after shift of one half of the culture (arrow) to 30° (—○—).

Isolation of a Gene carrying Essential Parts of the abt Synthetase

Even though the *in vitro* temperature sensitivity of abt synthetase is not very pronounced, it still seems to be significant enough to conclude that a somewhat temperature sensitive enzyme apparently causes a temperature sensitivity in sporulation. Therefore, it seems as if mutations in the structural gene for abt synthetase affect sporulation.

To test this possibility further, we isolated *B.subtilis* wild-type DNA, cleaved it with restriction endonuclease EcoR1 and Pst and separated the fragments by agarose gel electrophoresis. As can be seen in figure 5, many fragments were obtained both with EcoR1 and Pst. No single bands are visible. As molecular weight markers we used an EcoR1 cleaved octamer of phage λdv.

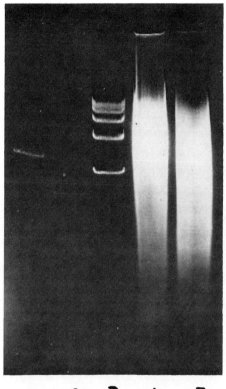

Figure 5. Agarose gel of Eco R1 (column 5) and Pst (column 4) cleaved B.subtilis DNA. Molecular weight markers (column 3) are partially Eco R1 cleaved λdv21 octamers (molecular weight of the monomer is 2.08 x 10^6 daltons). Column 2 shows the covalently closed plasmid pBS161-1 and column 1 shows the same plasmid cleaved with Eco R1.

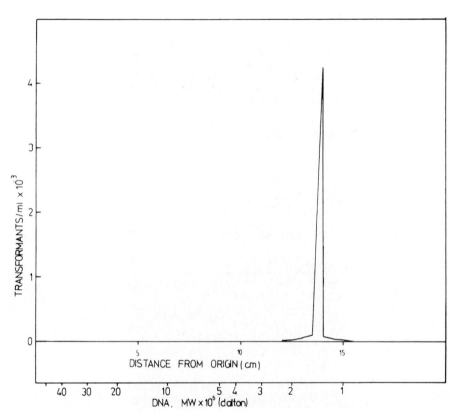

Figure 6. Transformation of strain JH649 to sporogeny by DNA fragments obtained after separation of Eco R1 cleaved DNA on agarose gels (see Figure 5). Sporogenous transformants were selected among asporogenous nontransformants by transferring the transformation culture to SYM, thus allowing one duplication and subsequent sporulation. Nonsporulating cells were killed by heat treatment (80° C, 20 minutes).

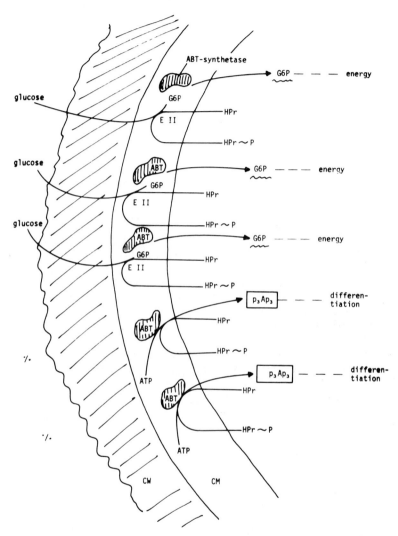

Figure 7. Sensor model of initiation of differentiation. The enzyme abt synthetase located in the cytoplasmic membrane of B.subtilis is inhibited by sugar phosphates, for exampe glucose-6-phosphate (G6P). Upon removal of sugars, like glucose, abt synthetase may be activated and able to synthesize p_3Ap_3 using HPr~P as possible phosphate donor.

The agarose gels were cut up in 0.5 cm wide strips, and the restricted *B.subtilis* DNA's eluted from the gel, and used to transform the asporogenous *spoOF* mutant JH649 to sporogeny. As can be seen in figure 6, no sporogenous transformants were observed with all fractions except one, which consists of fragments of the average molecular weight of 1.3 million daltons.

When we calculated the factor by which the *spoOF* gene was enriched, we found that the enrichment factor was 870; the purity of the *spoOF* gene is approximately 60 %.

We now treated this fragment, which is large enough to carry a maximum of two genes, with methylmethane sulfonate, transformed competent JH649 cells to sporogeny at 30° and selected for temperature sensitive asporogenous mutants. A temperature sensitive sporulation mutant was obtained, which has similar properties as the mutant obtained by treating competent cells of a sporogenous strain with unfractionated mutagenized DNA.

These results show that the mutated DNA fragment apparently contains the structural gene for abt synthetase, because mutations in this gene lead to altered abt synthetase. It shows furthermore, that mutations in this gene affect sporulation. Therefore it seems that p_3Ap_3 is essential for sporulation.

Together with our knowledge of the regulation of p_3Ap_3 synthesis and sporulation by environmental conditions we postulate that p_3Ap_3 is involved in initiation of sporulation.

The above described DNA fragment was also used to clone the *spoOF* gene in a *B.subtilis* plasmid pBS 161-1. After cutting the plasmid with the same restriction endonuclease by which the *spoOF* fragment was obtained, ligation with T4 DNA ligase was employed and the resulting recombinant plasmid transformed into a *rec E4* derivative of strain JH649 by the method of Chang and Cohen. A tetracyclin resistant clone was obtained which apparently contains a plasmid with parts of the *spoOF* gene (details will be published elsewhere).

This clone, however, was not able to sporulate nor was synthesis of p_3Ap_3 detectable. We assume that not the entire gene is contained in the fragment described above, but rather parts of it carrying regions which are apparently mutated in the *spoOF* mutant JH649.

DISCUSSION

The above described experiments seem to indicate that carbon source deprivation triggers the synthesis of a highly phosphorylated nucleotide which, according to data we have obtained so far, seems to be an adenosine-5',3'(2')bis-triphosphate. The previously described *in vivo* and *in vitro* inhibition of the enzyme abt synthetase, synthesizing this substance (Rhaese and Groscurth, 1976; Rhaese and Groscurth, 1979) and the apparent dependence of sporulation on the prior synthesis of this nucleotide allowed us to formulate the so called sensor model of initiation of sporulation in *B.subtilis*.

This model postulates that the enzyme abt synthetase is an integral part of the cytoplasmic membrane and present there at all times during the entire life cycle of *B.subtilis*. It is, however, inhibited as long as carbon sources, for example glucose, are transported as sugar phosphates through the cytoplasmic membrane by the phosphoenol pyruvate: sugar phosphoryltransferase system, PTS. These sugar phosphates are formed by enzyme II and HPr$\sim$P as phosphate donor as shown in figure 7. The inhibition of abt synthetase is mediated by phosphorylated metabolites of carbon sources (Rhaese and Groscurth, 1976; Rhaese and Groscurth, 1979). As long as sugars (for example glucose) are phosphorylated and transported through the membrane, abt synthetase cannot form adenosine-bis-triphosphate. However, as soon as sugars are removed from the growth medium, phosphorylation of ATP by uninhibited abt synthetase can take place, possibly by using the HPr$\sim$P as phosphate donor. Adenosine-bis-triphosphate is then formed, triggering sporulation in an as yet unknown manner.

ACKNOWLEDGEMENTS

This work was supported by the Deutsche Forschungsgemeinschaft.

REFERENCES

Kraft, J., Bernhard, K., and Goebel, W. (1978). Recombinant plasmids capable of replication in *Bacillus subtilis* and *Escherichia coli*. Molec.Gen. Genet.162, 59-67.

Rhaese, H.J. and Groscurth, R. (1974). Studies on the control of development. *In vitro* synthesis of HPN and MS nucleotides by ribosomes from either sporulating or vegetative cells of *Bacillus subtilis*. FEBS Letters 44, 87-93.

Rhaese, H.J., Dichtelmüller, H., and Grade, R. (1975). Studies on the control of development. Accumulation of guanosine tetraphosphate and pentaphosphate in response to inhibition of protein synthesis in *Bacillus subtilis*. Eur.J.Biochem.56, 385-392.

Rhaese, H.J., Grade, R., and Dichtelmüller, H. (1976). Studies on the control of development. Correlation of initiation of differentiation with synthesis of highly phosphorylated nucleotides in *Bacillus subtilis*. Eur.J.Biochem.64, 205-213.

Rhaese, H.J. and Groscurth, R. (1976). Control of development: Role of regulatory nucleotides synthesized by membranes of *Bacillus subtilis* in initiation of sporulation. Proc.Natl.Acad.Sci.USA 73, 331-335.

Rhaese, H.J., Hoch, J.A., and Groscurth, R. (1977). Studies on the control of development: Isolation of *Bacillus subtilis* mutants blocked early in sporulation and defective in synthesis of highly phosphorylated nucleotides. Proc.Natl.Acad.Sci.,USA,74,3 1125-1129.

Rhaese, H.J. (1978). Sporenbildung bei *Bacillus subtilis* - ein einfaches System zum Studium der Zelldifferenzierung. forum mikrobiologie 1, 24-29.

Rhaese, H.J. and Groscurth, R. (1979). Apparent dependence of sporulation on synthesis of highly phosphorylated nucleotides in *Bacillus subtilis*. Proc. Natl.Acad.Sci.,USA,76, 842-846.

Saito, H. and Miura, K.-J. (1963). Preparation of transforming desoxyribonucleic acid by phenol treatment. *Biophys.Acta 72*, 619-629.

Sterlini, J.M. and Mandelstam, J. (1969). Commitment to sporulation in *Bacillus subtilis* and its relationship to actinomycin resistance. *Biochem.J. 113*, 29-37.

Part III
HIGHLY PHOSPHORYLATED NUCLEOTIDES IN EUKARYOTES

Section 1
Isolation and Characterization of Novel Dinucleotides

Section 2
Effects of Dinucleotides and Adenosine Nucleotide Pools in Replication of DNA and Protein Synthesis

REGULATION OF MACROMOLECULAR SYNTHESIS
BY LOW MOLECULAR WEIGHT MEDIATORS

STUDIES ON THE BIOSYNTHESIS AND FUNCTION OF
DINUCLEOSIDE POLYPHOSPHATES IN *ARTEMIA* EMBRYOS[1]

A.H. Warner

Department of Biology
University of Windsor
Windsor, Ontario, Canada

The brine shrimp, Artemia salina, is a purine-requiring organism whose oocytes accumulate large amounts of Gp_4G during oogenesis. Following fertilization, large quantities of Gp_3G and small amounts of Gp_2G and Gp_3A are synthesized in the developing embryo prior to their entrance into dormancy. Collectively, the dinucleoside compounds represent about 2% of the dry weight of dormant Artemia embryos. In Artemia cysts Gp_2G, Gp_3G and Gp_4G are found predominantly in the yolk platelets, whereas Gp_3A is localized in the postmitochondrial fraction. The experimental data suggest that Gp_3G serves as a shuttle in the utilization of yolk platelet Gp_4G during early development. Although the reactions leading to the synthesis of Gp_2G and Gp_3A have yet to be characterized, in vitro experiments have shown that Gp_3G and Gp_4G are synthesized primarily in the yolk platelets and that the enzyme GTP:GTP guanylyltransferase in yolk platelets catalyzes the synthesis of these compounds. In Artemia Gp_4G is the primary source of all purine containing compounds for development. In addition, all dinucleoside compounds can be methylated. These data and others suggest that the dinucleoside compounds may play a regulatory role in protein synthesis during development in Artemia.

[1]Research supported by the N.R.C. of Canada.

INTRODUCTION

Embryos of the brine shrimp, *Artemia salina*, are rich in a group of compounds known as dinucleoside polyphosphates among which diguanosine tetraphosphate (Gp_4G)[2] is the most abundant. These nucleotides were first described 16 years ago(1), and since then nucleotides belonging to this group have been found in embryos of *Daphnia*(2), *Eubranchipus*(3), *Branchipus*(4) and in several mammalian cells(5). The physiology of these nucleotides has been studied most thoroughly in *Artemia*, but progress towards elucidation of their function has been hampered due to the inability of *Artemia* to synthesize purines *de novo* and the impermeability of encysted embryos and prenauplii to radiolabeled purines(3,6). Despite these constraints, *Artemia* embryos are a good system to study the function of dinucleoside polyphosphates.

The major morphological and metabolic changes which occur in encysted embryos of *Artemia* are outlined in Fig. 1. Near the end of gastrulation all macromolecular syntheses come to a halt and about 95% of the embryonic water is expelled as the embryo enters dormancy. Although dormancy appears to be an obligatory step in the development of the encysted embryo, it may be terminated at any time by immersion of the cyst in 0.5 M NaCl. Associated with the resumption of development is the onset of protein synthesis followed shortly by RNA synthesis. DNA synthesis does not resume until hatching begins; thus, development between the end of dormancy and the onset of hatching occurs in the absence of DNA synthesis and cell division.

The purpose of this work is to review the salient aspects of diguanosine nucleotide metabolism in brine shrimp embryos and to consider whether the diguanosine nucleotides serve mainly as a source of purines and high energy phosphate, or whether they and related nucleotides also play an important role in the regulation of macromolecular processes during development.

[2] *The following abbreviations have been used: Gp_4G, guanosine 5'-tetraphospho-5'-guanosine; Gp_3G, guanosine 5'-triphospho-5'-guanosine; Gp_3A, guanosine 5'-triphospho-5'-adenosine; and Gp_2G, guanosine 5'-diphospho-5'-guanosine.*

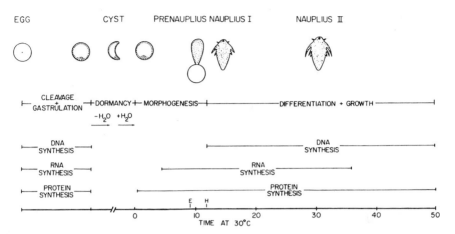

Fig. 1. Profile of the major morphological and metabolic changes in encysted embryos of Artemia salina following fertilization. E and H refer to the time of emergence and hatching, respectively.

METHODS

Yolk platelets and other subcellular fractions were prepared from *Artemia* embryos as described by Warner et al.(7). The isolation and purification of Gp_4G guanylohydrolase(8), GTP:GTP guanylyltransferase(9), and the acid soluble nucleotides have also been described previously(10,11).

For the elongation assays ribosomal wash proteins and unwashed polysomes were prepared from 0-h and 12-h *Artemia* embryos, respectively as described by Warner et al.(12). The assays contained the following in 0.15 ml: Tris-HCl, pH 7.4, 50 mM; KCl, 100 mM; $MgCl_2$, 8.1 mM; EDTA, 0.1 mM; DTT, 1 mM; ATP, 2.1 mM; GTP, 0.21 mM; creatine phosphate, 2 mM; creatine phosphokinase, 12.5 µg; deacylated *Artemia* cyst tRNA, 0.5 A_{260}; 0.25 µCi of a mixture of 15 ^{14}C-amino acids; 0.05 mM of each remaining ^{12}C-amino acids; 2 A_{260} units of 12-h polysomes; 0.14 mg 0-h ribosomal wash proteins; and dinucleoside polyphosphates, 0.26 mM. For the aminoacylation assays the reaction mixtures were as above except that polysomes and GTP were omitted and the amount of ribosomal wash was 28 µg. All incubations were conducted at 30°C and aliquots were processed for protein synthesis or aminoacylation as described previously(12).

The methylation study was conducted using a post-ribosomal fraction prepared from 0-h *Artemia* cysts as enzyme source(13). The reaction mixture was similar to that described by Ensinger and Moss(14) except that it contained 50 μM S-adenosyl [methyl-^{3}H] methionine (421 dpm/pmol), 50 μM nucleotide, and 6.5 mg protein in 1.0 ml. Incubations were conducted at 35°C and aliquants of 0.3 ml were removed at selected times and processed for nucleotide methylation using the organic-amine method of Warner and McClean(3).

RESULTS AND DISCUSSION

Nucleotide Content of Artemia Embryos

The acid soluble nucleotides in dormant cysts of *Artemia* comprise about 3% of the dry weight of the cysts. These nucleotides have been fractionated by ion-exchange chromatography on DEAE-cellulose(10) and Dowex-1-formate columns(11). Sixteen nucleotides have been identified in *Artemia* embryos including four dinucleoside compounds; at least eight others which are present in small amounts remain to be characterized. The contribution of each nucleotide to the total nucleotide pool is summarized in Table 1. These data show that the guanine nucleotides comprise about 90% of the total nucleotide pool. Of primary interest to this conference are the data which show that *Artemia* embryos contain four dinucleoside compounds. The most abundant of these are Gp_3G and Gp_4G and the least are Gp_2G and Gp_3A.

Distribution of Dinucleoside Compounds in Artemia

In 1968 Warner and McClean surgically removed the eggs from the ovisacs of adult brine shrimps and demonstrated, by direct chemical tests, that Gp_4G is present only in ovarian tissue of adult brine shrimp(3). These researchers also demonstrated that females lacking eggs or males (except for nauplii) do not incorporate ^{3}H-guanosine into either Gp_3G or Gp_4G during a 21-h incubation period despite a large intracellular pool of GTP. Other data showed that both non-encysted and encysted eggs contain similar amounts of Gp_4G. Collectively, these studies demonstrated that the developing egg of *Artemia* is the site of both synthesis and storage of Gp_4G.

TABLE 1. Nucleotide Composition of Artemia Cysts

Compound	μmoles/gram	Percent Dry Weight
Gua + Guo	5.38	0.152
CMP	0.525	0.017
AMP	0.950	0.033
UMP	0.638	0.021
GMP	5.18	0.188
IMP	0.445	0.015
sAMP	0.129	0.006
ADP	1.25	0.053
GDP	4.38	0.194
UDP	0.141	0.006
UDP-NAG	0.183	0.011
ATP	0.513	0.026
GTP	4.59	0.240
Gp_2G	0.188	0.013
Gp_3G	2.33	0.183
Gp_3A	0.108	0.008
Gp_4G	21.19	1.850
Totals	48.130	3.016

When the nucleotide composition of various fractions of encysted *Artemia* embryos was studied, we observed that the nucleotides are partitioned almost equally between the post-mitochondrial fraction and the yolk platelets(7). Moreover, the yolk platelets were found to be rich in the diguanosine compounds, especially Gp_4G which comprises 79% of the total nucleotide pool in the platelets. The distribution of the dinucleoside compounds among the various fractions is shown in Table 2. These results indicate that Gp_2G, Gp_3G and Gp_4G are localized primarily in yolk platelets of *Artemia* embryos, whereas Gp_3A is found exclusively in the post-mitochondrial fraction.

Biosynthesis of Metabolism of Gp_3G and Gp_4G

Dormant embryos of *Artemia* contain at least two enzymes which function in the metabolism of Gp_3G and Gp_4G. The first enzyme discovered was Gp_4G guanylohydrolase(8). This enzyme has been purified over 200-fold by Warner and Finamore and some of its properties have been studied by them(8) and others(15). The enzyme has a molecular weight of about 20,000(16) and it is localized mainly in the post-ribosomal fraction(15). At pH 8 and 4-5 mM $MgCl_2$ the enzyme catalyzes the hydrolysis of Gp_4G to give equimolar amounts of GMP and GTP. Under similar conditions the enzyme has no effect on Gp_3G, but at high $MgCl_2$ (10-30 mM) it catalyzes slowly the hydrolysis of Gp_3G to GMP and GDP(16). Although commercial preparations of GMP and GTP inhibit this enzyme(15), neither GMP nor GTP isolated from reaction mixtures has any effect on the enzyme(8). The inhibitory activity in commercial GTP appears to be due to guanosine tetraphosphate(15). After hatching of *Artemia* Gp_4G guanylohydrolase activity increases about 2-fold and remains at this level until the Gp_4G content is nearly exhausted(16).

In 1972 the second diguanosine nucleotide metabolizing enzyme was discovered(7). This enzyme was named GTP:GTP guanylyltransferase (Gp_4G synthetase) when its primary function was shown to catalyze the formation of Gp_4G from two equivalents of GTP(17). In dormant cysts of *Artemia* about 80% of the Gp_4G synthetase activity is present in the yolk platelets; the remainder is in the post-ribosomal fraction. Both enzymes have similar chromatographic properties on columns of Sepharose 6B except for an additional minor peak of activity in the cytoplasmic fraction (unpublished observation); therefore, they appear to be the same enzyme. The yolk platelet enzyme has been studied *in vitro* as a component of intact yolk platelets and as a partially purified protein. Both yolk platelets and purified enzyme catalyze the incorporation of ^3H-GTP into ^3H-Gp_4G and ^3H-GDP into ^3H-Gp_3G. However, the latter reaction requires the addition of either GTP or Gp_4G for maximal activity since Gp_4G is also a substrate in the reaction(9,17). When intact platelets are used as the source of enzyme, or when purified enzyme is supplemented with Gp_4G, the ratio of rates of synthesis of $Gp4G$ to $Gp3G$ is about 10 to 1 using ^3H-GTP and ^3H-GDP, respectively as substrates. This value is in good agreement with the relative composition of Gp_4G and Gp_3G in *Artemia* yolk platelets (see Table 2). These results show that the activity of yolk platelets *in vitro*, at least with respect to diguanosine nucleotide metabolism, reflects the activity of these organelles *in vivo*. Equations 1 and 2 describe the reactions catalyzed by GTP:GTP guanylyltransferase. This enzyme

(1) $2 \text{ GTP} \rightleftharpoons Gp_4G + PP_i$

(2) $GDP + Gp_4G \rightleftharpoons Gp_3G + GTP$

appears to be related to RNA guanylyltransferase which catalyzes "capping" of the 5'-end of mRNA(18), but it is not known whether it catalyzes guanylation of mRNA.

During oogenesis in *Artemia* Gp_4G is synthesized in the developing egg in close association with yolk platelet formation, and it accumulates in the platelets to about 3.5% of

TABLE 2. Distribution of Dinucleoside Polyphosphates Among Subcellular Functions of Artemia Cysts

	μmoles nucleotide/gram dormant Artemia cysts[a]			
Nucleotide	Yolk Platelets[b]	Mitochondria[c]	Post-Mitochondrial Supernatant[d]	Totals
Gp_2G	0.135	0	0.053	0.188
Gp_3G	1.84	0	0.483	2.33
Gp_3A	0	0	0.108	0.108
Gp_4G	19.7	0	1.49	21.19

[a] All nucleotides have been corrected for column losses.
[b] The 1,000 g sediment was washed 3x with homogenizing bugger and contains about 95% yolk platelets; the remainder are nuclear fragments.
[c] The mitochondrial-rich 15,000 g sediment was washed 3x with the homogenization buffer.
[d] The 15,000 g supernatant fluid and washes from c.

their dry weight(7). During this time no Gp_3G synthesis can be detected within the maturing egg. Thus the Gp_3G of dormant cysts must be synthesized between fertilization and the onset of dormancy. If we assume that Gp_4G synthetase catalyzes the synthesis of both Gp_3G and Gp_4G, how is the synthesis of these two nucleotides regulated? Since Gp_4G and Gp_3G are not found in large quantities in the cytosol (compared to the platelets), any attempt to answer this question must consider the GTP/GDP ratio in the cytosol and yolk platelets, as well as the structure and/or activity of the yolk platelets. The proposed cycle for the biosynthesis and metabolism of Gp_4G shown in Fig. 2 is based on these considerations and other data collected in this laboratory during the past several years. During oogenesis the cycle operates to allow for the accumulation of Gp_4G in yolk platelets but not in the cytosol. During early development Gp_3G serves as a shuttle for the utilization of yolk platelet Gp_4G to provide GTP to the cytosol. As GDP is generated, additional Gp_4G would be utilized concomitant with Gp_3G formation. After hatching Gp_4G may also diffuse into the cytosol where it could react with PP_i to give additional GTP. The cytosol level of Gp_4G may also be regulated by Gp_4G guanylohydrolase

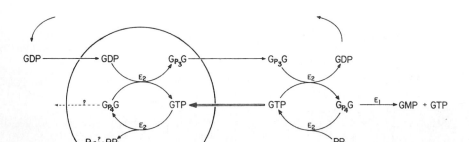

Fig. 2. *Proposed cycle for the biosynthesis and metabolism of Gp_4G and Gp_3G in Artemia prior to hatching. E_1 and E_2 are Gp_4G guanylohydrolase and GTP:GTP guanylyltransferase, respectively.*

and enzymes involved in the direct utilization of Gp_4G for ATP synthesis (see below).

The mechanism which permits Gp_4G to accumulate in the yolk platelets of the developing egg is poorly understood. Since PP_i inhibits Gp_4G synthesis, the accumulation of Gp_4G can only occur if the yolk platelets possess a mechanism for the removal of Gp_4G and/or PP_i from the soluble matrix of the

platelet(9). Our findings support the view that PP_i is
sequestered into vesicles in yolk platelets. However, some
of the PP_i may be exported to the cytosol or hydrolyzed to
P_i(17).

The mechanism which permits the initiation of Gp_3G synthesis following fertilization has not been studied.
However, our data suggest that the GTP/GDP ratio in the platelets and cytosol plays a vital role in the regulation of
Gp_3G synthesis. When the GTP/GDP ratio is high, insufficient
GDP is available to drive the Gp_3G reaction (equation 2), but
as the energy demand of the embryo increases, the ratio falls
and GDP becomes available to drive the reaction. We also
know that GDP and Gp_3G are more loosely associated with yolk
platelets than GTP and Gp_4G(7). Consequently, these nucleotides may diffuse across the platelet membrane readily during
development. As Gp_3G enters the cytosol it could be converted to cytosol Gp_4G or it may serve in a regulatory capacity
in protein synthesis.

At this point nothing is known about the biosynthesis of
Gp_2G and Gp_3A; therefore, their position in the cycle shown
in Fig. 2 has been omitted.

Gp_4G: The Source of Purine-Containing Compounds for Development

Encysted embryos of *Artemia* are cleidoic during the first
day of development and they lack the ability to synthesize
purines *de novo*(3,19). Despite this fact these embryos show
a remarkable constancy of purines during the first 3-4 days
of development(19). During this period there is an overall
increase in nucleic acid purines with a concomitant decrease
in acid-soluble purines. Based on an accounting of all acid
soluble nucleotides, it is clear that the primary source of
the nucleic acid purines is Gp_4G which accounts for about 55%
of the acid soluble purine-containing nucleotides of dormant
cysts(1).

In developing embryos the rate of Gp_4G utilization varies.
Prior to hatching Gp_4G is consumed at a rate of 0.2-0.4
pmol/h/embryo, but after hatching the rate increases to
between 1.3 and 1.8 pmol/h/embryo(1). During this latter
period there is a marked increase in adenosine nucleotides
(19). Although *Artemia* cannot synthesize purines *de novo*, at
least three other pathways have been proposed to account for
the synthesis of adenosine nucleotides in *Artemia*. One pathway involves the conversion of GMP to AMP by way of IMP, a
second involves the direct utilization of adenine in a Gp_4G
mediated reaction, and the third involves the formation of
dATP in a DNA synthesis linked metabolism of Gp_4G. The latter
two pathways may be restricted to adenosine nucleotide synthesis following hatching, whereas the first may provide the

source of embryo adenine for pre-hatch development. The main aspects of these pathways are discussed below.

Studies by Renart and co-workers(20) have demonstrated the existence of GMP reductase in dormant embryos of *Artemia*. The presence of this enzyme supports the view that AMP may be formed, at least in part, from GMP in the cytosol. The following pathway permits the utilization of GMP for AMP

$$GMP \longrightarrow IMP \longrightarrow sAMP \longrightarrow AMP$$

synthesis. The ability of nauplii to convert exogenous IMP to AMP(21), and the fact that dormant cysts contain IMP and sAMP support the existence of this pathway(22). Although XMP is a potent inhibitor of *Artemia* GMP reductase(20), Gilmour has not been able to detect XMP in extracts of *Artemia* cysts (22). However, since Gp_4G can reverse the inhibition by XMP, the above pathway should be operational in *Artemia* embryos.

In 1974 Van Denbos and Finamore reported that *Artemia* nauplii possess an unusual pathway for the synthesis of ATP involving Gp_4G(23). Their proposal was based on purine nucleotide labeling patterns following brief exposure of nauplii to ^{14}C-Guo. They showed that newly synthesized ATP has a ^{14}C-Ade to ^{14}C-ribose ratio that is distinctly different from newly synthesized AMP and ADP, and they suggested that ATP synthesis does not require either AMP or ADP as substrate. Other experiments showed that *Artemia* nauplii contain considerable adenine phosphoribosyltransferase (APRT) activity. From their data they proposed that Gp_4G plays a central role in ATP synthesis according to the following reactions. Initially, AMP would be derived from GMP as described

(3) $AMP + PP_i \rightleftharpoons Ade + PRPP$

(4) $Ade + Gp_4G \rightleftharpoons Gp_4A + Gua$

(5) $Gp_4A \longrightarrow GMP + ATP$

above and in the presence of APRT and PP_i, Ade would be produced (equation 3). Next, a base exchange reaction occurs between Ade and one Gua of Gp_4G to give Gp_4A (equation 4). Finally, Gp_4A would be hydrolyzed asymmetrically to GMP and ATP by Gp_4G guanylohydrolase (equation 5). Despite the fact that neither Gp_4A nor the enzyme required to catalyze the base exchange reaction has been identified in *Artemia*, such a pathway for the synthesis of ATP is consistent with the experimental data.

In 1969 Finamore and Clegg reported that Gp_4G metabolism in *Artemia* is closely linked to the resumption of DNA synthesis after hatching(21). On the basis of radiolabeled purine incorporation into nauplii, they suggested that two separate ribonucleotide pools exist in *Artemia*, one which supplies purines for ATP synthesis and another which supplies purines for dATP synthesis. Moreover, they noted that FUdR blocks Gp_4G utilization and that the incorporation pattern into nucleic acid purines, in the presence and absence of FUdR, is consistent with the hypothesis that Gp_4G functions as the principle source of dATP in a multistep process intimately linked to DNA synthesis. Although no intermediates or enzymes have been isolated to support their proposal, recent evidence that Ap_4A activates DNA synthesis in some mammalian cells suggests a possible role for nucleotides such as Gp_4G in DNA synthesis(24). It would appear fruitful to investigate further the mechanism of DNA synthesis in *Artemia* nauplii to ascertain whether Gp_4G or related nucleotides are linked to the resumption of DNA synthesis in these embryos.

Effect of Dinucleoside Compounds on Protein Synthesis in Artemia

Protein synthesis during development of *Artemia* may be regulated by several factors among which are the diguanosine nucleotides. Several studies have demonstrated that certain analogues of the 5'-end of capped mRNA inhibit translation of eukaryotic or capped mRNA *in vitro* at the initiation stage of protein synthesis(25,26). The most active of these analogues are $m^7GpppGm$, $m^7GpppAm$, m^7GTP and m^7GMP, but high concentrations (0.1-0.2 mM) of Gp_3G and Gp_3A such as that found in *Artemia* have also been shown to inhibit reovirus mRNA translation in a wheat germ system(25). Moreover, Gp_2G, Gp_3G and Gp_4G (at 0.1-0.2 mM) inhibit the binding of ^3H-$m^7GpppGpC$ to *Artemia* cap binding protein(25). Thus there exist in *Artemia* embryos the potential for protein synthesis regulation by endogenous dinucleoside compounds. In the encysted gastrula of *Artemia* the concentration of dinucleoside compounds is high. Based on the data in Table 2, and the fact that cysts imbibe 1.25 grams of water per gram dry cysts during normal hydration(27), *Artemia* gastrula cytosol contains 0.042 mM Gp_2G, 0.368 mM Gp_3G, 0.086 mM Gp_3A and 1.19 mM Gp_4G. Since concentrations of 0.1 mM Gp_3G and Gp_4G inhibit cap binding to *Artemia* cap binding protein by about 60% and 75%, respectively(25), it appears that these two nucleotides alone may play an important role in the regulation of protein synthesis in *Artemia* embryos.

Although Gp_3G, Gp_3A and Gp_4G have been shown to inhibit initiation of protein synthesis *in vitro*, their effects on other reactions in protein synthesis have not been studied. Thus we initiated a study on the effects of these compounds on polypeptide chain elongation and aminoacylation of tRNA using cell-free systems derived from *Artemia* embryos. Using polysomes from 12-h embryos which are unable to initiate new

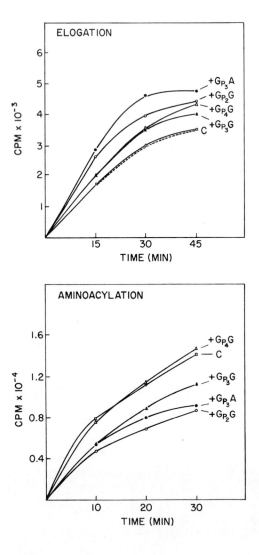

Fig. 3. Effect of dinucleoside compounds on polypeptide chain elongation and aminoacylation of tRNA. All nucleotides were tested at 0.26 mM and the dotted line in the top panel shows the effect of 0.1mM aurintricarboxylic acid on a control reaction. C represents control reactions lacking dinucleosides compounds.

complexes, we tested the effect of the dinucleoside compounds on elongation. The results of one series of experiments are shown in Figure 3 (top panel). At a concentration of 0.26 mM, Gp_2G and Gp_3A stimulate the rate of elongation by 30% and 40%, respectively whereas Gp_3G and Gp_4G have only a slight stimulatory effect (15-17%). In other experiments we observed that between 0.5-1.0 mM nucleotide all compounds show a more marked stimulatory effect. By comparison, high concentrations of GTP (1 mM) are slightly inhibitory. We also noted that the stimulatory effect of these compounds is greater at low $MgCl_2$ (3.4 mM) although the overall rates of elongation are reduced considerably. Although the dinucleoside polyphosphates inhibit initiation of protein synthesis at the concentrations tested, they promote chain elongation of preformed polysomes.

Since the stimulatory effect of these nucleotides on elongation may reflect changes in the rate of aminoacylation of tRNA, we tested these compounds in an aminoacylation reaction containing unfractionated cyst tRNA and enzymes and the same mixture of ^{14}C-amino acids used in the elongation experiments. The results shown in Figure 3 (bottom panel) indicate that, in general, Gp_3G, Gp_3A and Gp_2G inhibit the rate of aminoacylation, whereas Gp_4G has no effect. Thus stimulation of elongation by the dinucleoside compounds does not appear to be due to an increased rate of aminoacyl-tRNA formation but to some other factor(s). Furthermore, the effect of these compounds on aminoacylation of tRNA is more complex than revealed by these data alone. Other experiments have shown that lysyl-tRNA synthetase activity is inhibited by Gp_2G and Gp_3G, whereas Gp_3A and Gp_4G have no effect. In contrast, leucyl-tRNA synthetase activity is stimulated by all of the above compounds and phenylalanyl-tRNA synthetase is stimulated by Gp_3A only. None of the nucleotide effects, however, is apparent at high enzyme concentrations. Although the full meaning of these results is still unclear, they indicate that the dinucleoside polyphosphates, as a group, have a modifying effect on aminoacylation of tRNA.

The structural requirements for cap analogues to be effective inhibitors of protein synthesis *in vitro* are the presence of the 7-methyl and 5'-phosphate groups(28). Although we have not been able to demonstrate the presence of free methylated cap analogues in *Artemia* embryos, we have found that a post-ribosomal preparation from cysts catalyzes the methylation of dinucleoside compounds. The results in Table 3 show that the most active substrates for methylation are Gp_2G and Gp_3A but that all dinucleoside compounds can be methylated *in vitro*. The methylation that occurs with GTP may be due to the small amount of Gp_4G which is formed in this reaction. These data and those presented above suggest that initiation of protein synthesis in *Artemia* embryos may be controlled by changing levels of methylated and non-methylated dinucleoside polyphosphates. In addition, it seems likely that some of these compounds, expecially Gp_3G, play an important role in protein synthesis regulation during periods of environmental anoxia. When development of

Table 3. Methylation of Dinucleoside Compounds By Artemia Cytosol Enzymes

Nucleotide Substrates	Concentration μM	^3H-Methyl Incorporation[a] pmol/h
GTP	50	2.4
Gp_2G	50	14.8
Gp_3G	50	5.2
Gp_3A	50	15.2
Gp_4G	50	7.2

[a] *Enzyme alone showed no incorporation; these values represent total incorporation in 0.3 ml of the reaction mixture.*

encysted embryos is arrested by oxygen removal, protein synthesis stops abruptly and the level of Gp_3G increases nearly three-fold(29). Conceivably this compound or its 7-methyl derivative could be an important mechanism for blocking protein synthesis initiation until oxygen is restored to the medium.

CONCLUSIONS

In this paper I have attempted to integrate some of our recent findings on the metabolism and function of the dinucleoside polyphosphates in *Artemia* with previously published results. It is hoped that this work will generate further interest in these nucleotides as potential regulators of macromolecular processes.

ACKNOWLEDGEMENTS

The author wishes to thank V. Shridhar and S.J. Gilmour for their help and interest in completing this project and G. Sinclair for typing the manuscript.

REFERENCES

1. Finamore, F. J., and Warner, A. H. (1963). The occurrence of P^1,P^4-diguanosine 5'-tetraphosphate in brine shrimp eggs. *J. Biol. Chem. 238*, 344.
2. Oikawa, T. G., and Smith, M. (1966). Nucleotides in the encysted embryos of *Daphnia magna*. *Biochemistry 5*, 1517.
3. Warner, A. H., and McClean, D. K. (1968). Studies on the biosynthesis and role of diguanosine tetraphosphate during growth and development of *Artemia salina*. *Dev. Biol. 18*, 278.
4. Zagalksy, P. F., and Gilchrist, B. M. (1976). Isolation of a blue canthaxanthin-lipovitellin from the yolk platelets of *Branchipus stagnalis* (L.) (Crustacea: Anostraca). *Comp. Biochem. Physiol. 55B*, 195.
5. Rapaport, E., and Zamecnik, P.C. (1976). Presence of diadenosine 5',5'''-P^1,P^4-tetraphosphate (Ap_4A) in mammalian cells in levels varying widely with proliferative activity of the tissue: A possible positive "pleiotypic activator". *Proc. Natl. Acad. Sci. USA 73*, 3984.
6. McClean, D. K., and Warner, A. H. (1971). Aspects of nucleic acid metabolism during development of the brine shrimp *Artemia salina*. *Dev. Biol. 24*, 88.
7. Warner, A. H., Puodziukas, J. G., and Finamore, F. J. (1972). Yolk platelets in brine shrimp embryos. Site of biosynthesis and storage of the diguanosine nucleotides. *Exp. Cell Res. 70*, 365.
8. Warner, A. H., and Finamore, F. J. (1965). Isolation, purification, and characterization of P^1,P^4-diguanosine 5'-tetraphosphate *asymmetrical*-pyrophosphohydrolase from brine shrimp eggs. *Biochemistry 4*, 1568.
9. Warner, A. H., Beers, P. C., and Huang, F. L. (1974). Biosynthesis of the diguanosine nucleotides. I. Purification and properties of an enzyme from yolk platelets of brine shrimp embryos. *Can. J. Biochem. 52*, 231.
10. Warner, A. H., and Finamore, F. J. (1967). Nucleotide metabolism during brine shrimp embryogenesis. *J. Biol. Chem. 242*, 1933.
11. Gilmour, S. J., and Warner, A. H. (1978). The presence of guanosine 5'-diphospho-5'-guanosine and guanosine 5'-triphospho-5'-adenosine in brine shrimp embryos. *J. Biol. Chem. 253*, 4960.
12. Warner, A. H., Shridhar, V., and Finamore, F. J. (1977). Isolation and partial characterization of a protein synthesis inhibitor from brine shrimp embryos. *Can. J. Biochem. 55*, 965.

13. Muthukrishnan, S., Filipowicz, W., Sierra, J. M., Both, G. W., Shatkin, A. J., and Ochoa, S. (1975). mRNA methylation and protein synthesis in extracts from embryos of brine shrimp, *Artemia salina*. *J. Biol. Chem. 250*, 9336.
14. Ensinger, M. J., and Moss, B. (1976). Modification of the 5'-terminus of mRNA by an RNA (guanine-7-)-methyltransferase from HeLa cells. *J. Biol. Chem. 251*, 5283.
15. Vallejo, C. G., Sillero, M. A. G., and Sillero, A. (1974). Diguanosinetetraphosphate guanylohydrolase in *Artemia salina*. *Biochim. Biophys. Acta 358*, 117.
16. Beers, P. C. (1971). Diguanosine tetraphosphate pyrophosphohydrolase in the development of the brine shrimp, *Artemia salina*. Dissertation, Univ. of Windsor.
17. Warner, A. H., and Huang, F. L. (1974). Biosynthesis of the diguanosine nucleotides. II. Mechanism of action of GTP:GTP guanylyltransferase on nucleotide metabolism in brine shrimp embryos. *Can. J. Biochem. 52*, 241.
18. Wei, C. M., and Moss, B. (1977). 5'-Terminal capping of RNA by guanylyltransferase from HeLa cell nuclei. *Proc. Natl. Acad. Sci. USA 74*, 3758.
19. Clegg, J. S., Warner, A. H., and Finamore, F. J. (1967). Evidence for the function of P^1, P^4-diguanosine 5'-tetraphosphate in the development of *Artemia salina*. *J. Biol. Chem. 242*, 1938.
20. Renart, M. F., Renart, J., Sillero, M. A. G., and Sillero, A. (1976). Guanosine monophosphate reductase from *Artemia salina*: Inhibition by xanthine monophosphate and activation by diguanosine tetraphosphate. *Biochemistry 15*, 4962.
21. Finamore, F. J., and Clegg, J. S. (1969). Biochemical aspects of morphogenesis in the brine shrimp, *Artemia salina*. in "The Cell Cycle. Gene-Enzyme Interactions" (G. M. Padilla, G. L. Whitson, and I. L. Cameron, eds.), p. 249. Academic Press, New York.
22. Gilmour, S. J. (1978). The acid-soluble nucleotides of encysted embryos of the brine shrimp, *Artemia salina*. Thesis, Univ. of Windsor.
23. Van Denbos, G., and Finamore, F. J. (1974). An unusual pathway for the synthesis of adenosine triphosphate by the purine-requiring organism *Artemia salina*. *J. Biol. Chem. 249*, 2816.
24. Grummt, F. (1978). Diadenosine 5',5'''-P^1,P^4-tetraphosphate triggers initiation of *in vitro* DNA replication in baby hamster kidney cells. *Proc. Natl. Acad. Sci. USA 75*, 371.

25. Filipowicz, W., Furuichi, Y., Sierra, J. M., Muthukrishnan, S., Shatkin, A. J., and Ochoa, S. (1976). A protein binding the methylated 5'-terminal sequence, m^{7}GpppN, of eukaryotic messenger RNA. *Proc. Natl. Acad. Sci. USA 73*, 1559.
26. Hickey, E. D., Weber, L. A., and Baglioni, C. (1976). Inhibition of initiation of protein synthesis by 7-methylguanosine-5'-monophosphate. *Proc. Natl. Acad. Sci. USA 73*, 19.
27. Clegg, J. S. (1974). Interrelationships between water and metabolism in *Artemia salina* cysts: Hydration-dehydration from the liquid and vapour phases. *J. Exp. Biol. 61*, 291.
28. Filipowicz, W. (1978). Functions of the 5'-terminal m^{7}G cap in eukaryotic mRNA. *FEBS Letters 96*, 1.
29. Stocco, D. M., Beers, P. C., and Warner, A. H. (1972). Effect of anoxia on nucleotide metabolism in encysted embryos of the brine shrimp. *Dev. Biol. 27*, 479.

REGULATION OF MACROMOLECULAR SYNTHESIS
BY LOW MOLECULAR WEIGHT MEDIATORS

HISTORICAL BACKGROUND ON
DIADENOSINE 5',5'''-P^1,P^4-TETRAPHOSPHATE (A_p4A)
AND CURRENT DEVELOPMENTS

Paul C. Zamecnik
Eliezer Rapaport

Huntington Laboratories
Massachusetts General Hospital
Boston, Massachusetts

In 1966 we were studying the interaction of an aminoacyl tRNA synthetase and its cognate tRNA in the cell of an optical rotatory dispersion apparatus. Our hope was to observe a conformational change in either or both macromolecules upon their interaction at 37°. In a control preparation involving lysyl tRNA synthetase, ATP, lysine, Mg^{++}, and tris buffer, however, we found that an unusual Cotton effect centered around 267 nm gradually appeared over a 15 min incubation. This change in optical rotation was dependent on the presence of ATP. By means of PEI-cellulose chromatography and radioactive precursors we isolated and identified A_p4A as the new compound formed in abundance in this *in vitro* system (Randerath et al., 1966; Zamecnik et al., 1966). We found in further developments that numerous pyrophosphoryl groups (ppN) and triphosphoryl groups (pppN) could participate in this back reaction, as depicted in Figure 1. In general there appears to be an conformational barrier which interferes with easy reversibility of the reaction by which the dinucleoside polyphosphates are formed. No detailed kinetic studies of this point have as yet been made. We did identify a most unusual α-α ring stacking of the two adenine rings in A_p4A (Scott & Zamecnik, 1969). Recently, ^1H and ^{31}P magnetic resonance

Supported by Contracts 79-EV02403 and NO177001 of EDRA and the Dept. of HEW, respectively, and by Grants in aid ACSBC278 and CA-22904 from the American Cancer Society and the National Cancer Institute.

studies have shown that in Ap_5A and Ap_4A as well as Ap_3A and Ap_2A, the adenine rings are stacked at pH's above 5 with disruption of the stacking as the temperature is raised. At pH's 3-5, however, Ap_5A and Ap_4A (but not Ap_3A or Ap_2A) assume an unstacked, "folded" conformation with the two adenine rings interacting with the phosphate moieties. At low pH's (1-2) all the diadenosine 5',5'''-polyphosphates assume an extended conformation (Rapaport & Kolodny, unpublished data).

$$aa_1 + pppA + E_1 \qquad aa_1 - pA \cdot E_1 + pp$$

$$aa_1 + AppppN + E_1 \qquad aa_1 - pA \cdot E_1 + pppN$$

$$aa_1 + ApppN + E_1 \qquad aa_1 - pA \cdot E_1 + pcppA$$

$$E_1 + aa_1 + Appp\text{-}thiamine \qquad aa_1 - pA \cdot E_1 + pp\text{-}thiamine$$

$$E_1 + aa_1 + ApppRNA \qquad aa_1 - pA \cdot E_1 + ppp^5{'}RNA$$

$$E_1 + aa^1 + A^{5'}ppp^{5'}D^{3'}pp \qquad aa_1 - pA \cdot E_1 + pp^{5'}G^{3'}pp$$

FIGURE 1. *First step in protein synthesis, and mechanism of formation of Ap_4A and related compounds.*

We next looked for Ap_4A in living cells, and found it to be present in the small variety of living forms investigated, ranging from E. coli to rat liver (Zamecnik & Stephenson, 1978; Zamecnik & Stephenson, 1969).

The concentration hovered around 1 x 10^{-8} M, an uncomfortable level at which to work a decade ago in nucleotide biochemistry. In 1975, we instituted a search for a possible metabolic role for Ap_4A, and found it to be present around 10^{-8} M in eukaryotic cells arrested in G_0/G_1 phase and around 10^{-6} M in cells in early S phase of their cycle (Rapaport & Zamecnik, 1976). Ap_4A was shown by us to possess a higher metabolic lability than ATP (Rapaport & Zamcenik, 1976). Since the nucleotide is formed from ATP and aminoacyl adenylates, it reflects the cellular ATP level and the rate of protein synthesis. In cell-free protein synthesizing systems Ap_4A is readily produced. In *in vivo* systems, however, the levels of Ap_4A are low as compared to cell-free systems. The Ap_4A steady state levels (pool size) are invariably low in non-growing or G_1-arrested cells, as compared to actively non-proliferating cells. Liver cells, which posses an efficient protein synthesizing machinery and a high pool size of ATP, nevertheless contain low levels of Ap_4A. An enzyme has been isolated from liver capable of catalyzing the degradation of Ap_4A (diadenosine tetraphosphatase activity) with high substrate specificity (Vallejo et al., 1976). The Ap_4A pool size within the cell may therefore be controlled at the degradative level. We have suggested that Ap_4A is acting as a positive "pleiotypic activator" in mammalian cells, providing the link between protein synthesis and DNA replication (Rapaport & Zamecnik, 1976). Protein synthesis throughout all of the G_1 phase of the cell cycle is an obligatory requirement for the entry of cells into S phase, as has recently been demonstrated in two different systems (Brooks, 1977; Roufa, 1978).

There is a 50-100 fold increase in the pool size of Ap_4A at the beginning of the S phase of BHK cells as compared to cells arrested in the G_0/G_1 phase of their cycle. We therefore suspected that Ap_4A might play a role in triggering one of the enzymes poised at the initiation of DNA replication. We then began to use permeabilized cells, and to investigate the effect of added Ap_4A on DNA synthesis.

Following our study on the increase in Ap_4A concentrations related to cellular proliferation (Rapaport & Zamecnik, 1976), F. Grummt took up the quest of looking for an effect of Ap_4A on the events related to initiation of DNA synthesis in permeabilized mammalian cells. Grummt found that added Ap_4A stimulated DNA replication in G_0/G_1-arrested cells and dramatically increased the number of replication eyes seen by electron microscopy (Grummt, 1978). In a recent abstract (Grummt, 1979) and in unpublished work currently presented at this meeting (Grummt, 1979) Grummt et al. report a tight noncovalent binding of Ap_4A exclusively to one of the six subunits of DNA polymerase α. The mechanim of action of the bound Ap_4A is, however, still uncertain. The DNA polymerase reportedly is active without added Ap_4A.

We have found evidence (P.C. Zamecnik & E. Rapaport, unpublished data) for adenylylation of cytoplasmic protein(s) by $[α-^{32}P]Ap_4A$. Whether these are adenylylations of the same proteins with Ap_4A as a precursor as with ATP as a precursor, however, we have not yet determined. We also observed adenylylation of nuclear protein(s) which is exclusively from Ap_4A as a precursor (E. Rapaport & P.C. Zamecnik, unpublished data).

We have recently shown that isolated nuclei, without the presence of cytoplasmic material, engage only in elongation of DNA along existing replication forks (Rapaport et al., 1979). In isolated synchronized G_1 or S phase 3T6 nuclei, freed of cytoplasmic material, no effect on DNA synthesis (elongation) has been observed with varying Ap_4A levels. We are therefore investigating the possibility of Ap_4A acting in the initiation of replication units (replicons). Such initiation can possibly be achieved by adenylylation of a protein providing the 5'-end to the growing nascent DNA molecule. The attachment of proteins to the 5' end of nascent genomic RNA molecule of polio virus (Ambros & Baltimore, 1978; Rothberg et al., 1978) and the 5' end of nascent adenovirus DNA (Rekosh et al., 1977) have been reported.

An additional characteristic of $A_{p_4}A$ is its high metabolic lability (although its structure renders it stable to phosphomonoesterases). We have shown in a variety of cells in vivo, that the metabolic lability of $A_{p_4}A$ is higher than that of ATP (Rapaport & Zamecnik, 1976). Under conditions of ischemia or hypoxia the $A_{p_4}A$/ATP ratio always drops faster than the ATP/ADP ratio (Rapaport & Zamecnik, 1976). The high metabolic lability of $A_{p_4}A$ is suggestive of a role as an intracellular signal since its pool can fluctuate rapidly in response to environmental conditions which affect growth.

Thus, it remains to be determined whether the pathway of degradation and utilization of $A_{p_4}A$ is a reversal of that of its formation; and one's hunch is that it is not. Although an $A_{p_4}A$ has been found in liver (Vallejo et al., 1976), it's relationship ot metabolic events in the life cycle of $A_{p_4}A$ remains to be determined.

$A_{p_4}A$ thus presents itself as an intracellular "hormone" which is formed as part of the enzymatic reaction in the first step in protein synthesis, and which communicates with the initiation of the DNA synthesizing mechanism. When protein synthesis declines, due to the absence of an essential amino acid in a tissue culture system, the level of $A_{p_4}A$ subsides. However, when the rate of polypeptide polymerization is inhibited disproportionately to the amino acid starvation step, the level of $A_{p_4}A$ rises (Plesner et al., in press). More effort is needed to deepen our understanding of the signal $A_{p_4}A$ carried between the initiation of protein synthesis and the initiation of DNA synthesis.

REFERENCES

Ambros, V. & Baltimore, D. (1978), Protein is linked to a 5' end of poliovirus RNA by a phosphodiester linkage to tyrosine. J. Biol. Chem. 253, 5263.
Brooks, R.F. (1977), Continuous protein synthesis is required to maintain the probability of entry into S phase. Cell 12, 311.
Grummt, F. (1978), Diadenosine 5',5'''-P^1,P^4-tetraphosphate triggers initiation of in vitro DNA replication in baby hamster kidney cells. Proc. Nat. Acad. Sci. USA 75, 371.
Grummt, F., Waltl, G., Jantzen, M., Hamprecht, K., Hübscher, H. & Keunzle, C.C. (1979), Diadenosine tetraphosphate ($A_{p_4}A$)-

a ligand of DNA polymerase α and trigger of replication. *Hoppe-Seyer's Z. Phys. Chem. 360*, 272.

Grummt, F. (1979), Diadenosine tetraphosphate (Ap_4A)-a ligand of DNA polymerase α and trigger of replication. *this meeting.*

Plesner, P., Stephenson, M.L., Zamecnik, P.C. & Bucher, N.L. in Alfred Benzon Symposium 13, Diadenosine tetraphosphate (Ap_4A), an activator of gene function. ed. Engberg, J., Leick, V. & Klenow, H., Munksgaard and Academic Press (in press).

Randerath, K., Janeway, C.L., Stephenson, M.L. & Zamecnik, P.C. (1966), Isolation and characterization of dinucleoside tetra- and triphosphates formed in the presence of lysyl-sRNA synthetase. *Biochem. Biophys. Res. Commun. 24*, 98.

Rapaport, E. & Zamecnik, P.C. (1976), Presence of Ap_4A in mammalian cells in levels varying widely with proliferative activity of the tissue. A possible positive "pleiotypic activator". *Proc. Natl. Acad. Sci. USA 73*, 3984.

Rapaport, E., Garcia-Blanco, M.A. & Zamecnik, P.C. (1979), Regulation of DNA replication in S phase nuclei by ATP and ADP pools. *Proc. Natl. Acad. Sci. USA 76*, 1643.

Rapaport, E. & Kolodny, N., unpublished data

Rekosh, D.M.K., Russell, W.C., Bellett, A.J.D. & Robinson, A.J. (1977), Identification of a protein linked to the ends of adenovirus DNA. *Cell 11*, 283.

Rothberg, P.G., Harris, T.J.R., Nomoto, A. & Wimmer, E. (1978), O^4-(5'-uridylyl) tyrosine is the bond between the genome-linked protein and the RNA of poliovirus. *Proc. Natl. Acad. Sci. USA 75*, 4868.

Roufa, D.J. (1978), Replication of a mammalian genome: the role of *de novo* protein biosynthesis during S phase. *Cell 13*, 129.

Scott, J.F. & Zamecnik, P.C. (1969), Some optical properties of diadenosine-5'-phosphates. *Proc. Natl. Acad. Sci. USA 64*, 1308.

Vallejo, C.G., Lobaton, C.D., Quintanilla, M., Sillero, A. & Sillero, M.A.G. (1976), Dinucleosidetetraphosphatase in rat liver and artemia salina. *Biochim. Biophys. Acta 438*, 304.

Zamecnik, P.C., Stephenson, M.L., Janeway, C.L. & Randerath, K. (1966), Enzymatic synthesis of diadenosine tetraphosphate and diadenosine triphosphate with a purified lysyl-sRNA synthetase. *Biochem. Biophys. Res. Commun. 24*, 91.

Zamecni, P.C. & Stephenson, M.L. (1968), A possible regulatory site located at the gateway to protein synthesis. in Regulatory mechanism for protein synthesis in mammalian cells. ed. San Pietro, A., Lamborg, M.R. & Kenney, F.T., Academic Press, New York, p. 3.

Zamecnik, P.C. & Stephenson, M.L. (1969), Nucleoside pyrophosphate related to the first step in protein synthesis. *in* Alfred Benzon Symposium I. The role of nucleotides for the function and conformation of enzymes. ed. Kalckar, H.M., Klenow, H., Munch-Peterson, G., Ottesen, M. & Thuysen, J.M., Munksgaard , Copenhagen, p. 276.

**REGULATION OF MACROMOLECULAR SYNTHESIS
BY LOW MOLECULAR WEIGHT MEDIATORS**

THE ACCUMULATION OF HIGHLY PHOSPHORYLATED
COMPOUNDS IN <u>DROSOPHILA MELANOGASTER</u> CELLS AFTER
HEAT SHOCK

A.A. Travers

MRC Laboratory of Molecular Biology,
Hills Road, Cambridge, England.

When cultures of the K_{co} cell line of Drosophila melanogaster are shifted from 25° to 37° the cells accumulate a highly phosphorylated compound, DSS_1. The concentration of this compound increases by 5-fold during the first 20-30 min following the temperature shift, while the concentrations of the nucleoside triphosphate pools remain relatively unaffected. Additions of 10^{-3} M sodium azide or 10^{-3} M dinitrophenol to K_{co} cells, treatments known to induce similar changes in the pattern of transcription to heat shock (Leenders & Berendes, 1972) fail to elicit a similar accumulation of the phosphorylated compound. Instead the triphosphate pools fall to 20% of their original level. DSS_1 is absorbed to Norite and is alkali labile. After prolonged incubation at 25° or 37° DSS_1 is found in the extra-cellular medium.

Transfer of Drosophila melanogaster from 25° to 37° results in the rapid activation of transcription at a small number of loci with a concomitant reduction of activity at other loci (Ritossa, 1962; Ashburner, 1970). The activity of at least one other locus, that encoding histone mRNA, is maintained after the temperature shift (Spradling, Pardue & Penman, 1977). This response is characteristic both of the intact organism and of isolated organs and cells in tissue culture medium. The switch in the pattern of transcription after heat shock is, in some ways, formally analogous to the stringent response in the procaryote, Eschericha coli . In the latter situation aminoacid starvation results in a substantial reduction of synthesis of rRNA and also of many mRNA species e.g. r-protein messenger together with a stimulation of the production of a few characteristic proteins (Reeh, Pedersen & Friesen, 1976). This response is believed to be mediated, at least in part, by the nucleotide ppGpp (Cashel, 1969). Consequently it was of interest to determine whether related highly phosphorylated compounds accumulate in

Drosophila cells after heat shock. I report here that accumulation of highly phosphorylated compounds does indeed occur under these conditions. These compounds are apparently unrelated structurally to ppGpp and their role, if any, in the physiology of the heat shock response remains to be established.

MATERIALS AND METHODS

Cell culture. Eschalier's K_{co} cell line of *D. melanogaster* cells (Dolfini, 1971), adapted to grow in the absence of serum, was obtained from D. Glover. The cells were grown at 25° in a low phosphate medium D20-P (Rubin & Hogness, 1975) in monolayer or suspension culture. Cells were labelled with ^{32}P by the addition of 2 mCi inorganic phosphate (Radiochemical Centre, Amersham) to 5 ml cell culture.

Analysis of acid soluble ^{32}P-labelled compounds. Formic acid was added to a final concentration of 1 M to aliquots of the labelled culture. The cell debris was removed by centrifugation and aliquots of the supernatant were analysed by chromatography on PEI-cellulose thin layer sheets (Merck, Darmstadt). The developing solvents were either 1.5 M potassium phosphate pH 3.4 for one dimensional analysis or 3.3 M ammonium formate plus 4.2% boric acid, pH 7, followed by 1.5 M potassium phosphate pH 3.4 for two dimensional analysis (Cashel & Kalbacher, 1970). Preparation of ^{32}P-DSS_1 was by the method employed by Cashel and Kalbacher (1970) for the isolation of ^{32}P-ppGpp.

RESULTS

To investigate the levels of acid soluble phosphate containing compounds in *Drosophila* cells the K_{co} cell line was grown in a serumfree low phosphate medium described by Rubin and Hogness (1975). In this medium the specific activity of the triphosphate pools become constant within 90' of the addition of ^{32}P-phosphate. After this prelabelling period cultures were shifted from 25° to 37° and formic acid was added to aliquots from the culture at various times after the temperature shift. Analysis of the acid soluble extract by one dimensional chromatography on PEI-cellulose revealed that within 20-40' of incubation at 37° the amount of label in a compound with a mobility about 40% that of ATP increased about 5-fold to a level of 10% of the label in GTP. This compound we term DSS_1. During the same period the level of the ATP and GTP pools remained constant or fell by 20-30%.

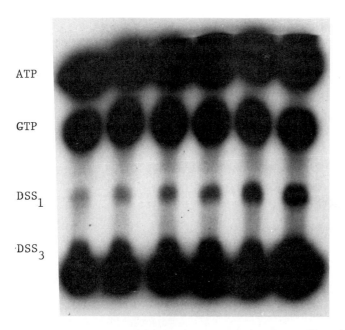

FIGURE 1. Kinetics of DSS_1 accumulation after heat-shock. 5 μl aliquots of samples taken at (from l. to r.) 0, 5, 10, 20, 40 and 60' after a temperature shift from 25° to 37°. DSS_2 is not visible in this chromatogram.

Two other compounds are also observed to be preferentially labelled after heat shock with approximately the same kinetics as DSS_1. The appearance of these compounds, termed DSS_2 and DSS_3, is apparently dependent on the state of growth of the cells prior to heat shock. Their mobilities relative to ATP are 0.20-0.25 and 0.10-0.15 respectively.

Analysis by two dimensional chromatography on PEI-cellulose with a formate-borate buffer as the developing solvent in the first dimension showed that DSS_1 in this solvent had approximately the same mobility as CTP while DSS_2 remained close to the origin.

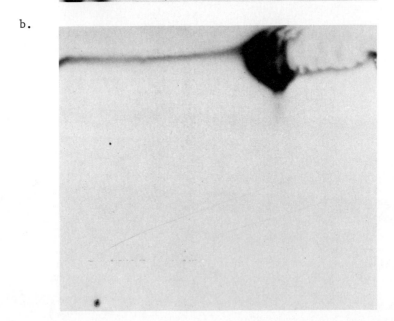

FIGURE 2. Two-dimensional analysis of acid soluble phosphate labelled compounds in K_{CO} cells. a. 5 μl aliquot from cells kept at $37°$ for 1 hr. A second aliquot was chromatographed in the potassium phosphate dimension only to serve as a marker. b. a parallel sample treated with Norite.

In some preparations another ^{32}P labelled compound was observed to accumulate after heat shock. The detection of this spot, DSS_{1a}, which has the same mobility in the first dimension as ATP and in the second demension as DSS_1, is extremely variable. It is invariably found as a contaminant of purified DSS_1 preparation and migrates identically to one of the alkaline digestion products of DSS_1.

The qualitative changes in the pattern of transcription elicited by heat shock can also be induced by certain inhibitors of energy metabolism, in particular dinitrophenol and azide (Leenders & Berendes, 1972). Cyanide alone however fails to stimulate transcription at the loci normally activated by heat shock. Accordingly K_{co} cells were exposed to 10^{-3} M sodium azide, 10^{-3} M dinitrophenol or 10^{-3} M potassium cyanide for 30'. The pools of acid soluble phosphate containing compounds were then analysed by one dimensional chromatography. All the metabolic inhibitors were without effect on the level of DSS_1

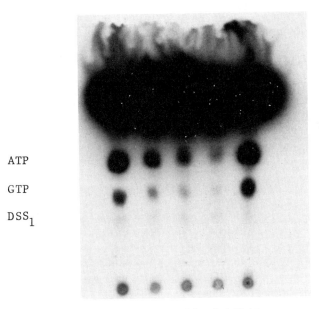

FIGURE 3. *Effect of metabolic inhibitors on acid soluble phosphate labelled compounds in K_{CO} cells. From l. to r.: control, $+10^{-3}$ M potassium cyanide, $+10^{-3}$ M dinitrophenol, $+10^{-3}$ M sodium azide, control.*

but all reduced the pools of ATP and GTP. This reduction was on average $\sim$ 5-fold for dinitrophenol and for azide and $\sim$ 20-fold for cyanide.

Although the intracellular concentration of DSS_1 increases 5-fold following a temperature shift from 25º to 37º no corresponding decrease in the level of DSS_1 is observed following the reverse shift. In fact the amount of label in DSS_1 in the total cell culture continues to increase, until at 24 hrs after the initial temperature shift the GTP and DSS_1 pools are labelled to approximately the same extent. In contrast to situation observed during heat shock 90-95% of this labelled DSS_1 is not intracellular but is present in the external medium.

What is the chemical nature of the phosphate containing compounds after heat shock? All are absorbed by Norite at acid pH and hence must be organic phosphates. Preliminary analysis of DSS_1 suggests that it is a hexaphosphate. Mild alkaline hydrolysis (0.4 M potassium hydroxide at 37º for 16 hrs) of this compound yields two products labelled in the ratio 5:1. The former has a mobility identical to DSS_{1a} when analysed by two dimensional chromatography. DSS_1 is sensitive to bacterial alkaline phosphate yielding inorganic phosphate and one other product. The compound is however resistant to snake venon phosphodiesterase, spleen phosphodiesterase and RNase T2. DSS_1 is thus probably structurally dissimilar to the highly phosphorylated nucleotides observed in bacteria and lower eukaryotes.

DISCUSSION

The observations reported here show that three highly phosphorylated compounds occur in the K_{co} cell line of *Drosophila melanogaster*. The kinetics of their accumulation following a temperature shift from 25º to 37º approximately parallel those of the heat shock induced activation of transcription at certain chromosomal loci (Ashburner, 1970). However dinitrophenol and azide which also activate transcription at the same loci, fail to induce similar accumulation of these phosphorylated compounds. Nevertheless in both situations the ratio of DSS_1 to the purine triphosphate pools is approximately the same. Thus if these phosphorylated compounds have any regulatory role in the physiology of the response to heat shock it appears probable that such regulation would depend on the relative levels of the triphosphate pools and the highly phosphorylated compounds. In this context it may be relevant that the transcriptional switch induced by dinitrophenol is suppressed by the addition of exogenous ATP (Leenders & Berendes, 1971). Clearly however it remains to be established whether these compounds have any significant role whatsoever in the changes in macromolecular synthesis following heat shock.

Also of interest is the preferential appearance of DSS_1 in the culture medium. Since the cell membrane is normally an effective barrier to the passage of highly phosphorylated compounds the extracellular location of DSS_1 implies either that it is actively excreted by cells or that it is released on cell death. This latter possibility appears unlikely since such extracellular accumulation occurs in cultures in which no visible cell disintegration has occurred. The alternative of excretion would imply that DSS_1 is either inimical to the normal growth of the cell or that it can function as an intercellular messenger.

REFERENCES

Ashburner, M. (1970), Patterns of puffing activity in the salivary gland chromosomes of Drosophila. V. Response to environmental treatments. Chromosoma 30, 356.

Cashel, M. (1969), The control of ribonucleic acid synthesis in Escherichia coli. IV. Relevance of unusual phosphorylated compounds from aminoacid starved stringent strains. J. Biol. Chem. 244, 3133.

Cashel, M. & Kalbacher, B. (1970), The control of ribonucleic acid synthesis in Escherichia coli. V. Characterisation of a nucleotide associated with the stringent response. J. Biol. Chem. 245, 2309.

Dolfini, S. (1971), Karyotype polymorphism in a cell population of Drosophila melanogaster cultured in vitro. Chromosoma 33, 196.

Leenders, H.J. & Berendes, H.D. (1972), The effect of changes in the respiratory metabolism upon genome activity in Drosophila. I. The induction of gene activity. Chromosoma 37, 433.

Reeh, S., Pedersen, S. & Friesen, J.D. (1976), Biosynthetic regulation of individual proteins in $relA^+$ and $relA$ strains of Escherichia coli during aminoacid starvation. Mol. Gen. Genet. 149, 279.

Ritossa, F. (1962), A new puffing pattern induced by temperature shock and DNP in Drosophila. Experientia 18, 571.

Rubin, G.M. & Hogness, D.S. (1975), Effect of heat shock on the synthesis if low molecular weight RNAs in Drosophila: accumulation of a novel form of 5S RNA. Cell 6, 207.

Spradling, A., Pardue, M.L. & Penman, S. (1977), Messenger RNA in heat-shocked Drosophila cells. J. Mol. Biol. 109, 559.

REGULATION OF MACROMOLECULAR SYNTHESIS BY LOW MOLECULAR WEIGHT MEDIATORS

HS3 - A BIZARRE DINUCLEOSIDE POLYPHOSPHATE AS POSSIBLE PLEIOTYPIC REGULATOR OF EUKARYOTES¶

Herb B. LéJohn, Glen R. Klassen & Swee Han Goh [*]

Department of Microbiology
University of Manitoba
Winnipeg, Manitoba
Canada R3T 2N2

A bizarre dinucleoside polyphosphate, HS3, accumulated rapidly in fungal and established mammalian (permanent) cells in culture when they were starved of L-glutamine. Restoration of glutamine caused HS3 to decay in minutes. The effect was specific for glutamine among all protein amino acids. This effect of glutamine was antagonised by P_i and, it then acted as a stimulant for HS3 synthesis. When glutamine induced HS3 decay in the fungal cell, sporogenesis was prevented. Ammonia and glucoseamine also inhibited HS3 synthesis and induced its decay. In these cases, only a combination of L-glutamate and P_i could reverse the inhibition.

Biochemical studies of HS3 biosynthesis supported the tentative structure of HS3 determined by chemical, physical and enzymic methods. We had shown (McNaughton, Klassen, Loewen & LéJohn, Can. J. Biochem. 56: 217-226, 1978) that HS3 is probably ADP-ribitol(glutamate)UDP-mannitoltetraphosphate. We were able to show that appropriately labelled precursors of de novo and salvage pathways of nucleotide biosynthesis were incorporated into HS3. In addition, both L-glutamate and D-mannitol were incorporated into it. Osmotic shock studies and labelling of fungal protoplasts and plasma membranes suggested that assembly of the molecule (phosphorylation ?) may occur at the level of the cell membrane.

The rate of HS3 synthesis in the fungal and mammalian cell varied inversely with the rate of DNA and RNA synthesis but not with protein synthesis. This may be physiologically significant as HS3 powerfully inhibited all eukaryotic DNA-dependent RNA polymerases and ribonucleoside diphosphate reductases in vitro. In vivo, transcription of permeabilised mammalian cells and isolated fungal nuclei were strongly inhibited by HS3. This molecule may be a pleiotypic regulator.

¶Supported by a grant from the National Research Council of Canada to H.B.L.

[*]Present address: Department of Biochemistry, Seattle, Wash.

INTRODUCTION

Generally, when cells that can sporulate are starved of certain nutrients, they do so (Grelet, 1951, 1957). Therefore starvation has come to be regarded as an environmental factor signalling sporogenesis (Szulmajster, 1973). Sporogenesis can be inhibited by nitrogen-containing metabolites (Schaeffer, 1969) or by limiting the supply of carbon, nitrogen or phosphate in bacteria (Schaeffer et al., 1965). From many physiological, biochemical and genetic data on *Bacillus* spp., many have thought, though not a consensus, that sporogenesis may be controlled by a single substance containing C, N and P atoms (Freese, 1977). But none of the experimental data implicating small molecular weight molecules such as ATP (Hanson, 1975), cyclic GMP (Clark & Bernlohr, 1972) and the ill-defined adenosine polyphosphate (Raese et al., 1977) in sporogenesis are convincing. The experimental findings recorded here may add other small molecular weight molecules to the growing list of presumptive sporogenesis regulators. One of three related molecules, HS3, is a dinucleoside polyphosphate containing 1 purine and 1 pyrimidine nucleotide, 1 glutamate, 2 polyhydric alcohols and a total of eight phosphates (McNaughton et al., 1978). When it accumulated in starving fungal cells, they sporulated; and when it was prevented from doing so by L-glutamine, ammonia or glucoseamine, they did not. Also, when HS compounds accumulated during starvation, DNA and RNA synthesis decreased dramatically and, *in vitro* and *in vivo* studies showed that the molecules powerfully inhibited RNA polymerases and ribonucleoside diphosphate reductases.

Growth control of mammalian cells in culture is poorly understood. Hershko et al., (1971) formulated the *pleiotypic response hypothesis* relating stringent control in bacteria to growth control in mammalian cells. They suggested that a *positive pleiotypic response* ensued when growth-promoting nutrients are added to cells and a *negative pleiotypic response* when they are withdrawn. The result is reduction of nucleic acid and protein synthesis in the negative state. They did not identify the *pleiotypic mediator*. Some of our findings on the physiological responses of mammalian cells to HS3 metabolism indicate that this molecule should be considered as a pleiotypic regulator.

MATERIALS AND METHODS

The biology and physiology of growth of the water-mold *Achlya*, have been described (Cameron & LéJohn, 1972). Starvation technique, method of isolating and quantitating HS compounds, osmotic shock treatment have been described in these reports (LéJohn et al., 1978; McNaughton et al., 1978; and Cameron & LéJohn, 1978).

Growth techniques, isolation of HS3 from mammalian cells labelling procedures and cell types used have all been recorded elsewhere (Goh & LéJohn, 1977; Goh, Wright & LéJohn, 1977; Goh, Doctoral thesis, Univ. Manitoba, 1979).

RESULTS

Structure and Biosynthesis of HS3

HS3 isolated from Chinese hamster ovary (CHO) cells and *Achlya* were structurally identical and displayed similar physiological and biochemical properties (Table 1). The downshift condition for *Achlya* refers to incubation in nutrient-free medium containing 0.1 mM $CaCl_2$ and 1 mM Tris-acetate, pH 6.5 and is really a starvation condition. Downshift for CHO cells is normal growth medium minus L-glutamine.

Biosynthesis during downshift. Through chemical, physical and enzymic analysis of HS3, we had determined that HS3 is a complex molecule of tentative structure: ADP-ribitol(glutamate)UDP-mannitoltetraphosphate (McNaughton et al., 1978). This rather bizarre structure prompted us to attempt *in vitro* labelling of the molecule with appropriate ^{14}C, ^{3}H and ^{32}P precursors.

When *Achlya* cells in downshift were supplied with labelled precursors of *de novo* nucleotide biosynthesis and then analysed for the presence of the labels in HS3 and other HS compounds which are synthesised at the same time, the results shown in Fig. 1 were obtained. Although labelling studies were done with ^{14}C and ^{3}H-labelled glycine, aspartate and orotate, ^{14}C-labelled formate and carbonate and $^{33}P_i$ and $^{32}P_i$, the results presented are for $^{33}P_i$ (panel (a)), L-(2,3-^{3}H)-aspartate (panel (b)) and (2-^{14}C)glycine (panel (c)). They show that while HS3 incorporated all three labels, HS2 and HS1 incorporated only $^{33}P_i$ and L-(2,3-^{3}H)aspartate. In the other labelling patterns not shown, the precursor of purine nucleotide (formate) was incorporated into HS3 only while orotate and sodium carbonate (pyrimidine nucleotide precursors) were incorporated into all three types of HS compounds. These observations lent biochemical support to our structural analysis data which had shown that HS3 contained both purine and pyrimidine nucleotides whereas HS2 and HS1 contained only pyrimidine nucleotides (McNaughton et al., 1978).

Further support for the structure of HS3 was obtained when we showed that D-(1-^{14}C)mannitol and L-(U-^{14}C)glutamate were incorporated into HS3 (Fig. 2, panels (a) and (b)).

Table I. A Comparison of the Properties of Achlya and CHO HS3

Property	HS3 Achlya	CHO [†]
R_f	0.22^a $(0.35)^b$	0.21^a $(0.36)^b$
A25-Sephadex elution at pH 8 (pH 3.6)	1.1M (0.28M)	1.1M (0.27M)
λ_{max} at pH 7	260 nm	260 nm
λ_{min} at pH 7	232 nm	231 nm
$A_{250:260}$ at pH 7	0.84	0.84
$A_{280:260}$ at pH 7	0.45	0.41
$A_{290:260}$ at pH 7	0.16	0.16
Action of BAP & SVPD[*]	resistant	resistant
Action of NP[*]	sensitive	sensitive
Components identified:		
adenosine	1 mole	1 mole
uridine	1 mole	1 mole
glutamate	1 mole	1 mole
mannitol	1 mole	1 mole
ribitol	1 mole	not determined
phosphate	8 moles	8 moles
Physiological responses:		
downshift (- glutamine)	induced	induced
upshift (+ glutamine)	reduced	reduced
purines/pyrimidines	reduced	reduced
inorganic phosphate	induced	induced
azaserine	induced	induced
sulphoximine	induced	not tested

[†] Lesch-Nyhan human fibroblast has similar chemical properties.
[a] First dimension (3.3 M ammonium formate in 4.2% borate, pH 7)
[b] Second dimension, 1.5M KH_2PO_4, pH 3.65
[*] BAP, bacterial phosphomonoesterase; SVPD, snake venom phosphodiesterase; NP, nucleotide pyrophosphatase.

Radioactively labelled purines and pyrimidines were also incorporated into HS3 (Fig. 2, panels (c) and (d)) indicating that the molecule could be synthesised via the salvage reactions of nucleotide synthesis. We also note that labelled glutamate and purines were incorporated into HS3 only while labelled mannitol and pyrimidines were incorporated into the three HS molecules.

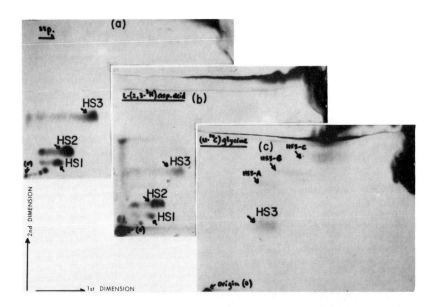

Fig. 1. Autoradiograms of two-dimensional PEI-cellulose chromatograms of osmotic shock fluids of Achlya cells labelled with (panel (a)), $^{33}P_i$; (panel (b)), L-$(2,3-^3H)$aspartate and (panel (c)), $(U-^{14}C)$glycine during downshift condition. Incubation was for 4h. See footnote to Table 1 for chromatographic condition. Medium had 1 mM each of glutamine and $\underline{PO_4}$.

The labelling patterns for the other three entities marked HS3-A, HS3-B and HS3-C are not discussed here as this is the subject of another study.

Preliminary results have shown that fungal protoplasts were capable of synthesising HS3 and HS2 when supplied with ^{32}P but the cell membranes and vesicles prepared from protoplasts required γ-^{32}P-ATP for their synthesis. Therefore, whether the recovery of HS compounds from osmotic shock fluids in the course of synthesis is related to the protoplast findings is yet to be studied further.

Metabolism of HS3

L-Glutamine inhibits HS synthesis. Both the fungal and mammalian cells stopped synthesising HS3 when downshifted cells were supplied with L-glutamine (Fig. 3). In fact, whatever HS3 had accumulated prior to the addition of glutamine decayed rapidly (within 10 min in CHO cells and within 1h in *Achlya* cells). Withdrawal of glutamine caused HS3 to accumulate again just as rapidly. In the mammalian cell, the level of HS3 increase was 5-6 fold whereas in the fungal cell

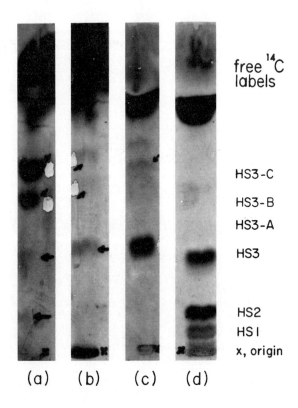

Fig. 2. Autoradiograms of one-dimensional PEI-cellulose chromatograms of osmotic shock fluids of Achlya cells labelled with (panel (a)), D-(1-^{14}C)mannitol; (panel (b)), L-(U-^{14}C)-glutamate; (panel (c)), (U-^{14}C)adenosine and (panel (d)), (U-^{14}C)uridine. The experimental conditions were as for the experiment of Fig. 1 and 1.5 M KH_2PO_4, pH 3.65 solvent used for chromatography.

it was more than 12-fold. That the inhibition was on synthesis not only phosphorylation was shown by ^{14}C-labelling studies (Table 2). Glutamine was unique in displaying this effect among all protein amino acids. A few amino acids such as L-histidine, L-glutamate, L-asparagine, L-proline, L-threonine and L-serine stimulated HS3 synthesis when inorganic phosphate was added to the culture, but this reflected an enhancement of phosphate uptake as well (Table 3).

Table 2. Inhibition of HS3 biosynthesis by L-glutamine (gln) in the absence of phosphate and stimulation of its synthesis in the presence of phosphate.

label supplied		HS3- Total cpm x 10^{-5}	
		plus gln (1 mM)	minus gln
NO PHOSPHATE ADDED			
sodium(^{14}C)carbonate	(25 µCi)	0.03	0.08
sodium(^{14}C)formate	(0.1 mCi)	0.12	0.36
(U-^{14}C)uridine	(20 µCi)	0.02	0.06
1 mM PHOSPHATE ADDED			
sodium(^{14}C)formate	(0.1 µCi)	2.44	0.51
L-(U-^{14}C)aspartate	(25 µCi)	0.22	0.04
(U-^{14}C)adenosine	(20 µCi)	1.51	0.29

HS compounds were recovered from Achlya cells in downshift after incubation for 4h in the presence of the radioactive precursor supplied at the specified activity. The cells were osmotically shocked and the shock fluid analysed by PEI-cellulose chromatography and autoradiography. HS3 areas were cut and radioactivity determined. The recorded counts represent the values of one-tenth the total isolate.

Ammonia inhibits HS synthesis. Ammonia inhibited HS3 synthesis more powerfully than glutamine (Table 3) for it could not be antagonised by phosphate. Only with the addition of L-glutamate as well as phosphate was there a partial relief of ammonia inhibition, but even so, the cells failed to sporulate (Table 3). An important feature is that HS1 synthesis was still inhibited under these conditions. The only other nitrogenous compound metabolised by the organism that was inhibitory was glucoseamine and phosphate could not overcome its inhibition.

Phosphate antagonism of glutamine. Although phosphate stimulated HS synthesis in a concentration dependent manner, this was probably due to a general enhancement of phosphate uptake for all other nucleotides showed similar increase in their cellular levels as did total phosphate uptake. What was of interest was the fact that, in the presence of a fixed

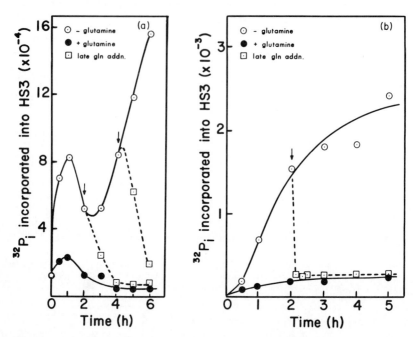

Fig. 3. Inhibition of HS3 synthesis and promotion of its decay by L-glutamine in fungal (a) and mammalian (b) cultures. Glutamine (1 mM) was added at the time signified by the arrow. HS3 was labelled with $^{32}P_i$, isolated and quantitated as described by LéJohn et al. (1978).

concentration of inorganic phosphate, glutamine stimulated the synthesis of HS compounds (Fig. 4) and the cells sporulated. It may be metabolically significant in that a combination of glutamate, ammonia and phosphate also stimulated HS synthesis (Table 3), much like glutamine and phosphate, while ammonia and phosphate could not. Therefore, glutamine is an important metabolite for HS synthesis acting in this capacity when phosphate is not limiting. Unavailability of phosphate initiates glutamine-induced decay of HS compounds. The results of Fig. 1 showing *de novo* synthesis of HS compounds which must require glutamine is therefore substantiated by these observations.

Table 3. Effect of various nitrogenous compounds on HS synthesis and sporogenesis in Achlya in the presence of phosphate.

Compound	conc. (mM)	% inc. ^{32}P up.	relative amnt. HS3	HS2	HS1	sporogen.
nil	–	100	1.0	1.0	1.0	+
glucoseamine	5	147	0.56	0.32	0.12	–
glutamine	5	212	2.35	1.37	0.92	+
ammonia	5	79	0.79	0.28	0.13	–
ammonia + glutamate	2 + 1	146	2.11	1.22	0.12	–
his, glu, asn, pro, thr and ser ¶	5		(1.6 – 3.5)			+

Abbreviations: up, uptake; amnt, amount; sporogen, sporogenesis.

¶ *The amino acids were supplied at 5 mM in each case and they enhanced phosphate uptake by 181 - 237%.*

+; *sporogenesis occurred;* -; *no sporogenesis.*

Phosphate was supplied at 1 mM in all assays.

HS Compounds Affect Nucleic Acid Metabolism

It has been shown that there is an inverse relationship between the rate of nucleic acid synthesis and the rate of HS synthesis in both fungal (LéJohn et al., 1978) and mammalian (Goh, Wright & LéJohn, 1977) cells. Protein synthesis was unaffected. We therefore studied the effects of these HS compounds on (i) the enzymes of nucleic acid synthesis and (ii) *in vivo* transcriptional activities.

RNA polymerases. Highly purified DNA-dependent RNA polymerases I, II and III isolated from *Achlya*, RNA polymerase II from wheat germ, RNA polymerase I from *Blastocladiella emersonii* and RNA polymerases I and II from CHO cells were powerfully inhibited by HS compounds obtained from *Achlya* and CHO cells as shown in Table 4. The prokaryotic polymerase from *Escherichia coli* was affected the least. Detailed kinetic studies showed that (i) HS compounds were bound by the enzyme(s); (ii) HS compounds did not interact with the enzyme at either the nucleotide or bivalent cation binding sites and (iii) HS compounds probably inhibited the initiation step of transcription (Table 5).

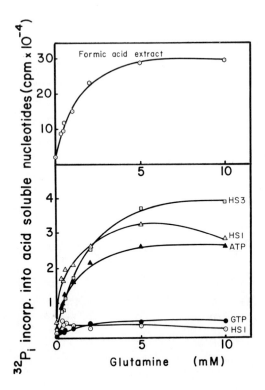

Fig. 4. Stimulation of phosphate uptake and incorporation into HS compounds and other nucleotides by L-glutamine. Phosphate was provided at a fixed concentration of 1 mM and glutamine was supplied at different concentrations as specified. Fungal cells were incubated with $^{32}P_i$ (1 mM) for 3 h under starvation condition then extracted by formic acid and analysed for HS compounds and nucleotides as described by LéJohn et al., (1978).

Ribonucleoside diphosphate reductases. HS3 inhibited CHO and *Achlya* ribonucleoside diphosphate reductases (Lewis et al., 1977) but had only a marginal effect on the enzyme from *E. coli*. The inhibition was noncompetitive with respect to the substrates CDP, ADP and GDP with K_i values of 23 µM, 14 µM and 16 µM respectively. The intracellular concentration of HS3 is uncertain but is estimated as greater than 0.1 mM.

Permeabilised CHO cells. Incorporation of ^{3}H-UMP into RNA of permeabilised CHO cells was linear for at least 1h. When HS3 was added to such transcribing system, incorporation was reduced by 48.3% (average of three determinations).

Table 4. Inhibition of RNA Polymerases by HS Compounds.

Enzyme	concentration of HS (μM) causing 50% inhibition			
	Ach-HS3	Ach-HS2	Ach-HS1	CHO-HS3
Achlya RNA-P I[a]	3.5	2.5	3.5	9
Achlya RNA-P II[a]	2.5	1.5	2.0	3
Achlya RNA-P III[a]	40.0	3.0	1.0	NT*
Blastocladiella RNA-P I[a]	3.0	2.5	NT	NT
Wheat germ RNA-P II[b]	NT	3.0	NT	NT
CHO RNA-P I[c]	2.0	NT	NT	2
CHO RNA-P II[c]	3.0	NT	NT	1
E. coli RNA-P[d]	50.0	4.5	6.0	NT

Assays performed according to the following:
[a] Horgen and Griffin (1971); [b] Jendrisak and Burgess (1975); [c] Ingles, Guialis, Lam and Siminovitch (1976); [d] Burgess (1969)
*NT, not tested.

Table 5. A Summary of the Effects of HS Compounds on the In Vitro Transcription Machinery of Eukaryotes.†

I. Binding assays*: HS3 binds to RNA-P II of Achlya and wheat germ.
II. In reaction kinetics, DNA and HS3 were competitive, but noncompetitive with ATP, GTP and UTP.
III. Pre-incubation of enzyme with substrates did not alter HS inhibition pattern; pre-incubation with HS3 or HS2 increased the inhibitory effect. Addition of HS after reaction had started resulted in delayed inhibition.

†All technical details can be found in the doctoral thesis of Glen R. Klassen (University of Manitoba, 1979).
*After Wyss & Wehrli (1976) and by DNA-cellulose column method.

Achlya nuclei. Because HS2 and HS1 were more powerful inhibitors of RNA polymerases (see Table 5), they were used to study their inhibitory effect on isolated nuclei of Achlya. HS2 inhibited ^{3}H-UMP incorporation by Achlya nuclei by 62% and HS1 inhibited by 32% at a concentration of 100 μg/ml.

Table 6. Relative amounts of ^{32}P-labelled HS3 in various mammalian cells in culture in the presence and absence of L-glutamine.†

Cell type	HS3	
	(+ glutamine)	(- glutamine)
Permanent lines		
CHO-wt	12.1 (0.045)†	100 (0.683)
CHO-GAT$^-$	96.0 (0.049)	100 (0.090)
CHO-HPRT$^-$	16.7	100
CHO-HPRT$^-$/APRT$^-$	14.8	100
Balb 3T3 (mouse)	15.8	100
Balb SV-3T3 (mouse)	10.5	100
BHK 21/Cl 13	15.1	100
HeLa	24.7	100
L5178Y	41.0	100
Primary cultures		
WI-38	90.5 (0.183)	100 (0.214)
Lesch-Nyhan human fibroblast	94.3	100
Normal human fibroblast	99.0	100
Normal human foreskin	95.0	100 (1.07)
African green monkey kidney	106.0	100

Determination of HS3 made after 2h of incubation with $^{32}P_i$ under the condition indicated. In the case of WI-38, determinations have been made from 2-5h with no change.
†Values in parenthesis are ratios of HS3/GTP.

HS3 - *Different Responses by Primary and Permanent Cell Cultures*

In a limited survey of various mammalian cell types, we observed that all permanent cell lines could only accumulate HS3 when glutamine was withdrawn from the growth medium. By contrast, primary cultures failed to alter an apparently high level of HS3 they produced whether or not glutamine was present (Table 6). Chemical analysis of HS3 from one primary culture, Lesch-Nyhan fibroblast, showed it to be identical to HS3 of CHO (see Table 1). Preliminary data suggest that both serum and glutamine were important factors in this phenomenon and work is currently in progress to resolve it.

DISCUSSION

These studies have defined specific developmental and biochemical events that occur during starvation of the coenocyte, *Achlya*. The most prominent changes are (a) synthesis of three bizarre dinucleotides, HS3, HS2 and HS1 accompanied by diversion from vegetative to sporogenic metabolism (Fig. 3). Synthesis of HS3 also accompanied the onset of nutrient (L-glutamine) starvation in mammalian cells. (b) These changes in HS3 levels were inversely correlated with the rate of nucleic acid synthesis in both fungal and mammalian cells. (c) *In vitro* and *in vivo* studies showed that eukaryotic RNA polymerases and ribonucleoside diphosphate reductases were strongly inhibited by HS compounds (Table 4). (d) In the absence of phosphate, glutamine inhibited the synthesis of HS3 promoted its decay and blocked sporogenesis (Fig. 3). Thus these cells appear to have been 'fooled' into continuing vegetative growth even though suitable nutrients were lacking. The glutamine effect was mimicked by ammonia and glucoseamine, both of which were not antagonised by phosphate. These cells also did not sporulate. By contrast, phosphate antagonised glutamine effect and the cells sporulated (Fig. 4). Thus, it seems that a proper balance between phosphate and nitrogen must exist for HS synthesis and sporulation to take place. It has not escaped our attention that HS3 is a carbon, nitrogen and phosphate-rich molecule whose metabolism is directly correlated with vegetative (low HS3) and sporulative (high HS3) growth activities.

Mammalian cells do not produce HS2 and HS1 under these conditions. These are non-differentiating cell types. Whether these compounds play any direct role in differentiation remains to be seen. A preliminary conjecture is that HS3 may be involved with growth control while HS2 and HS1 may have other cellular functions. These HS compounds appear to be good candidates for classification as low molecular weight mediators of eukaryotes and HS3 should be considered as a potential *pleiotypic* regulator.

ACKNOWLEDGEMENTS

We thank Renate Meuser for able technical assistance.

LITERATURE CITED

Burgess, R.R. (1969) A new method for the large scale purification of *Escherichia coli* deoxyribonucleic acid-dependent ribonucleic acid polymerase. *J. Biol. Chem.* 244: 6160-6167.

Cameron, L.E. & H.B. LéJohn. (1978) On the involvement of calcium in amino acid transport of the fungus *Achlya*. *J. Biol. Chem.* 247: 4729-4739.

Cameron, L.E. & H.B. LéJohn. (1978) Isolation and analysis of fungal phosphorylated proteoglycan. *Can. J. Biochem.* 56: 237-245.

Castellot, J.J., M.R. Miller & A.B. Pardee. (1978) Animal cells reversibly permeable to small molecules. *Proc. Nat. Acad. Sci. (U.S.A.)* 75: 351-355.

Clark, V.L. & R.W. Bernlohr. (1972) Catabolite repression and the enzymes regulating cyclic adenosine 3',5'-monophosphate and cyclic guanosine 3',5'-monophosphate levels in *Bacillus licheniformis*. *In "Spores V"* (H.O. Halvorson, R. Hanson & L.L. Campbell, eds.). pp. 167-173. American Society for Microbiology, Washington, D.C.

Freese, E. (1977) Metabolic control of sporulation. *In "Spores Research 1976"* (A.N. Barker, J. Wolf, D.J. Ellar, G. Dring & G.W. Gould, eds.). p. 1-32. Academic Press, London.

Goh, S.H. & H.B. LéJohn. (1977) Genetical and biochemical evidence that a novel dinucleoside polyphosphate coordinates salvage and *de novo* biosynthetic pathways in mammalian cells. *Biochem. Biophys. Res. Commun.* 74: 256-264.

Goh, S.H., J.A. Wright & H.B. LéJohn. (1977) Possible regulation of macromolecular biosynthesis in mammalian cells by a novel dinucleoside polyphosphate (HS3) produced during stepdown growth conditions. *J. Cell Physiol.* 93: 353-362.

Hanson, R.S. (1975) Role of small molecules in regulation of gene expression and sporogenesis in bacilli. *In "Spores VI"* (R.N. Costilow & H.L. Sadoff, eds.). pp. 318-326. American Society for Microbiology, Washington, D.C.

Hershko, A., P. Mamont., R. Shields & G.M. Tomkins (1971) Pleiotypic response. *Nature (New Biology)* 232: 206-211.

Horgen, P.A. & D.H. Griffin. (1971) Specific inhibitors of the three RNA polymerases from the aquatic fungus *Blastocladiella emersonii*. *Proc. Nat. Acad. Sci. (U.S.A.)* 68: 338-341.

Ingles, C.J.A., A. Guialis., J. Lam & L. Siminovitch (1976). α-Amanitin resistance of RNA polymerase II mutant Chinese hamster ovary cell lines. *J. Biol. Chem.* 251: 2729-2734.

Jendrisak, J.J. & R.R. Burgess. (1975) A new method for the large scale purification of wheat germ RNA polymerase II. *Biochemistry 14:* 4639-4645.

LéJohn, H.B., L.E. Cameron., G.R. Klassen & R.U. Meuser (1978) Effects of L-glutamine and HS compounds on growth and sporulation metabolism of *Achlya*. *Can. J. Biochem.* 56: 227-236.

Lewis, W.H., D.R. McNaughton., S.H. Goh., H.B. LéJohn & J.A. Wright. (1977) Inhibition of mammalian ribonucleotide reductase by a dinucleotide produced in eukaryotic cells. *J. Cell. Physiol.* 93: 345-352.

Rhaese, H.J., R. Groscurth & G. Rumpf. (1977) Molecular mechanism of initiation of differentiation in *Bacillus subtilis*. In "*Spores VI*" (G. Chambliss and J.C. Vary, eds.). pp. 286-292. American Society for Microbiology, Washington, D.C.

Schaeffer, P., J. Millet & J.P. Aubert.(1965) Catabolite repression of bacterial sporulation. *Proc. Nat. Acad. Sci. (U.S.A.) 31:* 129-137.

Schaeffer, P. (1969) Sporulation and production of antibiotics exoenzymes and exotoxins. *Bacteriol. Rev.* 33: 48-71.

Szulmajster, J. (1973) Initiation of bacterial sporogenesis. *Symp. Soc. Gen. Microbiol.* 23: 45-83.

Wyss, E. & W. Wehrli. (1976) The use of dextran-coated charcoal for kinetic measurements: interaction between rifampicin and DNA-dependent RNA polymerase of *E. coli*. *Anal. Biochem.* 70: 547-553.

REGULATION OF MACROMOLECULAR SYNTHESIS
BY LOW MOLECULAR WEIGHT MEDIATORS

DIADENOSINE TETRAPHOSPHATE ($A_{p4}A$) - A LIGAND OF DNA
POLYMERASE α AND TRIGGER OF REPLICATION

Friedrich Grummt
Gert Waltl
Hans-Michael Jantzen
Klaus Hamprecht

Max-Planck-Institut für Biochemie
D-8033 Martinsried bei München
Germany

Ulrich Huebscher
Clive C. Kuenzle

Institut für Pharmakalogie
Veterinärmedizinische Fakultät der Universität Zürich
CH-8057 Zürich
Switzerland

Diadenosine tetraphosphate ($A_{p4}A$) induces replicative DNA synthesis in quiescent mammalian cells. By equilibrium dialysis an $A_{p4}A$ binding activity is shown to be present in mammalian cells. The Ap_4A binding activity co-purifies with DNA polymerase α during the isolation procedure including chromatography on phospho-, DEAE- and DNA-cellulose, gel filtration, sucrose gradient centrifugation and electrophoresis in non-denaturing polyacrylamide gels. After these purification steps DNA polymerase α appears as a homogeneous protein complex with an apparent M_r of 404,000, consisting of 7 subunits with apparent M_r of 64,000, 63,000, 62,000, 60,000, 57,000, 55,000 and 52,000. By affinity labeling the protein with M_r of 57,000 has been shown to be the Ap_4A binding constituent of DNA polymerase α. The $A_{p4}A$ binding site is lost in neuronal cells during maturation of rat brains concomitantly with the loss of DNA polymerase α and mitotic activity in those cells. From these results DNA polymerase α *seems to be the intracellular target of $A_{p_4}A$. For the elucidation of the mechanism of $A_{p_4}A$ action during DNA synthesis a structural analog has been synthesized in this laboratory. This analog, methylene-bis-ADP, has been shown to act as an antagonist to $A_{p_4}A$ during binding to DNA polymerase α as well as in DNA replication in vivo and in vitro, and inhibits cell proliferation.*

INTRODUCTION

Diadenosine tetraphosphate ($A_{p_4}A$) was found to be distributed in various eu- and prokaryotic cells (Zamecnik, 1969; Rapaport, 1976; Plesner, 1979). In mammalian cells the intracellular concentration of this compound fluctuates drastically in response to the proliferation rate, i.e. between 10^{-8} M in resting or slowly growing and 10^{-6} M in rapidly growing cells.

We demonstrated previously that $A_{p_4}A$ is capable to stimulate DNA synthesis *in vitro* in baby hamster kidney (BHK) cells (Grummt, 1978; Grummt, 1979). $A_{p_4}A$ stimulates ^{3}H-dTTP incorporation only in G_1-arrested cells, not, however, in cells from exponentially growing cultures.

Both the growth rate-dependent fluctuation in the intracellular concentration of $A_{p_4}A$ as well as the induction of DNA replication in quiescent cells by $A_{p_4}A$ support the hypothesis of $A_{p_4}A$ being an intracellular (positive) signal molecule for the control of the transition from the resting to the growing state in animal cells (Rapaport, 1976).

To further substantiate this hypothesis and to approach a molecular understanding of the action of $A_{p_4}A$ in DNA replication we carried out experiments to characterize the intracellular target of $A_{p_4}A$. We describe here equilibrium dialysis experiments which demonstrate that DNA polymerase α specifically binds $A_{p_4}A$.

The elucidation of the mode of action of $A_{p_4}A$ during DNA replication would be facilitated by specific antagonists of this nucleotide. We synthesized a structural analog to $A_{p_4}A$, methylene-bis-ADP, and could demonstrate that this nucleotide competes with $A_{p_4}A$ for its binding site at DNA polymerase α and inhibits DNA replication *in vivo* and *in vitro* as well as cell proliferation.

RESULTS AND DISCUSSION

Ap_4A has previously been shown to induce replicative DNA synthesis in quiescent BHK cells (Grummt, 1978; Grummt, 1979). Since DNA polymerase α was shown to be involved in the replication of nuclear and viral DNA in mammalian cells (Spadari, 1975; Bertazzoni, 1976; Arens, 1977; Edenberg, 1978; Otto, 1978; Wagar, 1978; Huebscher, 1979) we investigated whether this enzyme is the intracellular target of Ap_4A.

This was approached by purifying DNA polymerase α from calf thymus to apparent homogeneity and studying both the catalytic and the Ap_4A binding activity of the enzyme fractions by measuring the ^3H-dTTP incorporation into acid-precipitable DNA and the binding of ^3H-Ap_4A in equilibrium dialysis, respectively. DNA polymerase α was purified from calf thymus extracts by means of chromatography on phospho-, DEAE- and DNA-cellulose, sucrose gradient centrifugation, gel filtration and electrophoresis in non-denaturing polyacrylamide gels. The comparison of catalytic activity and the Ap_4A binding activity revealed a co-purification of both ^3H-dTTP incorporation activity and an Ap_4A binding activity during all the isolation procedures applied (Fig. 1 A-F). DNA polymerase α of apparent homogeneity was isolated by this purification scheme as can be seen after staining of the enzyme separated in a polyacrylamide gel under non-denaturing conditions (Fig. 2). This DNA polymerase α preparation reveals seven protein bands of different M_r (64,000, 63,000, 62,000, 60,000, 57,000, 55,000 and 52,000) if separated under denaturing condition in sodium dodecylsulfate gels.

In order to decide which of those constituents of DNA polymerase α presents the Ap_4A binding subunit we carried out affinity labeling experiments with periodate-oxidized ^3H-Ap_4A. The result showed that predominantly the subunit with M_r 57.000 of DNA polymerase α can be labeled with oxidized ^3H-Ap_4A (Fig. 3).

Ap_4A Binding Activity is lost Concomitantly with DNA Polymerase α in Cerebral Neurones of Rats

If DNA polymerase α represents the intracellular target for Ap_4A in animal cells then the Ap_4A binding activity should be diminished in cell types having lost their polymerase α activity. We have studied whether this proves right by using neuronal cells of different developmental stages. Cerebral rat neurones develop from actively proliferating precursor cells at late fetal stage *via* non-dividing immature neurones at

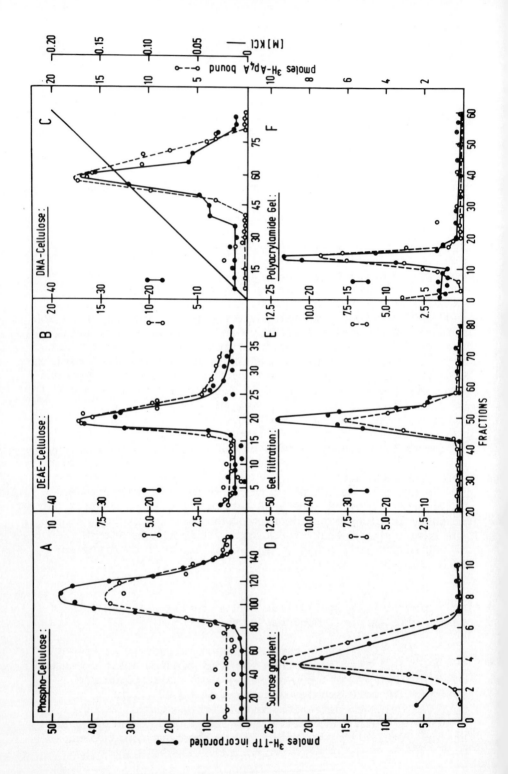

FIGURE 1. Co-purification of DNA polymerase α (●) and Ap_4A binding activity (○). (A), chromatography on phosphocellulose: calf thymus extract was adsorbed and finally eluted with 250 mM potassium phosphate, 10 µl of the eluate were tested in the ^{3}H-dTTP incorporation assay, 20 µl were assayed in equilibrium dialysis for Ap_4A binding activity. (B), chromatography on DEAE cellulose: the dialyzed phosphocellulose eluate was adsorbed and eluted with 250 mM KCl, 10 and 20 µl respectively, of the eluate were assayed for ^{3}H-dTTP incorporation and Ap_4A binding. (C), chromatography on DNA cellulose: the dialyzed DEAE eluate was adsorbed and eluted with a gradient of 0-200 mM KCl. (D), sucrose gradient centrifugation: 100 µl of the enzyme of the DEAE cellulose purification step was centrifuged in a 5 ml sucrose gradient (5-20%) in 10 mM Tris-HCl, pH 8.0, 100 mM KCl, 5 mM magnesium chloride, 7 mM 2-mercaptoethanol for 17 h at 42,000 rev/min and 4°C in a Spinco rotor SW 50.1.20 µl aliquots were assayed for both catalytic and Ap_4A binding activity. (E), gel filtration: the DNA cellulose eluate was concentrated by vacuum dialysis and applied on a Biogel AO.5 column (70 x 2.5 cm). 20 µl aliquots of the eluate were assayed for ^{3}H-dTTP incorporation and Ap_4A binding. (F), polyacrylamide gel electrophoresis: concentrated fractions of the gel filtrated enzyme were electrophoresed, eluted and assayed. Methods for equilibrium dialysis and enzyme assay were described previously (Grummt, 1979).

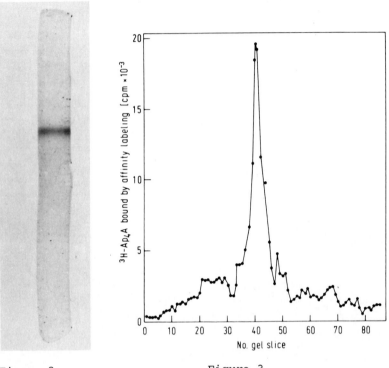

Figure 2   Figure 3

FIGURE 2. Electrophoresis of DNA polymerase α in polyacrylamide gels under non-denaturing conditions. 50 μl of the concentrated gel filtrated enzyme were applied on a tube gel (0.5 cm ∅) and processed as described elsewhere (Grummt, in press).

FIGURE 3. Electrophoresis of DNA polymerase α after affinity labeling with ^3H-$A_{p4}A$ in a sodium dodecylsulfate polyacrylamide gel. 50 μl of the gel filtrated enzyme were incubated for 6 h with 1 nmol (20 Ci/m Mol) ^3H-$A_{p4}A$ precedingly oxidized with 5 mM sodium peroxidase as described elsewhere. The gel was sliced into 1 mm sections and the radioactivity counted using a toluene-based scintillation fluid with 10% (v/v) Lumasolv ^{R}.

birth to terminally postmitotic neurones thereafter. These changes are accompanied by a specific and eventually complete loss of DNA polymerase α correlating with the decline of the *in vivo* rate of mitotic activity (Huebscher, 1979). In order to find out whether the cellular $A_{p4}A$ binding capacity declines concomitantly with the DNA polymerase α activity the binding of ^3H-Ap_4A by lysates of rat neuronal cells was analyzed. Figure 4 demonstrates that neurones from rat embryos at day 5 before birth have a significant Ap_4A binding activity. The capability to bind Ap_4A decreases sharply at the end of the fetal period resulting eventually in a complete loss in the early period. Figure 4 also shows the decrease of

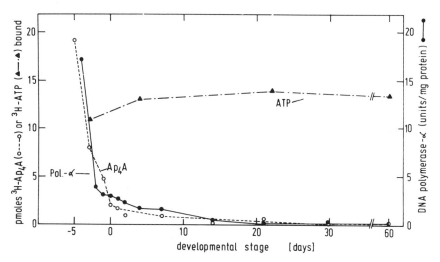

FIGURE 4. Loss of Ap_4A binding activity during development of rat brain neurones. (o), ^3H-Ap_4A binding activity, (▲), ^3H-ATP binding activity in neuronal extracts. The extracts were prepared as described elsewhere (Grummt, Proc.Natl.Acad. Sci., in press). (●), DNA polymerase α activity in neuronal extracts according to Huebscher (1977).

the DNA polymerase α activity as well as that of the mitotic rate during the development of rat neurones. The results clearly show that the capacity of neuronal lysates to bind ^3H-Ap_4A declines with a similar rate as the activity of the replicating enzyme in and the mitotic activity of neuronal cells. In control experiments the binding of ^3H-ATP by rat neuronal lysates was studied. In contrast to the Ap_4A binding capacity, no loss of ATP binding activity was observed in rat neuronal cells during the pre- and postnatal development (Fig. 4). Therefore, a specific correlation seems to exist between the level of replicating activity and the Ap_4A binding capacity in neuronal cells.

Methylene-bis-ADP as Antagonist of Ap_4A

Condensation of formaldehyde with adenosine or adenylates results in the production of the respective methylene-bridged dimers (Feldman, 1967; Alderson, 1973; Feldman, 1977) (Fig. 5). In those dimers the base moieties appear to be in a "stacking"

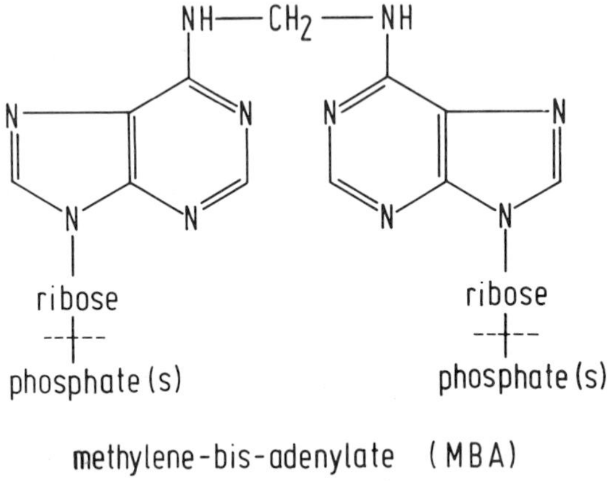

methylene-bis-adenylate (MBA)

FIGURE 5. *Structural formula of methylene-bis-adenylate*

conformation (Feldman, 1973) like that postulated for Ap_4A (Scott, 1969). Therefore, methylene-bis-ADP is a structural analog the special conformation of which is very similar to that of Ap_4A. We synthesized this analog by a method described by Feldman (Feldman, 1967) and studied its competition with Ap_4A for the binding site of DNA polymerase α, its effects on DNA synthesis *in vitro* and *in vivo* as well as its influence on cell proliferation in tissue cultures.

Figure 6 shows that methylene-bis-ADP competes at a 1:1 ratio with ^3H-Ap_4A for its binding site at DNA polymerase α. Methylene-bis-AMP is inactive in this respect, whereas ADP competes only at an 100-fold excess with Ap_4A for its site. DNA synthesis *in vitro* either in a system with purified DNA polymerase α and activated DNA or in a complex system containing cell lysates is strongly inhibited by methylene-bis-ADP (Fig. 7 A and B). At this concentration neither methylene-bis-AMP nor ADP have any effect on DNA replication *in vitro* (Fig. 7 A).

To study whether methylene-bis-adenylate derivatives can also inhibit DNA replication *in vivo* we synthesized methylene-bis-adenosine and -AMP, added these compounds to 3T3 and

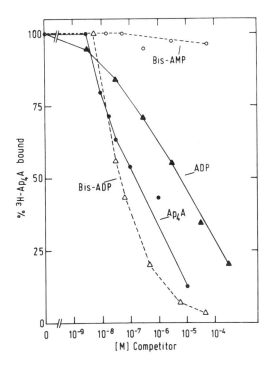

FIGURE 6. Binding of ^3H-$A_{p4}A$ to DNA polymerase α from calf thymus and its inhibition by unlabeled $A_{p4}A$, methylene-bis-ADP, methylene-bis-AMP, and ADP. 10 µl of calf thymus DNA polymerase α were used in equilibrium dialysis experiments as described previously (Grummt, 1979). Competition with unlabeled $A_{p4}A$ (●) methylene-bis-ADP (△), methylene-bis-AMP (○), ADP (▲).

Simian virus 40-transformed 3T3 (SV 3T3) cells and measured the incorporation of ^3H-thymidine into DNA. Figure 8 shows that methylene-bis-(5')AMP inhibits *in vivo* DNA synthesis by 50% at 0.05 mM and by 100% at 0.2 mM both in 3T3 and SV 3T3 fibroblasts. Similar effects had methylene-bis-adenosine and methylene-bis-(2' or 3')AMP (not shown). Since *in vitro* the DNA synthesis was exclusively inhibited by the ADP dimer we assume that *in vivo* the dimers are transported as adenosine derivatives through the cell membrane and are then phosphorylated intracellularily to the eventually active methylene-bis-ADP. Experiments to study whether this assumption proves right are now in progress in this laboratory.

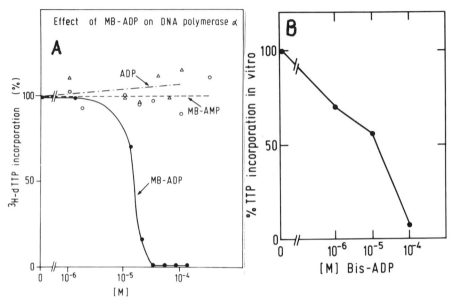

FIGURE 7. Effects on DNA polymerase α activity of methylene-bis-ADP, methylene-bis-AMP, and ADP. A, assay of purified DNA polymerase α from calf thymus (Grummt, 1979); B, ^{3}H-dTTP incorporation in vitro in a BHK cell lysate system (Grummt, 1978). Addition of methylene-bis-ADP (●), methylene-bis-AMP (o), and ADP (△).

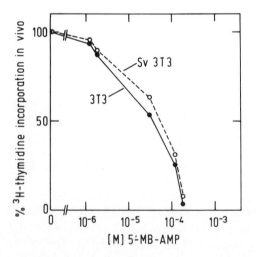

FIGURE 8. Effects of methylene-bis-(5')AMP on in vitro DNA synthesis of mouse 3T3 and SV 3T3 fibroblasts. Cells were grown and ^{3}H-thymidine incorporation assayed as described previously (Grummt, 1977). (o), 3T3 cells; (●), SV 3T3 cells.

Addition of methylene-bis-adenosine to the tissue culture cells does not only completely inhibit DNA replication but also the proliferation of those cells. Figure 9 shows the effect of 0.2 mM methylene-bis-adenosine on cell growth. BHK cells stop growing about 24 h after addition of the drug. However, most of the cells survive at least four days in the presence of methylene-bis-adenosine and remain attached to the plastics of the tissue culture dishes. These results let us hope that these Ap4A analogs might prove useful tools for the elucidation of the molecular mechanism of Ap_4A action as well as potential cytostatic drugs.

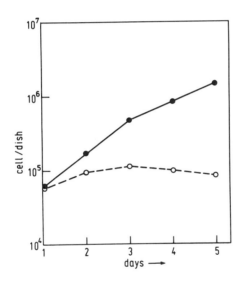

FIGURE 9. Growth of BHK cells in the presence and absence of methylene-bis-adenosine in the culture medium. Cell growth was measured without (●) or with 0.2 mM methylene-bis-adenosine (o) as described previously (Grummt, 1979).

REFERENCES

Alderson, T. (1973), Chemotherapy for an elective effect on mammalian tumor cells. Nature New Biol. 244, 3.
Arens, M., Yamashita, T., Padamanabhan, R., Tsuro, T. & Green, M. (1977), Adenovirus deoxyribonucleic acid replication. Characterization of the enzyme activities of a soluble replication system. J. Biol. Chem. 252, 7947.

Bertazzoni, U., Stefanini, M., Noy, G.P., Giulotto, E., Nuzzo, F., Falaschi, A. & Spadari, S. (1976), Variations of DNA polymerase-α and -β during prolonged stimulation of human lymphocytes. *Proc. Natl. Acad. Sci. USA 73*, 785.

Edenberg, H.J., Anderson, S. & DePamphilis, M.L. (1978), Involvement of DNA polymerase α in Simian virus 40 DNA replication. *J. Biol. Chem. 252*, 3273.

Feldman, M.Ya. (1973), Reactions of nucleic acids and nucleoproteins with formaldehyde. *Progr. Nucleic Acid Res. and Mol. Biol. 13*, 1.

Feldman, M.Ya. (1967), Reaction of formaldehyde with nucleotide and ribonucleic acid. *Biochim. Biophys, Acta 149*, 20.

Feldman, M., Ya., Balabanova, H., Bachrach, U. & Pyshnov, M. (1977), Effect of hydrolyzed formaldehyde-treated RNA on neoplastic and normal human cells. *Cancer Res. 37*, 501.

Grummt, F. (1978), Diadenosine 5',5'''-P^1,P^4-tetraphosphate triggers initiation of *in vitro* DNA replication in baby hamster kidney cells. *Proc. Natl. Acad. Sci. USA 75*, 371.

Grummt, F. (1979), Diadenosine tetraphosphate (Ap_4A) triggers *in vitro* DNA replication. *Cold Spring Harbor Symp. Quant. Biol. Vol. 43*, in press.

Grummt, F., Waltl, G., Jantzen, H.M., Hamprecht, K., Huebscher, U. & Kuenzle, C.C. *Proc. Natl. Acad. Sci. USA*, in press..

Grummt, F., Paul, D. & Grummt, I. (1977), Regulation of ATP pools, rRNA and DNA synthesis in 3T3 cells in response to serum or hypoxanthine. *Eur. J. Biochem. 76*, 7.

Grummt, F., Grummt, I. & Mayer, E. (1979), Ribosome biosynthesis is not necessary for initiation of DNA replication. *Eur. J. Biochem. 97*, 37.

Huebscher, U., Kuenzle, C.C., Limacher, W., Scherer, P. & Spadari, S. (1979), Functions of DNA polymerases α, β and γ in neurones during development. *Cold Spring Harbor Symp. Quant. Biol. Vol. 43*, in press.

Huebscher, U., Kuenzle, C.C. & Spadari, S. (1977), Variation of DNA polymerase during perinatal tissue growth and differentiation. *Nucleic Acid Res. 8*, 2917.

Otto, B. & Fanning, E. (1978), DNA polymerase α is associated with replicating SV 40 nucleoprotein complexes. *Nucleic Acid Res. 5*, 1715.

Plesner, P. (1979), Diadenosine tetraphosphate (Ap_4A), an activator of gene function. *in* Alfred Benzon Symp. 13, Munksgaard, Copenhagen, in press.

Rapaport, E. & Zamecnik, P.C. (1976), Presence of diadenosine 5',5'''-P^1,P^4-tetraphosphate (Ap_4A) in mammalian cells in levels varying widely with proliferative activity of the tissue: A possible positive "pleiotypic activator". *Proc. Natl. Acad. Sci. USA 73*, 3984.

Scott, J.F. & Zamecnik, P.C. (1969), Optical properties of diadenosine-5'-phosphates. *Proc. Natl. Acad. Sci. USA 64*, 1308.

Spadari, S. & Weissbach, A. (1975), RNA primed DNA synthesis: specific catalysis by HeLa cell DNA polymerase α. *Proc. Natl. Acad. Sci. USA 72*, 507.

Wagar, M.A., Evans, M.J. & Huberman, J.A. (1978), Effect of 2',3'-dideoxythymidine-5'-triphosphate on HeLa cell *in vitro* DNA synthesis: evidence that DNA polymerase α is the only polymerase required for cellular DNA replication. *Nucleic Acid Res. 5*, 1933.

Zamecnik, P.C. & Stephenson, M.L. (1969), The role of nucleotides for the function and conformation of enzymes. *in* Alfred Benzon Symp. (Eds. H.M. Kalckar et al.), Munksgaard, Copenhagen, *1*, 276.

**REGULATION OF MACROMOLECULAR SYNTHESIS
BY LOW MOLECULAR WEIGHT MEDIATORS**

ELEVATED NUCLEAR ATP POOLS AND ATP/ADP
RATIOS MEDIATE ADENOSINE TOXICITY IN FIBROBLASTS[1]

*Eliezer Rapaport
Sandra K. Svihovec*

The John Collins Warren Laboratories of the
Huntington Memorial Hospital of
Harvard University
at the Massachusetts General Hospital
Boston, Massachusetts 02114

We have recently suggested that the nuclear compartment pools of ATP and nuclear ATP/ADP ratios act as regulators of DNA replication in S phase 3T6 (mouse fibroblast) cells. A decrease in nuclear ATP/ADP ratios has been observed upon entry of 3T6 cells into the S phase of their cycle. High ATP/ADP ratios were shown to be inhibitory to DNA replication in isolated S phase 3T6 nuclei. The decrease in nuclear ATP/ADP ratios upon entry into S phase is probably produced by DNA-dependent ATPases which have been associated with increases in the proliferative activities of a variety of mammalian cells. This report indicates that inhibition of DNA synthesis observed in log phase 3T6 cells incubated in the presence of adenosine, is a result of an increase in nuclear ATP pools and ATP/ADP ratios. Adenine or inosine which yield increases in total cellular ATP pools and ATP/ADP ratios similar to those promoted by adenosine, do not produce similar increases in the nuclear compartment and consequently do not inhibit DNA synthesis in log phase 3T6 cells. An inverse relation is demonstrated between the incorporation of adenosine (at physiological levels) and the proliferative activity of cells. The data reported here suggest that adenosine incorporation, which was shown to yield compartmentalized nuclear ATP pools, may be used as an intracellular

[1]*Supported by a National Institute of Health Grant CA-22904*

growth regulatory mechanism. This mechanism would be mediated by the nuclear compartment's ATP pools and ATP/ADP ratios.

INTRODUCTION

The toxicity of adenosine to various mammalian cells in culture (1-3), has been shown to depend on the presence of cellular adenosine kinase (EC 2.7.1.20) and was thus attributed to the formation of phosphorylated derivatives in mouse and hamster fibroblasts (1,4). Toxicity of adenosine to lymphoid cells, however, was reported not to be dependent on initial phosphorylation of adenosine (5). In earlier studies we have demonstrated that the incorporation of adenosine into adenine nucleotide pools is inversely related to the proliferative activity of the cells (6) and yields compartmentalized (segregated) ATP pools which are kinetically or physically separated from *de novo* synthesized ATP pools or the ATP pools produced via salvage pathways from adenine, inosine or hypoxanthine (7). Other groups of investigators have recently shown that some acid soluble nucleotide pools in Novikoff hepatoma cells are compartmentalized into nuclear and cytoplasmic pools (8,9).

In recent studies we have demonstrated that high ATP levels as well as high ATP/ADP ratios are inhibitory to *in vitro* DNA replication in isolated S phase 3T6 nuclei (which exhibit only elongation of DNA along existing replication forks) (10). We have also determined the cellular and nuclear pools of ATP and ADP in intact synchronized 3T6 cells by high-pressure liquid chromatography. A good correlation with the studies on isolated nuclei has been observed. Whereas total cellular ATP pools increase during the progression of 3T6 cells from G_1 to S phase of the cell cycle, nuclear ATP pools do not increase, and the nuclear ATP/ADP ratios decrease once the cells enter the S phase of their cycle. These experiments suggest that nuclear ATP pools and ATP/ADP ratios act as S phase controls, regulating DNA elongation at sites where its synthesis has previously been initiated by cytoplasmic factors (10).

This report indicates that the inhibition of DNA synthesis observed in 3T6 cells (mouse fibroblasts) incubated in the presence of adenosine, is a result of an increase in nuclear ATP pools and ATP/ADP ratios. Adenine or inosine which yield increases in total cellular ATP pools and ATP/ADP ratios similar to those promoted by adenosine, do not produce similar increases in the nuclear compartment and consequently do not inhibit DNA synthesis in log phase 3T6 cells.

MATERIALS AND METHODS

Mouse fibroblasts line 3T6 was obtained from the American Tissue Culture Collection. Each new batch of cells was demonstrated to be free of mycoplasma contamination by two methods (11,12). 3T6 cells were inoculated in 35 mm petri dishes in Dulbecco modified Eagle's medium supplemented with 15% fetal calf serum. Following 44 hours of culturing, log phase cells (0.1 x 10^6 cells/cm^2) were washed twice in DME without serum and incubated with the corresponding precursor in DME without serum for 4 hours. [^{3}H]Thymidine (2 µCi/ml) was incorporated for the last 30 minutes of the incubation. Extraction of total cellular and nuclear acid soluble nucleotides pools was performed as described earlier (10). High pressure liquid chromatographic analysis of acid soluble nucleotides was performed on Whatman's Partisil-10 SAX columns with a Waters Associates ALC 202 instrument equipped with a programmer and a second pump for gradient elution. Peaks were detected by the instrument's UV detector at 254 nm and were integrated electronically with a Hewlett-Packard HP3380A integrator. Gradient elution of acid soluble nucleotides was performed according to a published procedure (13). [^{3}H]dTTP fractions were collected and total [^{3}H]dTTP radioactivity determined.

Rates of incorporation of [^{3}H]adenosine, [^{3}H]adenine and [^{3}H]inosine into ATP pools of chick embryo fibroblasts and Rous sarcoma virus transformed chick embryo fibroblasts were determined in semiconfluent, log phase cultures in 35 mm petri dishes (about 0.6 x 10^6 cells per dish, cultured in DME supplemented with 5 percent fetal calf serum and 10 percent tryptose phosphate broth). Preparation of primary CEF cultures and their infection with RSV were performed according to published procedures (14). Cells were prelabeled with $^{32}P_i$ in DME without serum (2 µCi/ml) for five hours followed by the incorporation of ^{3}H-labeled precursors in DME without serum containing the same ^{32}P-specific radioactivity. All operations were performed at 37°. The pool sizes of ATP as well as the rate of incorporation of ^{3}H-labeled precursors into the ATP pools were determined by utilization of $^{32}P_i$-labeling and chromatographic techniques (6,7). The use of high pressure liquid chromatographic analysis for determining the specific activity of [^{3}H]ATP in CEF and RSV-CEF cells, following incorporation of labeled precursors, yielded similar results.

RESULTS

The inverse relation between the rate of incorporation of adenosine into adenine nucleotide pools and the proliferative activity of cells is demonstrated in Figure 1. Primary chick embryo fibroblasts (CEF) when transformed with Rous sarcoma virus (RSV-CEF) show a much larger decrease in the rate of incorporation of adenosine, compared to adenine or inosine, into adenine nucleotide pools. The incorporation pattern of salvage precursors (Figure 1) as well as the increase in *de novo* synthesis of adenine nucleotides (data not shown) are observed as soon as the first morphological changes which accompany viral transformation can be observed. The recently reported substantial reduction in the level of adenosine deaminase following transformation of CEF cells with RSV (15) does not have any bearing on this study since at an adenosine concentration of 0.5 µM (physiological levels) phosphorylation to adenine nucleotides rather than deamination accounts for over 90 percent of adenosine metabolism (16).

Incubations of log phase 3T6 cells with increasing concentrations of adenosine, adenine or inosine for 4 hrs in DME without serum yielded the following results; only adenosine, at concentrations which produced increases in nuclear ATP pools and ATP/ADP ratios, markedly inhibited DNA synthesis (Table 1). While total cellular ATP pools and ATP/ADP ratios increased following incubations with all three precursors, only adenosine produced an increase in nuclear ATP pools and ATP/ADP ratios. Both adenosine and adenine inhibited *de novo* pyrimidine biosynthesis (1) as judged from the total cellular UTP pool size (Table 1). These results suggest that the decrease in *de novo* pyrimidine nucleotides biosynthesis is not responsible for the inhibition of DNA synthesis by adenosine under these conditions.

DISCUSSION

We have developed a procedure by which the *in vivo* adenine nucleotide pools of the nuclear compartment can accurately be determined in 3T6 cells (10). The technique is based on the property of 3T6 nuclei which adhere to the

Table 1. Total cellular and nuclear ADP, ATP and UTP pools in 3T6 cells incubated with adenosine, adenine and inosine. The effects of these precursors on DNA synthesis.

Precursor(mM)	Total Cellular Pools				[³H]Thymidine Incorporation (percent of control) [³H]dTTP Acid Insoluble		Nuclear Pools		
	ADP	ATP	ATP/ADP	UTP			ADP	ATP	ATP/ADP
None	1.1+0.2	8.0+0.6	7.4+0.6	2.1+0.3	100	100	0.9+0.2	2.0+0.3	2.1+0.3
Adenosine(0.01)	1.4+0.2	10.2+1.4	7.4+0.3	0.9+0.2	105	113	1.0+0.2	2.2+0.4	2.2+0.4
Adenosine(0.05)	1.6+0.2	13.8+0.7	8.7+1.0	0.7+0.2	129	98	1.0+0.2	2.5+0.4	2.5+0.4
Adenosine (0.1)	1.7+0.1	15.2+1.1	9.0+0.7	0.3+0.1	132	84	1.1+0.3	2.9+0.4	2.6+0.4
Adenosine(0.5)	1.6+0.1	13.9+0.6	8.7+0.7	0.3+0.1	71	40	1.1+0.3	3.1+0.4	2.8+0.4
Adenine (0.01)	1.5+0.2	11.7+0.5	8.1+0.7	1.7+0.4	123	108	1.0+0.1	2.3+0.4	2.2+0.3
Adenine (0.05)	1.4+0.3	12.6+1.2	8.9+1.0	0.7+0.3	122	112	1.0+0.2	2.4+0.4	2.3+0.3
Adenine (0.1)	1.5+0.2	13.0+0.7	9.0+0.8	0.6+0.1	122	115	1.0+0.1	2.3+0.2	2.2+0.3
Adenine (0.5)	1.6+0.1	15.2+1.3	9.7+1.3	0.4+0.1	125	116	1.2+0.2	2.5+0.4	2.2+0.3
Inosine(0.01)	1.3+0.3	11.3+1.2	8.6+0.8	1.9+0.3	118	112	1.1+0.3	2.2+0.4	1.9+0.3
Inosine(0.05)	1.3+0.2	10.7+0.9	8.6+0.6	1.8+0.4	130	104	0.9+0.1	2.0+0.4	2.1+0.4
Inosine(0.1)	1.3+0.2	11.5+0.7	9.4+0.9	1.8+0.3	141	105	0.9+0.2	2.2+0.4	2.3+0.3
Inosine(0.5)	1.3+0.2	11.4+0.9	9.1+0.7	1.4+0.2	107	91	1.0+0.2	2.2+0.3	2.2+0.3

All values represent the average of five experiments. Pool sizes are expressed as nmols/10^6 cells or nmols/10^6 nuclei and are shown as mean value ± standard deviation.

dish following destruction of cellular membrane and extraction of cytoplasmic compartments with Nonidet P-40 (17). Following a 10 second extraction with cold 1% NP-40 in buffer, soluble cytoplasmic components were completely removed with minimized diffusion from the nucleus due to the short treatment with cold buffer (10). Nuclear acid soluble nucleotide extracts were then analyzed by high pressure liquid chroma-

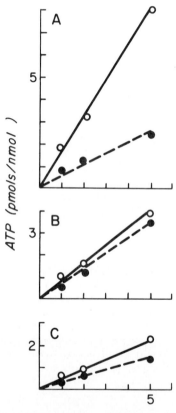

FIGURE 1. The rates of incorporation of 0.5 µM of [^{3}H]adenosine (A), 0.5 µM of [^{3}H]adenine (B), and 0.5 µM of [^{3}H]inosine (C) into the ATP pools of chick embryo fibroblasts (―――) and Rous sarcoma virus transformed chick embryo fibroblasts (-------). The ordinate denotes pmols of labeled precursor incorporated per nmol of existing cellular ATP pool.

tography (13). Nonaqueous fractionation techniques (18,19) involve removal of the cells from the dish followed by several manipulations including pelleting the cells, which was shown to widely affect adenine nucleotide pools in mammalian cells (unpublished data). Applying the aqueous extraction procedure to synchronized 3T6 cells we have shown that *in vivo* nuclear adenine nucleotide pools, but not total cellular pools, are consistent with the *in vitro* results dealing with the effects of ATP levels and ATP/ADP ratios on DNA replication in isolated S phase 3T6 nuclei. Namely, whereas total cellular ATP pools and ATP/ADP ratios increase upon progression of 3T6 cells from the G_1 to S phase of their cycle, a decrease in nuclear ATP/ADP ratios has been observed (10). These results were consistent with the previous reports demonstrating marked increases in DNA-dependent ATPase activities upon progression from resting to proliferating states of cells (20,21).

The data reported in Table 1, in conjunction with the earlier studies (10), suggest that adenosine promoted increases in the nuclear compartment pools of ATP and ATP/ADP ratios mediate adenosine toxicity in 3T6 mouse fibroblasts. Recently, increases in 2'-dATP were reported in erythrocytes of immunodeficient, adenosine deaminase-deficient patients (22). The high dATP levels were proposed as mediators of the toxic effects (22). Utilizing similar high pressure liquid chromatographic techniques we could not detect any substantial increases in total cellular or nuclear dATP pools upon incubations of log phase 3T6 cells with adenosine.

The incorporation of adenosine into adenine nucleotide pools of 3T6 cells indicate that at least part of the compartmentalized adenine nucleotide pools produced from adenosine (7) are located in the nuclear compartment. We have demonstrated an inverse relation between adenosine incorporation into adenine nucleotide pools and the proliferative activities of the cells (6) (Figure 1). The nuclear compartmentalization of adenine nucleotide pools produced from adenosine and the inhibitory effects of elevated nuclear ATP pools and ATP/ADP ratios on DNA synthesis in log phase 3T6 cells are indicated in Table 1. These data suggest that at physiological levels (Figure 1) adenosine incorporation may be used as an intracellular growth regulatory mechanism. It is indicated that the adenosine-promoted increases in ATP pools and ATP/ADP ratios of the nuclear compartment are the mediators of adenosine toxicity (inhibition of DNA synthesis) in mouse fibroblasts.

ACKNOWLEDGMENTS

We are grateful to Dr. Paul C. Zamecnik for his encouragement and many helpful discussions. We acknowledge the expert technical assistance of Mrs. B. Radner. This is publication No. 1563 of the Cancer Commission of Harvard University.

REFERENCES

1. Ishii, K., Green, H. (1973) Lethality of adenosine for cultured mammalian cells by interference with pyrimidine biosynthesis, *J. Cell Sci. 13,* 429.
2. Green, H., Chan, T. (1973) Pyrimidine starvation induced by adenosine in fibroblasts and lymphoid cells: role of adenosine deaminase, *Science 82,* 836.
3. Bynum, J.W., Volkin, E. (1976) Wasting of 18s ribosomal RNA by human myeloma cells cultured in adenosine, *J. Cell. Physiol. 88,* 197.
4. McBurney, M.W., Whitmore, G.F. (1974) Mutants of chinese hamster cells resistant to adenosine, *J. Cell. Physiol. 85,* 87.
5. Hershfield, M.S., Snyder, F.F., Seegmiller, J.E. (1977) Adenine and adenosine are toxic to human lymphoblast mutants defective in purine salvage enzymes, *Science 197,* 1284.
6. Rapaport, E., Zamecnik, P.C. (1976) Increased incorporation of adenosine into adenine nucleotide pools in serum-deprived mammalian cells, *Proc. Natl. Acad. Sci. USA 75,* 1145.
7. Rapaport, E., Zamecnik, P.C. (1976) Incorporation of adenosine into ATP: formation of compartmentalized ATP, *Proc. Natl. Acad. Sci. USA 73,* 3122.
8. Plagemann, P.G.W. (1972) Nucleotide pools in Novikoff rat hepatoma cells growing in suspension culture, *J. Cell Biol. 52,* 131.
9. Khym, J.X., Jones, M.H., Lee, W.H., Regan, J.D., Volkin, E. (1978) On the question of compartmentalization of the nucleotide pool, *J. Biol. Chem. 253,* 8741.
10. Rapaport, E., Garcia-Blanco, M.A., Zamecnik, P.C. (1979) Regulation of DNA replication in S phase nuclei by ATP and ADP pools. *Proc. Natl. Acad. Sci. USA 76,* 1643.
11. Schneider, E.L., Stanbridge, E.J., Epstein, C.J. (1974) Incorporation of ^{3}H-uridine and ^{3}H-uracel into RNA. *Exp. Cell Res. 84,* 311.
12. Hatanaka, M., DelGuidice, R., Long, C. (1975) Adenine formation from adenosine by mycoplasmas: adenosine phos-

phorylase activity, *Proc. Natl. Acad. Sci. USA 72*, 1401.
13. Chen, S-C., Brown, P.R., Rosie, D.M. (1977) Extraction procedures for use prior to HPLC nucleotide analysis using microparticle chemically bonded packings, *J. Chromatog. Sci. 15*, 217.
14. Zamecnik, P.C., Stephenson, M.L. (1978) Inhibition of Rous sarcoma virus replication and cell transformation by a specific oligodeoxynucleotide, *Proc. Natl. Acad. Sci. USA 75*, 280.
15. Chiang, P.K., Cantini, G.Z., Ray, D.A., Bader, J.P. (1977) Reduced levels of adenosine deaminase in chick embryo fibroblasts transformed by Rous sarcoma virus, *Biochem. Biophys. Res. Commun. 78*, 336.
16. Perrett, D., Dean, B. (1977) The function of adenosine deaminase in the human erythiocyte, *Biochem. Biophys. Res. Commun. 77*, 374.
17. Tsai, R.L., Green, H. (1973) Rate of RNA synthesis in ghost monolayers obtained from fibroblasts preparing for division, *Nature New Biol. 243*, 168.
18. Kirsch, W.M., Lectner, J.W., Gainey, M., Schulz, D., Lasher, R., Nakane, P. (1970) Bulk isolation in nonaqueous media of nuclei from lyophilized cells, *Science 168*, 1592.
19. Gurney, Jr., T., Foster, D.N. (1977) Nonaqueous isolation of nuclei from cultured cells, *Methods in Cell Biology*, ed. *Prescott, D.M., (Academic Press) 16*, 45.
20. Otto, B. (1977) DNA-dependent ATPases in concanavalin A stimulated lymphocytes, *FEBS Letters 79*, 175.
21. Hachmann, H.J., Lezius, A.G. (1976) An ATPase depending on the presence of single-stranded DNA from mouse myeloma, *Eur. J. Biochem. 61*, 325.
22. Cohen, A., Hirschhorn, R., Horowitz, S.D., Rubinstein, A., Polmar, S.H., Hong, R., Martin, Jr. D.W. (1978) Deoxyadenosine triphosphate as a potentially toxic metabolite in adenosine deaminase deficiency, *Proc. Natl. Acad. Sci. USA 75*, 472.

REGULATION OF MACROMOLECULAR SYNTHESIS
BY LOW MOLECULAR WEIGHT MEDIATORS

RELATION OF PROTEIN SYNTHESIS TO THE CONTENT OF
ADENOSINE POLYPHOSPHATES

Paul Plesner
Karsten Kristiansen

Department of Chemistry
Carlsberg Laboratory
Copenhagen, Denmark

and

Department of Molecular Biology
University of Odense
Odense, Denmark

The interdependence of changes in the concentration of adenosine polyphosphates and changes in the pattern of protein synthesis has been investigated in an in vivo unbalanced system, Tetrahymena pyriformis cells that were shifted from growth to starvation conditions. The following events were observed during the first 1-2 hours of starvation:
1. an abrupt inhibition of initiation of protein synthesis (disaggregation of polyribosomes); 2. a preferential synthesis of 7 proteins starting 15-30 min after the shift-down; 3. an unchanged pattern of proteins synthesized by in vitro translation of limiting amounts of poly(A) containing RNA isolated during the first hour of starvation; 4. a transition of

This work was supported by grants from the Danish Cancer Society, the Danish Medical Research Council, and the Danish Natural Science Research Council.

The following abbreviations are used: HPLC, high pressure liquid chromatography; Ap_4A, diadenosine $5',5'''-P^1,P^4$-tetraphosphate; Gp_4G, diguanosine $5',5'''-P^1,P^4$-tetraphosphate; p_3Ap_3, adenosine $5',3'(2')$-bis(triphosphate); DNP, 2,4-dinitrophenol.

ribosomal protein S6 from the non-phosphorylated to a highly phosphorylated state which is not impaired by DNP treatment and complete after about 1 h of starvation; 5. an increase in ATP content from 14 to 21 nmol·mg protein^{-1} during the first 30 min of starvation, followed by a decrease; 6. a steady increase in Ap_4A content from an unmeasurable level immediately after the shift to about 40 pmol·mg protein^{-1} at 1 h of starvation; 7. an immediate increase in $A_{p4}A$ content to about 40 pmol·mg protein^{-1} in DNP treated cells. The effects of ATP can pragmatically be considered to be turnover effects (hydrolysis of ATP) or concentration effects (binding of ATP and allosteric effects). In the present case, ATP is believed to accumulate passively as the result of a decreased utilization and then actively act as a regulator by virtue of the high concentration. The phosphorylation of the ribosomal protein S6 is considered to part of a mechanism that ensures preferential translation of de novo synthesized mRNA.

INTRODUCTION

At the Cold Spring Harbor Meeting in 1961 it was suggested that ATP might exert a regulatory control on the process of translation. ATP was thought to exert its regulatory effect by virtue of binding to other molecules; this implies that variations in the ATP concentration (rather than in turnover) are decisive (Plesner, 1961). The evidence came from in vivo experiments on *Tetrahymena pyriformis* synchronized for cell division; it was found that ATP concentration increased markedly prior to cell division (Plesner, 1958, 1964). It was difficult to accept that a doubling in the concentration of ATP would not have any effect on cellular metabolism; Lardy and Parks (1956) had stressed the importance of the ATP/Mg^{2+} ratio for the phosphofructokinase reaction and the effects of variations in Mg^{2+} on ribosome functions were being published (e.g. Tissières et al., 1960). In addition, Siekevitz and Palade (1960) had shown that variations in ATP within the physiological range could release active pancreatic enzymes from microsomes *in vitro*. These findings led to the idea that changes in turnover of ATP during unbalanced situations cause fluctuations in the concentration of cellular ATP, which in turn serve as regulatory signals. Thus, with regard to macromolecular synthesis, we distinguish between two types of ATP dependent reactions; one type which depends on ATP hydrolysis (turnover effect of ATP), and one in which ATP acts by virtue of its binding to other low molecular weight compounds or macromolecules without being cleaved (concentration effect of ATP).

A compound related to ATP, diadenosine tetraphosphate (Ap_4A) seems also to function as a pleiotypic signal compound (see Zamecnik; Grummt - this vol.). A similar compound, Gp_4G, is the most abundant energy rich nucleotide in *Artemia salina* cysts (Finamore & Warner, 1963). We have examined several unbalanced systems (nutrient shift-up in *E. coli* and *T. pyriformis*, regenerating rat liver, and cold-shocked Hodgkin's disease cells in tissue culture) and found characteristic changes in the Ap_4A content (Plesner et al., 1979).

Much of our present rather detailed knowledge on the process of protein synthesis in eukaryotes originates from studies on fractionated protein synthesizing systems, especially reticulocyte cell-free systems. However, recent analyses of the mechanisms behind the various types of translational control that operate in eukaryotic cells have also clearly emphasized the need for analyzing these phenomena in intact cells (for a review see Revel & Groner, 1978). In the present work we have used cultures of the ciliated protozoan *T. pyriformis* to analyze how a physiological change from growth to starvation conditions influences: a) the pattern of protein synthesis; b) ribosome structure and the accumulation of modified ribosome in polyribosomes; c) the cellular concentration of ATP and Ap_4A and attempted to correlate these parameters.

MATERIALS AND METHODS

Tetrahymena pyriformis, strain GL, was used in all experiments. The conditions for growth and starvation, isolation of ribosomes and extraction of ribosomal proteins, analysis of ribosomal proteins by two-dimensional polyacrylamide gel electrophoresis and determination of [^{32}P]phosphate incorporation into ribosomal proteins were as previously described (Kristiansen et al., 1978a, b; Kristiansen & Krüger, 1978). Protein was determined according to the procedure of Lowry et al. (1951) using bovine serum albumin as a standard. The details of fractionation of polyribosomes, labelling of cells, extraction and analysis by one-dimensional sodium dodecyl sulphate gel electrophoresis of total cellular protein, isolation of poly(A) containing RNA and its translation in a wheat germ derived cell-free system will be reported elsewhere.

The extraction and quantitative estimation of the cellular free nucleotides followed the HPLC methods developed by Phyllis Brown and co-workers (Hartwick & Brown, 1975; Chen et al., 1977). Figure 1 shows how the method works in our hands. XDP added to the trichloroacetic acid that was used for the

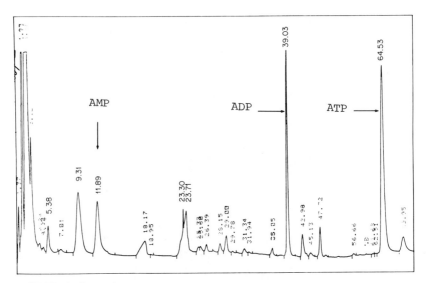

FIGURE 1. Chromatogram of a nucleotide extract of centrifuged T. pyriformis cells. Extraction and chromatography was performed as described by Hartwick & Brown (1975) (reproduced from Plesner et al. (1979) by permission).

extraction served as a standard for the quantitative determination of ATP, ADP and AMP. Ap$_4$A was measured by HPLC after digestion with alkaline phosphatase (Plesner et al., 1979), only was o.1 M NaHCO$_3$ used and not Tris, and the enzyme incubation mixture was injected without any foregoing acidification.

RESULTS
Changes in the Patterns of Protein Synthesis During the Adaptation from Growth to Starvation Conditions

The pattern of protein synthesis in T. pyriformis changes markedly during the transition from growth to starvation conditions. A general decrease in the rate of protein synthesis as determined by the disaggregation of polyribosomes is accompanied by a preferential incorporation of [^{35}S]methionine into 7 protein bands, designated 71K, 58K, 48K, 43K, 40K, 36K and 27K, according to molecular weights (Fig. 2). Pronounced labelling of protein 71K is only observed after 1 h of starvation in contrast to proteins 58K-27K, where preferential labelling is apparent after 15-30 min of starvation. Furthermore, protein 71K is a rather stable protein that accumulates in the starved cells, whereas proteins 58K-27K turn over rapidly (unpublished).

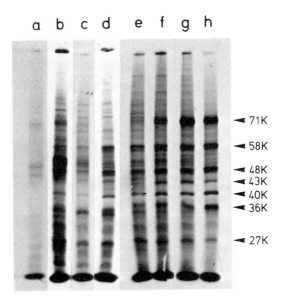

FIGURE 2. Autoradiograms of pulse-labelled proteins synthesized in exponentially growing and starved T. pyriformis cells. The cells were pulse-labelled with [^{35}S]methionine for 10 min; total cellular protein was solubilized with sodium dodecyl sulphate and analyzed on one-dimensional 10% polyacrylamide gel containing sodium dodecyl sulphate. (a) exponentially growing cells; (b-h) cells starved for 0, 15, 30, 60, 90, 120, 180 min, respectively.

Poly(A) containing RNA isolated from growing and starved cells directs the synthesis of distinct polypeptides in a wheat germ derived cell-free system. When this in vitro protein synthesizing system was programmed with limiting amounts of the poly(A) contained RNA isolated from exponentially growing cells and cells starved for 1 h, respectively, virtually identical patterns of protein synthesis were observed in striking contrast to the pronounced differences in the corresponding patterns of the proteins synthesized in vivo (Fig. 3).

Changes in Ribosome Structure and Accumulation of Phosphorylated 40-S Ribosomal Subunits in Polyribosomes during the Adaptation from Growth to Starvation Conditions

When exponentially growing cells of T. pyriformis are transferred into a non-nutrient medium pronounced phosphorylation of a single small subunit ribosomal protein S6 is in-

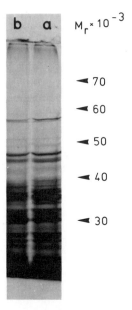

FIGURE 3. Autoradiograms of [^{35}S]methionine labelled proteins synthesized in a wheat germ cell-free system programmed with total poly(A) containing RNA from exponentially growing and starved T. pyriformis cells. Limiting amounts (1.5 μg/50 μl incubation mixture) of total poly(A) containing RNA from exponentially growing cells (a) and from cells starved for 1 h (b) were translated. Incubation was for 1 h at 25°C. The synthesized products were analyzed as in Fig. 2 except that gels containing 12.5% acrylamide were used.

duced. After 1 h of starvation only highly phosphorylated S6 is present on the ribosomes (Kristiansen & Krüger, 1978). During this process when the cells contain a heterogenous population of 40-S ribosomal subunits (Kristiansen & Krüger, 1979), a determination of the kinetics of S6 phosphorylation in 80-S ribosomes and polyribosomes may serve as a measure of the relative efficiency with which the different subclasses of 40-S ribosomal subunits participate in the initiation of protein synthesis, provided that the phosphorylation *per se* does not depend on protein synthesis, or that the overall rate of S6 phosphorylation in 80-S ribosomes and polyribosomes does not differ. Figure 4 shows that complete inhibition of protein synthesis with cycloheximide (5 μg/ml) does not impair the phosphorylation. Figure 5 demonstrates that phosphorylation of S6 in 80-S ribosomes and polyribosomes proceeds at virtually identical rates after 1 h of starvation. However,

Figure 6 and **Tab**le I show that there is a preferential
accumulation of 40-S ribosomal subunits containing highly
phosphorylated S6 in the polyribosomes after 30-40 min of
starvation. The enhanced accumulation of 40-S ribosomal subunits with highly phosphorylated S6 in the polyribosomes may,
therefore, signify a preferential utilization of these 40-S
ribosomal subunits for protein synthesis.

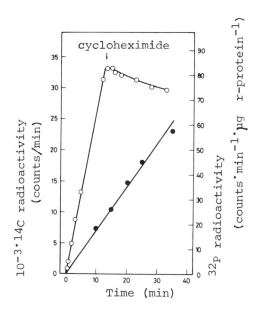

*FIGURE 4. Protein synthesis and S6 phosphorylation in
starved T. pyriformis cells following the addition of cycloheximide. The cells were starved for 80 min and then received
[^{14}C]proline (1 µCi/ml) and [^{32}P]phosphate (12 µCi/ml).
Cycloheximide (5 µg/ml) was added as indicated. Samples were
collected for determination of [^{14}C]proline incorporation into
hot trichloroacetic acid resistent materials, and for the isolation of ribosomes. The ribosomal proteins were extracted
with acetic acid and the incorporation of [^{32}P]phosphate into
ribosomal proteins was determined. Open circles: ^{14}C-radioactivity; solid circles: ^{32}P-radioactivity.*

Adenosine Polyphosphates

When *T. pyriformis* cells are transferred to a non-nutrient
medium Hamburger & Zeuthen (1957) found that the respiratory
rate dropped to 25% in division synchronized *Tetrahymena*; we
have found that the adenylate energy charge as defined by Atkinson (1969) increases to 0.95-0.98 under these conditions (un-

TABLE I. *The Extent of S6 Phosporylation in 80-S Ribosomes and Polyribosomes in T. pyriformis Cells During the First 1.5 h of Starvation.*

Duration of starvation (min)	Derivatives of S6 found on 80-S ribosomes			Derivatives of S6 found on polyribosomes		
	S6	S6b	S6a	S6	S6b	S6a
10	+	(+)		+	(+)	
20	+	+		+	+	
30	+	+		+	+	+
40	+	+		+	+	+
60		(+)	+			+
90			+			+

Exponentially growing cells were transferred into a nonnutrient medium. 80-S ribosomes and polyribosomes were isolated at intervals by sucrose gradient centrifugation; their proteins were extracted with acetic acid and analyzed by two-dimensional polyacrylamide gel electrophoresis. S6 denotes the non-phosphorylated form; S6b the intermediate phosphorylated and S6a the highly phosphorylated form of S6.

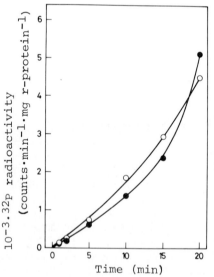

FIGURE 5. *The kinetics of S6 phosphorylation in 80-S ribosomes and polyribosomes in starved T. pyriformis cells. ^{32}P phosphate (32 µCi/ml) was added to cells starved for 1 h. Samples were collected at intervals and the ribosomes were fractionated by sucrose gradient centrifugation. The ribosomal proteins in the 80-S ribosomes and the polyribosomes were extracted with acetic acid and their ^{32}P-radioactivity was determined. Open circles: 80-S ribosomes; solid circles: polyribosomes.*

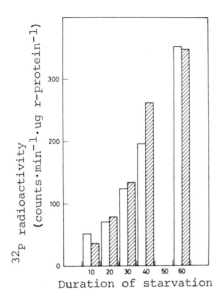

FIGURE 6. The kinetics of S6 phosphorylation in 80-S ribosomes and polyribosomes in T. pyriformis cells during the first hour of starvation. Exponentially growing cells were transferred into a non-nutrient medium containing [^{32}P]phosphate (33 µCi/ml). Samples were collected at intervals and the incorporation of [^{32}P]phosphate into ribosomal proteins of 80-S ribosomes and polyribosomes was determined.
Light bars: 80-S ribosomes; cross-hatched bars: polyribosomes.

published); assuming a constant P/O ratio, this must mean that a decrease in the turnover of ATP leads to an increase in concentration. The parallel situation for exponentially growing cells is shown in Figure 7. After the transfer to the starvation medium a steady increase in the ATP content is recorded until a peak content is reached at 30 min; the content then decreases. When XDP is used as an internal standard, the ATP determinations are quite exact (Fig. 1), especially when calculated over the protein content of the cells from which the ATP is extracted. Data from three experiments were in complete agreement. Analysis of cells subjected to anaerobiosis by centrifugation shows a constant level at half the concentration as that found in non-centrifuged cells. Addition of 2,4-dinitrophenol at the time of transfer (final concentration 0.5 mM) leads to an immediate decrease in ATP. Phosphorylation of S6, however, is not impaired by the addition of DNP (unpublished). This suggests a certain compartmentilization of ATP and that glycolytic generated ATP may serve as the phosphoryl donor in the phosphorylation of S6.

Figure 8 shows the content of Ap_4A in the same experiment as that in Figure 7. Immediately after the shift to starvation medium only trace amounts of Ap_4A are found; it then increases and reaches a value of 0.1-0.2% of the ATP content after 90 min of starvation. At that time only highly phosphorylated S6 is present and the starvation induced changes in the pattern of protein synthesis are fully expressed. DNP immediately brings Ap_4A to the level that is reached in the non-treated cells only after 1-2 hours.

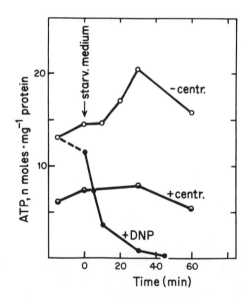

FIGURE 7. *ATP in T. pyriformis cells during the adaptation from growth to starvation conditions. 1 ml samples were removed at the times indicated and analyzed for free nucleotides by high pressure anion exchange chromatography. The protein in the acid precipitable residue was determined by the method of Lowry et al. (1951). Open circles: cells were not centrifuged prior to extraction; semi-solid circles: cells were collected by centrifugation prior to extraction; solid circles: 2,4-dinitrophenol (final conc. 0.5 mM) was added to a subculture at the time of resuspension in the starvation medium, the cells were not centrifuged prior to extraction.*

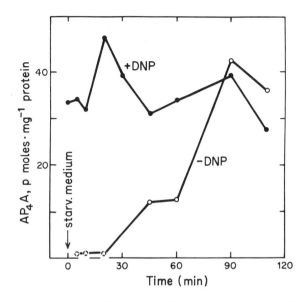

FIGURE 8. Ap_4A in T. pyriformis cells during the adaptation from growth to starvation conditions. 1 ml samples were removed at the times indicated and analyzed for Ap_4A by high pressure anion exchange chromatography after treatment of the nucleotide extract with alkaline phosphatase. The cells were not centrifuged prior to extraction. Open circles: no additions; broken circles: a peak could be detected but not quantitated; solid circles: 2,4-dinitrophenol was added as described in Fig. 7.

DISCUSSION

When T. pyriformis cells are transferred to a non-nutrient medium they respond by a series of rapid adjustments in ultrastructure and metabolism. These adjustments enable the cells to survive for even prolonged periods (several days) of starvation. In micronucleate strains of Tetrahymena starvation also induces pair formation, one of the initial events in the sexual process of conjugation (Orias & Bruns, 1976). One of the conspicuous biochemical events, the induced phosphorylation of the ribosomal protein S6 (Kristiansen et al., 1978b) may serve as a marker for at least some of these necessary adjustments, since it appears that those cells which adapt successfully to the starvation conditions always contain phosphorylated S6 during the transition period.

Pronounced increases in the levels of phosphorylation of the ribosomal protein S6 in higher eukaryotes were observed in regenerating rat liver (Gressner & Wool, 1974a) and in HeLa cells following serum stimulation (Lastick et al., 1977); conditions that apparently bear very little resemblance to the starvation conditions that induce S6 phosphorylation in *T. pyriformis*. However, common to all these unbalanced systems is an increase in the intracellular level of Ap_4A (and presumably ATP) (cf. Rapaport & Zamecnik, 1976; Grummt et al., 1977; Plesner et al., 1979). The correlation between Ap_4A accumulation and S6 phosphorylation may well be fortuitous, but it should be noted that a link between Ap_4A and S6 phosphorylation might explain, for example, the rather odd increase in S6 phosphorylation in rat liver *in vivo* following administration of the protein synthesis inhibitors cycloheximide and puromycin (Gressner & Wool, 1974b). That such a link exists may also be indicated by the parallel enhancement of phosphorylation of S6 and accumulation of Ap_4A in DNP treated cells.

Our results show that the pattern of protein synthesis changes markedly during the adaptation period (Fig. 2). However, these changes in the patterns of protein synthesis *in vivo* are not reflected when total poly(A) containing RNAs from growing and starved cells are translated in a wheat germ cell-free system (Fig. 3). Among various possible explanations these results may be reconciled in accordance with the concept that a non-specific inhibition of protein synthesis leads to a preferential translation of those mRNAs that have the highest rate constants for ribosome attachment (Lodish, 1976). On the other hand, a more specific translational control mechanism may operate. Thus, although it has been questioned whether S6 phosphorylation influences the process of protein synthesis *in vivo*, our finding that 40-S ribosomal subunits which carry highly phosphorylated S6 accumulate in the polyribosomes during the transition period (Fig. 6 & Table I), may point to a possible role for S6 phosphorylation during initiation. In fact, we envision that S6 phosphorylation is part of a mechanism that ensures preferential translation of *de novo* synthesized mRNA. Experiments to prove (or disprove) this hypothesis are in progress.

The results presented show that the ATP concentration increases during the first 30 min of the adaptation period and then decreases. We have found a pattern of changes in ATP concentration during the cell division cycle which resembles the present in the sense that ATP increases when the rate of cellular metabolism decreases (Plesner, 1964). In both cases the increase in ATP concentration is parallelled by an increase in adenylate energy charge (unpublished). Although the maintenance of a high energy charge seems essential for the formation of initiation complexes (Van Venrooij et al., 1972; Swedes et al., 1975; Walton & Gill, 1976), little is published about the effect of very high ATP concentrations (energy charges), e.g. in the same way as described for phosphofructokinase by Lardy & Parks (1956) and Uyeda (1979). It is, however, well known that an optimal Mg^{2+} concentration for protein synthesis exists beyond which the rate decreases (Tissières et al., 1960). In addition, from in vivo experiments on E. coli, Swedes et al. (1975) concluded that beyond a concentration of 5 nmol ATP·mg protein^{-1} (corresponding to an energy charge of 0.90) the rate of protein synthesis decreases. Even if a closer analysis of the data mentioned may lead to a hen and egg problem, it might mean that we can begin to understand a puzzling situation which exists after T. pyriformis cells have been transferred to a non-nutrient medium: a) the cellular ATP content (and the adenylate energy charge) increases (Fig. 7), b) the endogenous supply of precursors for protein synthesis appears plentiful immediately following the shift (cf. Nilsson, 1976) and, c) the cellular content of poly(A) containing RNA does not decrease significantly during the first hour of starvation (unpublished). Yet, the initiation of protein synthesis is abruptly inhibited. Thus, the adaptation of T. pyriformis to starvation conditions is reminiscent of sporulation of bacilli in that a considerable number of biochemical and morphological changes are induced by adverse environmental conditions and proceed while the cells still contain ample supplies of nutrients and energy rich compounds. In Bacillus subtilis the production of a highly phosphorylated nucleotide, p_3Ap_3, seems to be crucially involved in sporulation (Rhaese & Groscurth, 1979). In analogy, starvation of T. pyriformis cells leads to accumulation of Ap_4A and ATP (Fig. 7 and 8). It is assumed that Ap_4A is formed in the back reaction of the synthesis of aminoacyl adenylate, and that this back reaction is favored by an increase in the concentration of aminoacyl-tRNA and ATP in response to a decrease in the rate of protein synthesis (Zamecnik & Stephenson, 1969; cf. Aspen & Hoagland, 1978). These assumptions provide a plausible explanation of the increase in Ap_4A in starved T. pyriformis cells. When S6 becomes dephosphorylated after prolonged starvation (Kristiansen & Krüger, 1979), no Ap_4A is found (unpublished). The situation with a decreased protein synthesis when energy and precursors are abundant becomes meaningful if it succeeds a cellular event that reduces the ATP consumption by a few per cent; with a constant P/O ratio, this will effect an increase in ATP concentration in the order of magnitude as that in Figure 7. That high ATP concentrations could act inhibitory on protein synthesis by binding of ATP is supported by recent work, which indicate that glucose-6-phosphate, NAD^+ and GDP participate directly in the regulation of initiation of protein synthesis in reticulocyte lysates (Ernst et al., 1978; Wu et al., 1978; Safer & Jagus, 1979). In keeping with this, we believe that the changes in adenosine polyphosphate concentration are not solely passive events in the wake of changes in macromolecular synthesis. They may well act as decisive regulatory effectors.

ACKNOWLEDGMENTS

We thank Professor Martin Ottesen of the Carlsberg Laboratory warmly for his hospitality and kind help, and we thank Anna Krüger, Vibeke Ekholdt and Søren Andersen for expert technical assistance and Ulla Pedersen for typing the manuscript. We are thankful to Dr. Eliezer Rapaport for the Ap_4A, we used as standards and the staff of the Carlsberg Laboratory for the kind help.

REFERENCES

Aspen, A.J. & Hoagland, M.B. (1978), Uncoupling of amino acid turnover on transfer RNA from protein synthesis in HeLa cells. *Biochim. Biophys. Acta 518*, 482.
Atkinson, D.E. (1969), Regulation of enzyme function. *Annu. Rev. Microbiol. 23*, 47.
Chen, S., Brown, P.R. & Rosie, D.M. (1977), Extraction procedures for use prior to HPLC nucleotide analysis using microparticle chemically bounded packings. *J. Chrom. Sci. 15*, 218.
Ernst, V., Levin, D. & London, I.M. (1978), Evidence that glucose 6-phosphate regulates protein synthesis initiation in reticulocyte lysates. *J. Biol. Chem. 253*, 7163.
Finamore, F.J. & Warner, A.H. (1963), The occurrence of P^1, P^4-diguanosine 5'-tetraphosphate in brine shrimp eggs. *J. Biol. Chem. 238*, 344.
Gressner, A.M. & Wool, I.G. (1974a), The phosphorylation of liver ribosomal proteins in vivo. Evidence that only a single small subunit protein (S6) is phosphorylated. *J. Biol. Chem. 249*, 6917.
Gressner, A.M. & Wool, I.G. (1974b), The stimulation of the phosphorylation of ribosomal protein S6 by cycloheximide and puromycin. *Biochem. Biophys. Res. Commun. 60*, 1482.
Grummt, F., Paul, D. & Grummt, I. (1977), Regulation of ATP

pools, rRNA and DNA synthesis in 3T3 cells in response to serum or hypoxanthine. *Eur. J. Biochem. 76*, 7.

Hamburger, K. & Zeuthen, E. (1957), Synchronous divisions in *Tetrahymena pyriformis* as studied in an inorganic medium. The effect of 2,4-dinitrophenol. *Exptl. Cell Res. 13*, 443.

Hartwick, R.A. & Brown, P.R. (1975), The performance of microparticle chemically-bounded anion-exchange resins in the analysis of nucleotides. *J. Chrom. 112*, 651.

Kristiansen, K. & Krüger, A. (1978), Ribosomal proteins in growing and starved *Tetrahymena pyriformis*. Starvation-induced phosphorylation of ribosomal proteins. *Biochim. Biophys. Acta 521*, 435.

Kristiansen, K. & Krüger, A. (1979), Phosphorylation and degradation of ribosomes in starved *Tetrahymena pyriformis*. *Exptl. Cell Res. 118*, 159.

Kristiansen, K., Hartmann, B., Plesner, P. & Krüger, A. (1978a), Conservation of active ribosomes in acetone-treated cells of *Tetrahymena pyriformis*. *Eur. J. Biochem. 83*, 389.

Kristiansen, K., Plesner, P. & Krüger, A. (1978b), Phosphorylation in vivo of ribosomes in *Tetrahymena pyriformis*. *Eur. J. Biochem. 83*, 395.

Lardy, H.A. & Parks, R.E. Jr. (1956), Influence of ATP concentration on rates of some phosphorylation reactions. in "Enzymes: Units of Biological Structure and Function" (O.H. Gaebler), p. 584. Academic Press, New York.

Lastick, S.M., Nielsen, P.J. & McConkey, E.H. (1977), Phosphorylation of ribosomal protein S6 in suspension cultured HeLa cells. *Molec. Gen. Genet. 152*, 223.

Lodish, H.F. (1976), Translational control of protein synthesis. *Annu. Rev. Biochem. 45*, 39.

Lowry, O.H., Rosebrough, N.J., Farr, A.L. & Randall, R.J. (1951), Protein measurement with the Folin phenol reagent. *J. Biol. Chem. 193*, 265.

Nilsson, J.R. (1976), Physiological and structural studies on *Tetrahymena pyriformis*$_{GL}$. *Compt. Rend. Trav. Lab. Carlsberg 40*, 215.

Orias, E. & Bruns, P.J. (1976), Induction and isolation of mutants in *Tetrahymena*. in "Methods in Cell Biology" (D.M. Prescott), Vol. 13, p. 247. Academic Press, New York.

Plesner, P. (1958), The nucleoside triphosphate content of *Tetrahymena pyriformis* during the division cycle in synchronously dividing mass cultures. *Biochim. Biophys. Acta 29*, 462.

Plesner, P. (1961), Changes in ribosome structure and function during synchronized cell division. *Cold Spring Harbor Symp. Quant. Biol. XXVI*, 159.

Plesner, P. (1964), Nucleotide metabolism during synchronized cell division in *Tetrahymena pyriformis*. *Compt. Rend. Trav.*

Lab. Carlsberg 34, 1.
Plesner, P., Stephenson, M.L., Zamecnik, P.L. & Bucher, N.R.L. (1979), Diadenosine tetraphosphate (Ap$_4$A), an activator of gene function. in "Specific Eukaryotic Genes" (J. Engberg, V. Leick, H. Klenow & J. Hess Thaysen), p. 383. Munksgaard, Copenhagen.
Rapaport, E. & Zamecnik, P.C. (1976), Presence of diadenosine 5',5'''-P^1,P^4-tetraphosphate (Ap$_4$A) in mammalian cells in levels varying widely with proliferative activity of the tissue: A possible positive "pleiotypic activator". Proc. Natl. Acad. Sci. USA 73, 3984.
Revel, M. & Groner, Y. (1978), Post-transcriptional and translational controls of gene expression in eukaryotes. Annu. Rev. Biochem. 47, 1079.
Rhaese,H.J. & Groscurth, R. (1979). Apparent dependence of sporulation on synthesis of highly phosphorylated nucleotides in Bacillus subtilis. Proc. Natl. Acad. Sci. USA 76, 842.
Safer, B. & Jagus, R. (1979), Control of eIF-2 phosphatase activity in rabbit reticulocyte lysate. Proc. Natl. Acad. Sci. USA 76, 1094.
Siekevitz, P., Palade, G.E. (1960), A cytochemical study on the pancreas of the guinea pig. VI. Release of enzymes and ribonucleic acid from ribonuclein particles. J. Biophys. Biochem. Cytol. 7, 631.
Swedes, J.S., Sedo, R.J. & Atkinson, D.E. (1975), Relation of growth and protein synthesis to the adenylate energy charge in an adenine-requiring mutant of Escherichia coli. J. Biol. Chem. 250, 6930.
Tissières, A., Schlessinger, D. & Gross, F. (1960), Amino acid incorporation into proteins by Escherichia coli ribosomes. Proc. Natl. Acad. Sci. USA 46, 1450.
Uyeda, K. (1979), Phosphofructokinase. Adv. Enzymol. 48, 193.
Van Venrooij, W.J.W., Henshaw, E.C. & Hirsch, C.A. (1972), Effects of deprival of glucose or individual amino acids on polyribosome distribution and rate of protein synthesis in cultured mammalian cells. Biochim. Biophys. Acta 259, 127.
Walton, G.M. & Gill, G.N. (1976), Preferential regulation of protein synthesis initiation complex formation by purine nucleotides. Biochim. Biophys. Acta 447, 11.
Wu, J.M., Cheung, C.P. & Suhadolnik (1978), Stimulation and inhibition of the protein synthetic process by NAD$^+$ in lysed rabbit reticulocytes. J. Biol. Chem. 253, 7295.
Zamecnik, P.C. & Stephenson, M.L. (1969), Nucleoside pyrophosphate compounds related to the first step in protein synthesis. in "The Role of Nucleotides for the Function and Conformation of Enzymes" (H.M. Kalckar, H. Klenow, A. Munch-Petersen, M. Ottesen & J. Hess Thaysen), p. 276. Munksgaard, Copenhagen.

Part IV
ALTERATION OF TRANSLATIONAL MECHANISMS

REGULATION OF MACROMOLECULAR SYNTHESIS BY LOW MOLECULAR WEIGHT MEDIATORS

NUCLEOSIDETRIPHOSPHATE MEDIATED DISCRIMINATION
OF GENE EXPRESSION IN T1-INFECTED E.coli

Manfred Schweiger, Erwin F. Wagner

Institut für Biochemie
Universität Innsbruck
Innsbruck, Austria

T1-infected E.coli, a suitable system for gene controls:

T1-virus infected E.coli appears to be an advantageous system to study control mechanisms of gene expression. As the other E.coli viruses of the T-group, T1 exerts a pronounced temporal and quantitative regulation of synthesis of proteins (fig.1;1,2) However, T1 does not possess the complications of coupling of late viral protein synthesis to DNA replication as do the T-evens and T1 does not induce a viral specific RNA polymerase as do T3 and T7 (3). T1 is dissimilar in its mode of DNA injection to T5, which infects in a complex two step process (4). Additionally, T1 is a rather small DNA virus. Its DNA of $31 \cdot 10^6$ m.w. (4) codes for about 31 characterized proteins (1,2). Redirection of gene expression from host specificity toward viral synthesis takes place during the very first minutes of infection (1,2). Host protein synthesis is dramatically reduced and concomitantly production of "early T1-proteins" is started. Whereas the synthesis of the early gene products is turned down 8 minutes (at 30°C) after start of infection the early-late proteins are continuously formed until the cell is lysed (fig.1). The majority of viral structural proteins is synthesized late in the infectious cycle. The corresponding mechanisms of control of

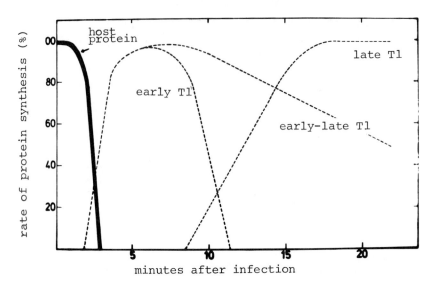

Figure 1: Protein synthesis after T1-infection, the schematic pattern.

gene expression are unknown. Of special interest is the T1-induced restriction of host gene expression (T1 - host repression) since a whole group of several hundreds of genes is commonly inactivated while another group, the early T1-genes, is activated.

Time course of T1-host repression:

Formation of individual enzymes as for instance ß-galactosidase (fig.2) as well as amino acid incorporation in host proteins is discontinued very soon after addition of the virus to the culture (2). Chloramphenicol exerts a faster action than T1, whereas rifampicin inhibits host protein synthesis after a considerable delay. This time course indicates: i) an action of T1 on the translational level rather than on transcription; ii) an effect of T1 on initiation of translation, since the lag period of establishment of repression takes longer than the one of the inhibitor of elongation of translation (CAM). Comparison of the time courses of T1 - host repression with the action of a drug acting on initiation of translation (Nitrofurantoin (5)) supports this presumption (2). In contrast to the T1-host repression the comparable T7-induced inhibition of host gene expression requires a longer lag period (6) (5 to 6 minutes). In this period T7-specific

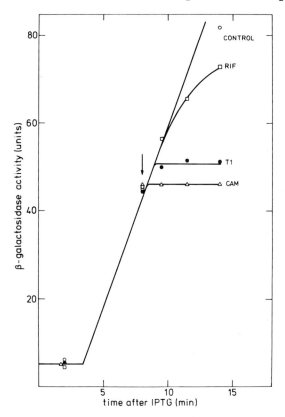

Figure 2: Time course of inhibition of host protein synthesis by T1. A culture of E.coli B_{S-1} (5×10^8 cells/ml) was induced by isopropyl thiogalactoside (IPTG, 1 mM) to synthesize ß-galactosidase (time 0 min.). To aliquotes of the culture were added at 8 min. either T1 (moi 10), or chloramphenicol (100 µg/ml), or rifampicin (200 µg/ml). ß-galactosidase was assayed at the given times.

control proteins are induced which act on transcription and on tranlation of host genes (3): T7-protein kinase, which acts on the ß'subunit of host RNA polymerase, T7-transcriptional inhibitor, which forms an inactive complex with host RNA polymerase and the T7-translational repressor, which discriminates in the initiation of translation against host specific translation. The brief period for the induction of T1 - host repression suggests that no viral repressor protein is induced.

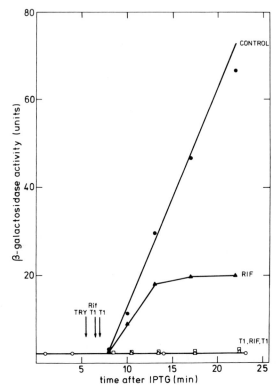

Figure 3: T1 inhibits the initiation of translation of host mRNA: An exponentially growing culture of E.coli B_{S-1} was induced with IPTG in the presence of tryptophan analogues (7-azatryptophan (50 µg/ml) and 5-methyltryptophan (5 µg/ml)). At the time indicated tryptophan was added (1.25 µg/ml). To aliquots were added: either rifampicin (200 µg/ml) or rifampicin and T1 (moi 10) or T1. At the given times ß-galactosidase was assayed.

T1 - host repression is exerted in translation:

To elucidate the level at which T1 - host repression is exerted, transcription and translation were separated by the following experimental approach (2):

ß-galactosidase synthesized in the presence of amino acid analogues is inactive. The ß-gal mRNA formation is not affected. Replacement of the analogues by an excess of natural tryptophan leads to synthesis of active enzyme. If the shift from analogues to natural amino acids is accompanied by the addition of an inhibitor of transcription, only translation takes place in the subsequent period. The now formed enzyme activity is a measure of the amount of specific mRNA present at the time of the shift. Any effect on translation would be manifested in this period. This technique revealed that T1 prevented any translation of ß-gal mRNA (fig.3). Thus T1 - host repression is exerted at the level of translation. Furthermore, detailed analysis of the time course in this translation system showed that T1 - host repression affected the initiation step. Essentially, the time course of establishment of T1 - host repression is identical to the action of nitrofurantoin, a specific inhibitor of initiation of translation (5).

T1-host repression is membrane mediated:

All our attempts to establish the T1 - host repression in the cell-free in vitro protein synthesizing system were unsuccessful (2). Systems prepared from cells infected with T1 for various periods were all capable to synthesize host enzymes as ß-galactosidase or galactokinase (2). In contrast, T7-infected cells lost the capability to form these host enzymes in cell-free systems (6). T7 induces the T7-translational repressor. However, in vitro systems from T1-infected cells were active under a wide range of conditions. Even gently lysed T1-infected cells synthesized ß-galactosidase in vitro. This indicates that T1 unlike T7 does not induce a control protein. How then is T1 - host repression established? This control is mediated by the cell membrane as is supported by the finding that establishment of T1 - host repression depends on the multiplicity of infection (fig.4). This is an indication for the involvement of the cell membrane.

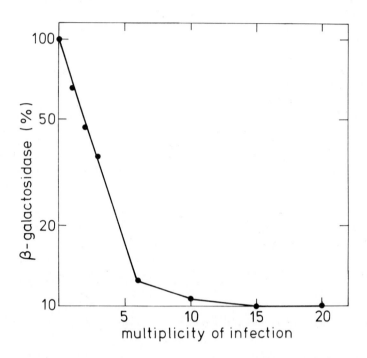

Figure 4: Multiplicity of infection (moi) dependence of T1-host repression: E.coli B_{S-1} was induced with IPTG. Immediately after induction, aliquotes were infected with the indicated multiplicities of T1. ß-galactosidase was determined after 10 minutes.

T1 alters the cell membrane:

Profound membrane alterations are a prerequisite for the central role of the membrane in this gene control. Alterations can be observed in the physicochemical state of the membrane and in its functions: Emission at 480 nm by the membrane fluorescence probe ANS is dramatically stimulated after T1-infection (E.Reider, unpublished). The fluidity of the membrane measured by the pyrene laser pulse relaxation technique (7) (Wagner, unpublished) is increased similarly. Concomitantly basic functions of the cell membrane are inhibited (fig.5). PMF-driven transports are inhibited, ATP-dependent uptakes are reduced and PTS-sugar accumulation is greatly stimulated (PTS is controlled by the PMF of the membrane (E.Reider, unpublished)). The detailed analysis of these alterations revealed that T1-infection reduced the electrochemical proton gradient of the cell membrane. ATPase deficient E.coli cells like wild type

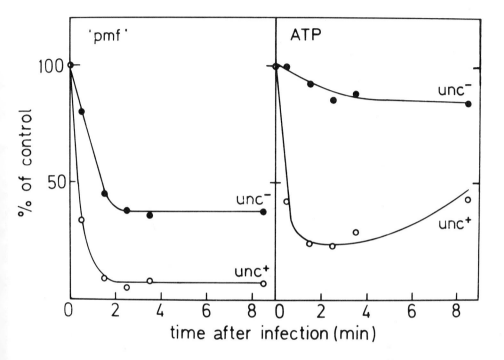

Figure 5: T1 affects "pmf"-and ATP-transports: E.coli B_{s-1} (unc^+) and E.coli uncAB (unc^-) grown in M9/glycerol were infected with T1 (moi 10) at zero time. Aliquotes of 0.5 ml were allowed to accumulate labeled L-(U-^{14}C) proline (PMF) (10 µCi/0.115 mg, 20 µM) or L-(^{3}H) glutamine (ATP) (21 µCi/nmol,10 µM) for one minute. The cells were collected by filtration and washed.

cells are inhibited in their PMF-driven transports after T1-infection, as elaborated in figure 5 by proline uptake. Whereas ATP-dependent transports are reduced only little in the unc$^-$ mutants, they are strongly affected in ATPase wild type cells (fig.5). Thus, T1 affects primarily the proton gradient and secondarily the ATP concentrations via the functionally active ATPase.

T1 dissipates the membrane potential:

The proton motive force consists of two components, the concentration gradient ΔpH and the electrical potential $\Delta\psi$. To elucidate which part of the PMF was affected mainly by T1-infection we determined the membrane potential by the distribution of the triphenylmethylphosphonium ion (TPMP$^+$) (8), a compound which can freely penetrate the lipophilic membrane bilayer due to its lipophilic property.

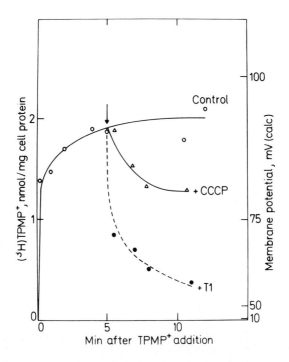

Figure 6: T1 reduces the membrane potential:
E.coli B_{S-1}, grown in M9/glycerol, were loaded with triphenylmethylphosphonium ion.
After 5 min. CCCP or T1 (moi 5) were added to aliquotes. (^{3}H)TPMP$^+$ was determined in the cells after collection.

Since it carries a positive charge, TPMP$^+$ enters the cell along the electrical potential (internal negative). From the in to out ratio of TPMP$^+$, the potential can be determined. Figure 6 shows that, as expected,

TPMP$^+$ penetrates into the cell along the potential (internal negative). The uncoupler CCCP reduces the electrical potential by dissipation of the proton gradient. It is remarkable that T1-infection causes a rapid, dramatic depolarization of the membrane. This is especially interesting, since concomitantly K$^+$ is exported, transporting positive charges (Wagner, unpublished). This would increase the potential. The net depolarization indicates an active import of protons.

ATP mediates the T1 - host repression:

It was demonstrated here that T1-infection redu-

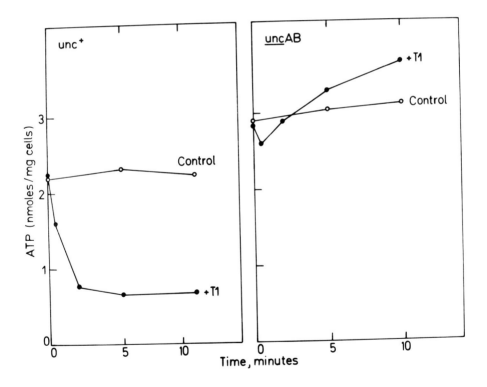

Figure 7: Intracellular ATP levels after T1-infection: Exponentially growing E.coli B_{S-1} and uncAB were infected at zero time with T1 (moi 10). After various perios, ATP was determined in aliquotes as described (11).

ćes the potential part of the PMF. Consequently, the ATPase converts ATP to electrochemical proton gradient. This was concluded from the ATP-driven transports in unc$^-$ and unc$^+$ E.coli. The direct determinations of the ATP concentrations proved this conclusion (fig.7). T1-infection causes a pronounced reduction of the ATP concentration in E.coli wild type (unc$^+$). In absence of an active ATPase (unc$^-$) this reduction did not take place (fig.7b). The observed elevation of the ATP concentration in unc$^-$ mutants reflects a control of substrate phosphorylation by the membrane energy.

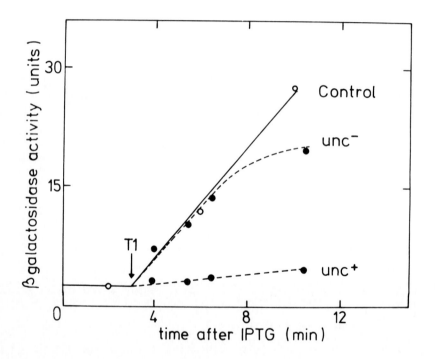

Figure 8: ATPase is essential in T1 - host repression: E.coli B_{S-1} and uncAB were induced with IPTG and infected with T1 (moi 10) at 3 minutes. ß-galactosidase was determined in aliquotes.

In the unc$^-$ cells the connection between the electrochemical proton gradient and the cell metabolism via the ATPase is disconnected. In these mutants T1-infection cannot repress host protein synthesis (fig.8), whereas normal T1 progenies are developed. Therefore ist must be concluded that ATP is the mediator of T1 - host repression. Its role can be either directly or indirectly, for instance via GTP. GTP is known to be cofactor in the initiation of translation. Then, it can be postulated that host mRNA has a higher requirement for GTP in initiation of translation than the viral mRNAs.

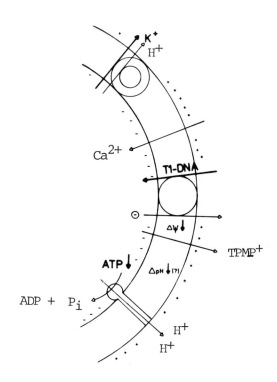

Figure 9: *T1-infection reduces the cell energy: The process of infection dissipates the proton gradient and reduces the membrane potential. As an indicator, TPMP$^+$ is exported. A proton gradient is reestablished on the expense of ATP by the ATPase or on the expense of the potassium concentration. The reduced ATP concentration mediates (via GTP) the T1-host repression.*

Why does T1 reduce the cell energy? (fig.9)

T1 attaches to the cellular ion transport system (10). An active proton gradient is required to transport the viral DNA into the cell (Wagner, thesis, 1978). The proton symport and the linked dissipation of the electrochemical proton gradient has several consequences: The membrane potential drops (fig.6). The ATPase is translocating protons on the expense of the ATP pool (and the nucleosidtriphosphate pool). The potassium gradient is utilized to reestablish a proton gradient. Concomitantly, Ca^{++} enters the cell and is not exported. The lowered concentrations of ATP and of the coupled nucleosidetriphosphates control the macromolecular synthesis. GTP is cofactor of the initiation of translation and thus a likely candidate for the mediator of T1-host repression.

ATP a general mediator of membrane control on macromolecular synthesis?

The control of protein synthesis by the membrane via the triphosphate concentration as demonstrated here might be a general mechanism. Membrane alterations are involved in many biological phenomena. Several of them could be mediated by the triphosphate concentrations. For instance: Aggregation and subsequent differentiation of Dictyostelium discoideum is triggered by starvation. Since starvation is accompanied by an increase in the concentration of cyclic AMP, most attention was concentrated on cAMP as the main signal for the triggering of the developmental program. However, a reasonable alternative would be that the decreased ATP concentration, caused by starvation, controls differentially gene expression.

Fertilization of sea-urchin eggs is a much studied developmental system. One of the very earliest events is a profound alteration of the cell membranes. Concomitantly, oxygen consumption is dramatically stimulated, a finding which was the concern of Otto Warburg's first biochemical publication in 1908 (9). The elevated oxydative phosphorylation leads to an increase of the ATP concentration. The elevated triphophate concentration can now turn on translation of mRNAs which were silent in the unfertilized egg. No additional repressor protein is required in this context. Indeed, even if there have been many attempts to find such a control protein in unfertilized sea-urchin eggs, no convincing success was reported yet.

The examples of biological developments, which could be easily understood by introducing the triphosphate concentration as mediator of membrane control on protein synthesis can be expanded.

REFERENCES

(1) Wagner,E.F., Ponta,H., Schweiger,M. (1977). Development of E.coli Virus T1: The Pattern of Gene Expression. *Mol.gen.Genet. 150, 21.*

(2) Wagner,E.F., Ponta,H., Schweiger,M. (1977). Development of *Escherichia coli* Virus T1: Repression of Host Gene Expression. *Eur.J.Biochem. 80, 255.*

(3) Schweiger,M., Wagner,E.F., Hirsch-Kauffmann,M., Ponta,H., Herrlich,P. (1978). Biochemistry of Development of E.coli Viruses T7 and T1 *in "Gene expression" (Clark,B.F.C., Klenow,H., Zeuthen,J. eds), pp.171-186, Pergamon Press.*

(4) Bresler,S.E., Kiselev,N.A., Manjakov,V.F., Mosevitsky,M.I., Timkovsky,A.L. (1967). Isolation and Physicochemical Investigation of T1 Bacteriophage DNA. *Virology 33, 1.*

(5) Herrlich,P., Schweiger,M. (1976). Nitrofurans, a group of synthetic antibiotics, with a new mode of action: Discrimination of specific messenger RNA classes. *Proc.Natl.Acad.Sci.U.S.A. 73, 3386.*

(6) Herrlich,P., Rahmsdorf,H.J., Pai,S.H., Schweiger,M. (1974). Translational Control Induced by Bacteriophage T7. *Proc.Natl.Acad.Sci.U.S.A. 71, 1088.*

(7) Ponta,H., Grätzel,M., Pfennig-Yeh,M., Hirsch-Kauffmann,M., Schweiger,M. (1977). Membrane Alteration Induced by T7 Virus Infection. *FEBS letters 73, 207.*

(8) Schuldiner,S., Kaback,H.R. (1975). Membrane Potential and Active Transport in Membrane Vesicles from *Escherichia coli*. *Biochemistry 14,* 5451.

(9) Warburg,O. (1908). Beobachtungen über die Oxydationsprozesse im Seeigelei. *Hoppe-Seyler's Z. physiol.Chem.57,1.*

(10) Braun,V., Hancock,R.E.W., Hantke,K., Hartmann, A. (1976). Functional Organization of the Outer Membrane of Escherichia Coli: Phage and Colicin Receptors as Components of Iron Uptake Systems. *J.Supramol.Struct.5, 37.*

(11) Klein,W.L., Boyer,P.D. (1972). Energization of Active Transport by *Escherichia coli*. *J.Biol. Chem.247, 7257.*

REGULATION OF MACROMOLECULAR SYNTHESIS BY LOW MOLECULAR WEIGHT MEDIATORS

COMPARTMENTALIZATION OF PHOSPHATE DONOR POOL FOR RIBOSOMAL PROTEIN S6 BEFORE AND AFTER STIMULATION OF QUIESCENT MOUSE FIBROBLAST CELLS WITH SERUM, AND NON-DEPENDENCE OF THE PHOSPHORYLATION ON cAMP POOLS

Julian Gordon, Luis Jimenez de Asua, Michel Siegmann, Anne-Marie Kübler and George Thomas

Friedrich Miescher-Institut
P.O.Box 273
CH-4002 Basel, Switzerland

The kinetics of ribosomal protein S6 phosphorylation have been investigated in quiescent, serum stimulated and serum withdrawn 3T3 mouse fibroblast cells. The kinetics of ^{32}P-labelling of soluble phosphate pools was also investigated, to facilitate interpretation of the data. In pulse-chase experiments, a Pi pool was found which rapidly exchanged with medium Pi. The exchange of this pool was unaffected by serum stimulation. On the other hand, the ATP-γ-phosphate pool exchanged relatively slowly, and its labelling was stimulated by serum. The labelling of the S6 phosphate, which was greatly stimulated by serum addition, was apparently via a rapidly exchanging donor, distinct from the total ATP pool in stimulated cells; but by a more slowly exchanging donor in quiescent cells. The phosphoryl of S6 was metabolically stable during the first 60 min of stimulation, but was rendered labile by serum withdrawal. In experiments with agents which affect cAMP pools, methylxanthines led to increased total cAMP, and decreased S6 phosphorylation and polysome assembly, while prostaglandin El led to increase in cAMP but no change in the other two parameters. It is concluded that changes in total measurable cAMP pools are not responsible for regulation of S6 phosphorylation and polysome formation.

INTRODUCTION

When quiescent mouse fibroblasts arrested in the Go/G1 phase of the cell cycle are stimulated with serum, a number of biochemical events take place which precede the activation of protein synthesis (reviewed in refs. 1 and 2). These changes include stimulation of phosphate, glucose, uridine, monovalent ion transport, decrease in cAMP levels (1,2), and more recently, phosphorylation of a protein of the small ribosomal subunit, namely S6 (3,4,5). Serum stimulation of protein synthesis was evidenced by increased total rate of translational and mobilization of pre-existing monosomes and mRNA into polysomes (6). We have confirmed this (4,5). As we discussed in detail earlier (2), any one of the above parameters may be directly or indirectly involved in the increased protein synthesis. It will only be possible to determine this when it is known which metabolic changes are relevant to the biochemical pathways involved. It is the purpose of this article to evaluate the effect of the state of the cell on the exchange of the phosphate in the medium with that in the protein S6, and thus determine whether the process is controlled by protein kinases and/or protein phosphatases; and also to determine whether the observed changes in total intracellular cAMP pools are the determining signals.

RESULTS

In general, labelling kinetics with phosphate has been considered difficult to carry out in eukaryotic systems because of the slow exchange of the internal phosphate pools with the phosphate in the medium (7). We have carefully re-examined the exchange kinetics of the external and internal Pi pools, using slightly modified procedures. We tried to avoid the possibility of release of labile bound phosphorus into Pi by using a rapid extraction procedure; and also established methodology for the analysis of the phosphate of ATP, which may be more relevant than the total ATP pool (4). We have already used this to show the kinetics of exchange of these two compartments with the medium Pi in serum stimulated 3T3 mouse fibroblast cells (4). Here we extend this study to investigate the effect of the serum stimulation on the exchange of these pools. The results are shown in Fig. 1. There was no effect of the serum stimulation on the exchange kinetics of the internal Pi pool. This is in apparent contrast to the data on uptake of ^{32}Pi into the internal acid soluble pool, which was stimulated by serum (1). Furthermore, both stimulated and quiescent cells showed an initial rapid exchange, followed by a second more slowly exchanging phase (Fig. 1). The internal Pi pool may therefore be composed of kinetically distinguishable compartments. On the other hand, the ATP-γ-phosphate pools exchanged relatively slowly. The results established

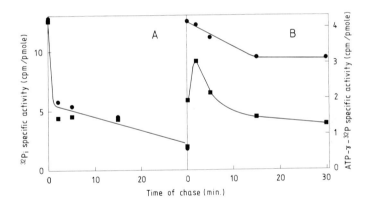

Figure 1. Pulse-chase label of internal (A) Pi and (B) ATP-γ-phosphate pools. Experimental protocol is described in detail elsewhere (4) with the additional data on quiescent cells. Cells were labelled for 30 min, with ^{32}P, either during quiescence (■——■), or following serum addition (●——●). The chase was carried out in the absence of serum for thequiescent cells, and in the same medium for the stimulated cells. Identical results were obtained if the quiescent cells were chased in medium containing serum, but conditioned by growth of cells to quiescence.

by others for total ATP phosphate in resting and growing cells (7) is essentially confirmed here. The exchange was slower in resting cells, since the initial specific activity reached was lower (0 time, Fig. 1B), and the subsequent kinetics is at least consistent with this: the initial increase in counts may not be significant. However, our observation of a previously undetected rapidly exchanging Pi compartment in both quiescent and stimulated cells raises the possibility that other internal phosphate pools may be rapidly exchangeable. This was in fact borne out by our finding the phosphate donor for S6 being in such a pool (4). We attempted to extend this observation for stimulated cells to resting cells, and the results are given in Fig. 2.

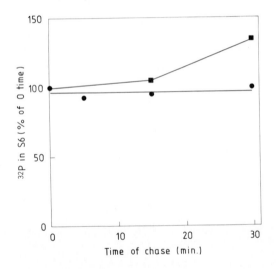

Figure 2. Pulse-chase kinetics of ^{32}P in protein S6 in quiescent and stimulated cells. The stimulated cells were as in ref. 4 (●——●), and the quiescent cells were first labeled 2 hr and then chased in non-radioactive conditioned medium (■——■). Both curves were normalized to 100% at 0 time of chase.

As was already shown (4), when cells were labelled with ^{32}P in the first 30 min of stimulation, and transferred to an identical non-radioactive medium, there was an immediate cessation of incorporation of further radio-activity into protein S6, during a period when the physical phosphorylation was still increasing. The only reasonable explanation for this cessation was that the donor pool was immediately chased out by non-radioactive phosphate (4). However, in the corresponding experiment with quiescent cells (Fig. 2) the labelling continued to increase. We can therefore only conclude that the phosphate donor pool does not exchange so rapidly with the external Pi as it does in stimulated cells. It should be noted that the initial incorporation (0 time of chase) is far lower in the quiescent cells (3), but the scales have been normalized to facilitate comparison. As noted above, the presence or absence of serum during chase made no difference to the fate of the ^{32}P label in the Pi and ATP pools in the quiescent cells. In contrast, we found that the presence of serum made a dramatic difference to the fate of ^{32}P label in S6 in both quiescent and stimulated cells. This is shown in Fig. 3.

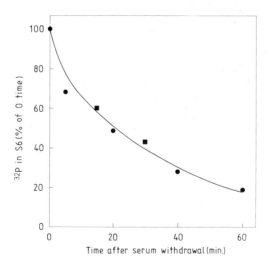

Figure 3. Effect of serum withdrawal on the fate of ^{32}P in S6, prelabelled for 30 min following serum addition (●———●), or during 2 hr of quiescence (■———■).

Withdrawal of serum led to identical kinetics for both cases with a rapid dephosphorylation reaction. The net dephosphorylation has also been confirmed for the stimulated cells from the shift in the mobility of the protein spot in two-dimensional gels (to be published in detail elsewhere: ref. 5).
These results show that, although the S6 phosphate is metabolically stable in stimulated cells (4), an active phosphatase must also appear following serum withdrawal, which was inactive in the presence of the serum. Although the experiments do not allow us to make any conclusions concerning such an activity in quiescent cells, because of the apparent inability to chase out the phosphate donor pool (Fig.2). an active phosphatase must be present during the serum withdrawal.

Since addition of serum causes a transient drop in cAMP levels, and serum withdrawal causes it to increase (reviewed in 2), we were interested to see whether, more generally, parameters which affect total cAMP in the cells also affect S6 phosphorylation, and whether this in turn is always coupled to increased protein synthesis.

Addition of the phosphodiesterase inhibitors theophylline or SQ20006 increases the cAMP pools (8). We have confirmed this (data in summary form in Table I, and to be published in detail in ref. 5), and furthermore found that both bring about decreased S6 phosphorylation and polysome formation. The decreased phosphorylation was determined both from ^{32}P incorporation in S6 and its altered mobility in two-dimensional gel electrophoresis (5). Since prostaglandin E_1 causes a greater change in cAMP pools than the above methylxanthines (5,8), we investigated its effects on the same parameters. These results are also summarized in Table I. In the case of PGE_1, there was neither change in S6 phosphorylation, nor in polysome formation.

Table I. Effect of various additions on total cAMP pools, polysome formation and S6 phosphorylation

Addition	cAMP	S6-P	Polysomes
+ serum	-	+	+
+ serum + theophylline	+	-	-
+ serum + SQ20006	+	-	-
+ serum + PGE_1	++	o	o
- serum	+	-	-

The data in this table is to be published in detail elsewhere (5), and where this is confirmatory of literature data see ref. 2 for review and ref. 8. +, - or o indicates change brought about by addition mentioned, except in the case of additions over serum, where the change is over and above the effect of serum itself.

DISCUSSION

The results described here show that in spite of difficulty in correlating intracellular pools with biochemical parameters, it is possible to draw some conclusions. Thus, while it is well known that the total phosphate pools are only sluggishly labelled, it is possible to detect compartments which are much more rapidly equilibrated. Further, by manipulation of the external conditions, it is possible to show that the total internal cAMP pool can be made to increase without measurable effect on polysome formation. It must therefore be considered that some other intracellular component must be the relevant signal.

The fact that we have discovered rapidly exchanging phosphate pools which are apparently unaffected by whether the cells are quiescent or stimulated is not in direct contradiction of the published data (1). We have measured the specific activity of the intracellular Pi pool, following chase with non-radioactive medium, whereas previous published data was on the uptake of ^{32}P into the total acid soluble pool. The lack of effect of serum stimulation may be due to the fact that this compartment is in such rapid equilibrium that no detectable change occurs during the time scale of the experiment. On the other hand, the labelling kinetics of the ATP pools is not very different from that found by others (7). But as suggested above, even careful measurement of the phosphate label does not reveal the true phosphate donor for S6, which must be in a separate compartment, not readily accessible to direct measurement. The donor may not even be ATP. Others have presented evidence for compartmentalization of ATP pools from either kinetic evidence or direct determination (9-12).

The data also permits us to make some deductions concerning the kinds of pathways responsible for defining the level of the ribosomal protein S6, and hence perhaps of polysome formation. During the first 60 min of stimulation, the S6 phosphate is metabolically stable (4). Thus, the phosphatase, if present, must have been inactivated by the serum addition, or was not present at all. When serum was withdrawn, a phosphatase must have been active, both in resting and quiescent cells. The observation leads to the idea that the phosphorylation can respond up or down, depending on the immediate demands on the cell, with a corresponding response in polysome formation, or dis-assembly.

Since the phosphorylation of S6 was not blocked by cycloheximide (5), we can also deduce that a *de novo* synthesis of kinase is not required for the stimulation. This may well account for the extreme responsiveness of the phosphorylation of ribosomal protein S6 in a variety of systems. However, the data summarized in Fig. 1 suggests that we must continue to seek the relevant internal signal.

REFERENCES

1) Rudland, P.S. and Jimenez de Asua, L. (1979). Action of growth factors in the cell cycle. *Biochim. Biophys. Acta 560*, 91-133.

2) Thomas, G. and Gordon, J. (1979). Regulation of protein synthesis during the shift of quiescent animal cells into the proliferative state. *Molecular and Cell Biology Reports,* in press.

3) Haselbacher, G.K., Humbel, R.E. and Thomas, G. (1979). Insulin-like growth factor: insulin or serum increase the phosphorylation of ribosomal protein S6 during transition of the stationary chick embryo fibroblasts into early G1 phase of the cell cycle. *FEBS Letters 100*, 185-190.

4) Thomas, G., Siegmann, M. and Gordon, J. (1979). Multiple phosphorylation of ribosomal protein S6 during the transition of quiescent cells into early G1, and cellular compartmentalization of the phosphate donor. *Proc. Nat. Acad. Sci. USA*, in press.

5) Thomas, G., Siegmann, M., Kubler, A.-M., Gordon, J. and Jimenez de Asua, L. (1979). Possible role of ribosomal protein S6 phosphorylation in activation of protein synthesis and its relationship with intracellular cyclic AMP levels. In preparation.

6) Rudland, P.S. (1974). Control of translation in cultured cells: continued synthesis and accumulation of messenger RNA in non-dividing cultures. *Proc. Nat. Acad. Sci. USA 71*, 750-754.

7) Weber, M.J. and Edlin, P. (1971). Phosphate transport, nucleotide pools, and ribonucleic acid synthesis in growing and density inhibited 3T3 cells. *J. Biol. Chem. 246*, 1828-1833.

8) Chasin, M., Harris, J.A., Phillips, M.B. and Hess, S.M. (1972). I-Ethyl-4(isopropylidenehydrazino)-1 H-pyrazolo-(3,4-6)-pyridine-5'-

carboxylic acid, ethyl ester, hydrochloride (SQ20009) - a potent new inhibitor of cyclic 3',5'-nucleotide phosphodiesterases. *Proc. Nat. Acad. Sci. USA 73*, 3122-3125.

9) Rapoport, E. and Zamecnik, P.C. (1976). Presence of diadenosine 5',5'''-P',P^4-tetraphosphate (AP$_4$A) in mammalian cells in levels varying with the proliferative activity of the tissue: a possible positive "pleiotypic activator". *Proc. Nat. Acad. Sci. USA 73*, 3122-3125.

10) Grummt, I. and Grummt, F. (1976). Control of nucleolar RNA synthesis by the intracellular pool size of ATP and GTP. *Cell 7*, 447-453.

11) Sobol, S., Scholz, R. and Heldt, H. (1978). Subcellular metabolite concentrations. Dependence of mitochondrial and cytosolic ATP systems on the metabolic state of perfused rat liver. *Eur. J. Biochem. 87*, 377-390.

12) Rapoport, E-, Garcia-Blanco, M.A. and Zamecnik, P.C. (1979). Regulation of DNA replication in S-phase nuclei by ATP and ADP pools. *Proc. Nat. Acad. Sci. USA 76*, 1643-1647.

REGULATION OF MACROMOLECULAR SYNTHESIS BY LOW MOLECULAR WEIGHT MEDIATORS

MEMBRANE MEDIATED AMPLIFICATION OF TRANSLATIONAL CONTROL IN EUKARYOTES: A PLEIOTROPIC EFFECT [1]

Gebhard Koch
Patricia Bilello
Joachim Kruppa

Physiologisch-Chemisches Institut der Universität Hamburg
Abteilung Molekularbiologie
Hamburg, W.Germany

Tissue culture cells respond to amino acid starvation, virus infection and exposure to hypertonic medium with a reduced rate of polypeptide chain initiation. This response is accompanied by changes in the phosphorylation state of certain ribosomal proteins and by the activation and/or release of a low molecular weight mediator which inhibits protein synthesis when added to cell-free extracts. The inhibitor is reversibly bound to ribosomes and decreases the ability of ribosomes to participate in the formation of functional initiation complexes. A preliminary list of some of the biochemical properties of this low molecular weight mediator is given below.

INTRODUCTION

Animal cells in culture grow optimally only within a narrow range of medium osmolarity. Variations in the tonicity of the growth medium are accompanied by a rapid flow of water into or out of the cell thereby leading to alterations in the intracellular osmolarity. Intracellular levels of specific ions change more slowly. A rapid elevation of the intracellular osmolarity of tissue culture cells by increasing the

---

[1] *Supported in part by Deutsche Forschungsgemeinschaft.*

tonicity of the growth medium results in a specific inhibition of polypeptide chain initiation (Hypertonic Initiation Block = HIB) (Koch et al., 1976). The observed inhibition is independent of the salt used to increase the osmolarity of the medium. NaCl, KCl, NH_4Cl or even sucrose are equally well suited for this purpose (Oppermann et al., 1973).

Translation of host cell mRNAs is also rapidly inhibited after infection by several viruses (Lodish; 1976; Revel & Groner, 1978; Koch et al., 1979a). This inhibition is triggered well before the onset of viral protein synthesis. Protein synthesis in virus infected cells is less sensitive to inhibitors of polypeptide chain initiation than that of uninfected cells. On the other hand, virus infected and uninfected cells respond equally to inhibitors of polypeptide chain elongation such as puromycin and cycloheximide (Koch et al., 1976). These observations could be interpreted in several ways: Virus infection could change the properties of cells resulting in a higher resistance of all messengers to inhibitors of polypeptide chain initiation. Alternatively, viral mRNA translation selectively proceeds under conditions where host mRNA translation is severely inhibited. Our experimental findings indicate that the latter alternative appears to be the more likely one. It remains to be determined *why* this inhibition of cellular protein synthesis enhances the expression of viral genes and *how* this inhibition is brought about.

RESULTS AND DISCUSSION

Differential Effect of HIB on Synthesis of Individual Cellular Proteins

The mouse plasmacytoma cell line MPC-11 was used as a model to study the differential effect of HIB on protein synthesis in an uninfected cell line. MPC-11 cells synthesize an immunoglobulin gamma heavy (H) chain of approximately 55 k daltons and a 23 k molecular weight kappa light (L) chain (Laskov & Scharff, 1970). The L and H chains are reported to account for as much as 20% of the newly synthesized polypeptides (Laskov & Scharff, 1970). Since the two IgG polypeptides can be easily identified and quantitated, MPC-11 cells provide a useful system to investigate the effect of HIB on the synthesis of several individual cellular polypeptides (Nuss & Koch, 1976b). The percentage of total ^{35}S-methionine incorporation which is associated with the L chain increases from a value of 6.9% at isotonic conditions to a value of 27.2% when cells are pulse labeled under hypertonic conditions. Similarly, the percentage of total ^{35}S-methionine incorporation increases for the H protein from 8.8% to 12.8%. There is approximately a 3.5 to 4.0 fold increase in the relative incorporation into the L chain under hypertonic conditions when compared to that observed under isotonic conditions. Likewise, there is an increase in the relative incorporation into the H chain polypeptide of approximately 1.5 fold. Comparable results were obtained by Sonenshein & Brawerman (1976) using amino acid starvation for inhibition of polypeptide chain initiation on L and H chain synthesis. These results suggest that the mRNAs coding for these specialized polypeptides are (similar to viral mRNAs, see below) more efficient messengers than the mRNA species coding for the other cellular proteins. The ratio in the synthesis of light and heavy chains is 1.6 under isotonic conditions. This ratio is changed to a value above 6 when total protein synthesis is inhibited by more than 98% by inhibitors of polypeptide chain initiation, but remains essentially unaltered when protein synthesis decreases due to inhibitors of polypeptide chain elongation (Nuss & Koch, 1976b). Only the quantitative decrease in the amount of protein synthesis following inhibition at the level of polypeptide chain initiation is accompanied by an extensive alteration in the protein pattern of treated and untreated cells. We suggested (Nuss, Oppermann & Koch, 1975) that under conditions which result in a reduction in the overall rate of peptide chain initiation, each mRNA is translated with its own characteristic relative translational efficiency (RTE). Our working hypothesis is that an mRNA with a high relative translational efficiency possesses a higher binding affinity for ribosomes and/or initiation factors (Lawrence & Thach, 1974) and thereby outcompetes those with lower RTE. Such competition would be considerably amplified when ribosomes or initiation factors are limited.

When overall protein synthesis in tissue culture cells is inhibited by more than 90% by the hypertonic initiation block, the synthesis of one major cellular protein, actin, is reduced to a level of 1%, that is 10 times more than overall protein synthesis. In contrast, the synthesis of other host proteins and viral proteins in particular is unchanged indicating that the translation of some species of mRNAs is more sensitive to HIB (i.e. actin) than others (i.e. viral mRNA) RTEs for individual mRNAs can be determined by analysis of protein synthesis under optimal and restricted conditions. RTEs for viral and cellular mRNAs can differ over a wide range, and as such mRNAs can be classified in a hierarchical order according to translation efficiencies. Based on the example of actin stated above, we would assign an RTE of 0.1 to actin and an RTE of 1 to mRNAs from which the relative synthesis of the

TABLE I. Relative Translational Efficiencies of mRNAs in BCS-1 Cells

		Incorporation under HIB % of control	Relative translational efficiency
Cellular	average	5.0	1.0
	p55	5.5	1.1
	p54	6.0	1.2
	p50	7.0	1.4
	p41 (actin)	1.0	0.2
	p34	5.0	1.0
	p13	20.0	4.0
SV 40 VPI		17.8	3.6
Polio NCVP 2		27.6	5.5
NCVP X, Y		15.0	3.0
VP.0,1,3		6.6-7.4	1.4
VSV L G		20.0	4.0
NS		16.0	3.2
N		27.6	5.5
'p42'		25.4	5.1
M		30.8	6.2
Vaccinia	average	30.4	6.1
Major early 28K		31.4	6.3
Major early 28K 13K		37.6	7.5
26K 30K		25.6	5.1
33K		22.4	4.5
Semliki F.V. all proteins		50.0	10.0
Reovirus μg1,x,y,z		40.0	8.0
μg2		1.0	0.2

Polyacrylamide gel electrophoresis of ^{35}S-methionine pulse labeled mock and HIB treated cellular proteins was performed. Relative translational efficiencies were estimated from densitometer tracing of the resulting autoradiographs.

corresponding protein remains unaltered by HIB conditions.
This led us to the proposal that alterations in the pattern of protein synthesis are not dependent in the availability of mRNA specific factors. An indiscriminative change in the binding affinity of ribosomes and/or initiation factors for mRNAs will influence differentially the translation of

every mRNA species (Nuss et al., 1975; Koch et al., 1976). Thus, translational control enables animal cells to respond rapidly to altered nutritional conditions or to other external signals with a pleiotropic effect on the synthesis of many proteins.

In further studies with this technique, the relative translational efficiencies of individual cellular mRNAs and a number of viral mRNAs were estimated. Table 1 summarizes these results. A HIB experiment was performed with a number of cell lines to compare the translational efficiencies of major mRNA species in different cells (Fig. 1). Inspection of the autoradiographs reveal a number of HIB resistant mRNAs translational products. In MPC-11 cells (channels A and B), these include in addition to the L (23 k) and H (55 k) chains, a polypeptide of 45 k molecular weight and the histones (11 k and 14 k). The translation of these non-IgG mRNAs in HeLa, L-929 and BHK-21 cell lines is also resistant to HIB (HeLa: channels D-F; L-929: channels G-I; and BHK-21: channels J-L). In contrast, the synthesis of two major cellular proteins, actin (41 k) and myosin (200 k) are highly sensitive to HIB. A closer inspection of the autoradiograph of Figure 1 reveals a number of other differences in the labeling pattern of proteins in the various cell lines. A HIB resistant translation product of 73 k is present mainly in HeLa and L-cells. A corresponding band is present in myeloma cells but it is not labeled under HIB conditions. Similarly, a 12 k and 13 k protein is clearly visible in the autoradiographs.

Characteristics of Virus-Induced Inhibition of Synthesis of Cellular Proteins

In spite of intensive study over the last two decades, the mechanism(s) of virus induced inhibition of host macromolecular synthesis has not been elucidated. *A priori* there is no reason to expect only one unique mechanism for virus induced event(s). On the contrary, it is known that phages kill their host cells in several independent ways (Schweiger et al., 1978). Viral proteins might interfere directly with the protein synthesizing machinery (Racevskis et al., 1976) or indirectly by altering membrane functions (activating pre-existing cellular mediators - Koch et al., 1976), or by increasing membrane permeability for ions (Carrasco & Smith, 1976). Matthews et al. (1973) and Wright & Cooper (1974) proposed that virus induced suppression of host mRNA translation might be exerted by alterations of the protein synthesizing mechanism which allow ribosomes to specifically interact only with viral mRNA. Virus induced suppression of host mRNA trans-

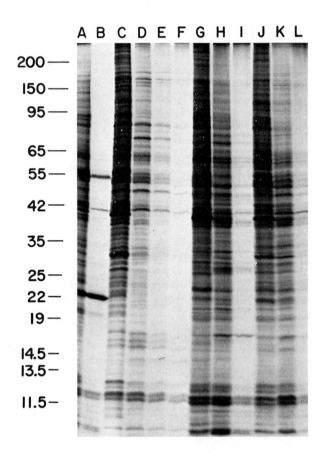

FIGURE 1. HIB induced changes in the relative synthesis of individual cellular proteins. Protein synthesis in various tissue culture cells after isotopic labeling under control and hypertonic conditions was analyzed by polyacrylamide gel electrophoresis. Myeloma cells, channels A and B, HeLa cells (C-F), L929 cells (G-I) and BHK21 cells (J-L). The first channel shown for each cell type (A, C, G, and J) are polypeptides from cells labeled under isotonic conditions. Extra NaCl was added to all other cell cultures in the following amounts 100 mM (D), 120 mM (E, H, K) and 140 mM (B, F, I, and L). All cultures with extra NaCl received 2x the amount of ^{35}S-methionine added to the isotonic control. The migration of polypeptides in the discontinuous buffer system does not accurately reflect their molecular weight. The assigned molecular weights serve merely for orientation on the autoradiographs.

lation (Leibowitz & Penman, 1971), like HIB, acts at the level of peptide chain initiation by inducing an indiscriminate reduction on the rate of polypeptide chain initiation. Accordingly virus infection and HIB should cause comparable alterations in the pattern of protein synthesis. In order to test this hypothesis we compared the effects of HIB and virus infection on the relative synthesis of IgG and non-IgG proteins in myeloma cells. Infection of myeloma cells with VSV resulted in a rapid inhibition of total protein synthesis (Nuss & Koch, 1976a) and alterations in the distribution of labeled amino acids in L, H and non-IgG polypeptides which were similar to those observed following exposure of uninfected cells to HIB. Thus in VSV infected myeloma cells host mRNAs with high RTE - i.e. those for heavy and light IgG - are more resistant to virus directed suppression than other cellular mRNAs.

This conclusion was further substantiated by studies on poliovirus infected HeLa cells (Oppermann & Koch, unpublished, Fig. 2) and SV 40 infected BSC-1 cells which demonstrated that the synthesis of actin is preferentially suppressed by virus infection. These results strengthen our proposal (Nuss et al., 1975) that virus directed suppression of host protein synthesis need not involve virus specific or virus induced factor(s) which possess the capacity to actively discriminate between viral and host mRNA. A virus may favor its own replication by lowering the overall rate of polypeptide chain initiation for the translation of all mRNAs, in a fashion similar to HIB.

Changes in the Phosphorylation State of Ribosomal Proteins

Ribosomes and especially ribosomal proteins might provide another site in the translational machinery where regulation of polypeptide synthesis could be exerted. Therefore, we examined the ribosomal protein patterns of MPC-11 cells before and after HIB (Fig. 3) as well as the pattern of the Ehrlich Ascites cells before and after glucose starvation. Specifically we qualitatively and quantitatively analyzed the incorporation of ^{32}P into individual ribosomal proteins. In MPC-11 cells we observed a positive correlation between the phosphorylation of the small subunit protein S3 and the amount of ribosomes engaged in protein synthesis (Martini & Kruppa, 1979). Raising the tonicity of the growth medium by addition of 100 mM excess NaCl results in a large increase of single ribosomes with a concomitant reduction in the phosphorylation of protein S3 in single ribosomes and in residual polysomes (Fig. 4) (Kruppa & Martini, 1978). Phosphoprotein S3

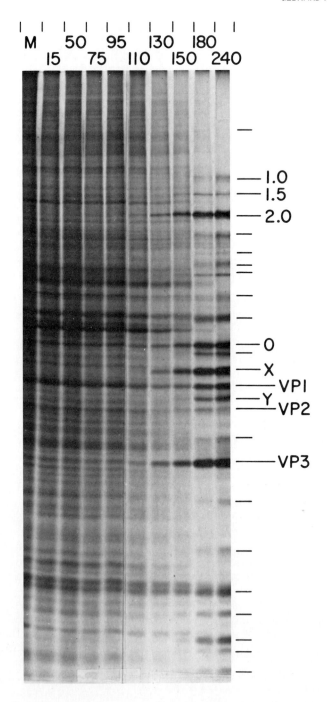

Figure 2 A

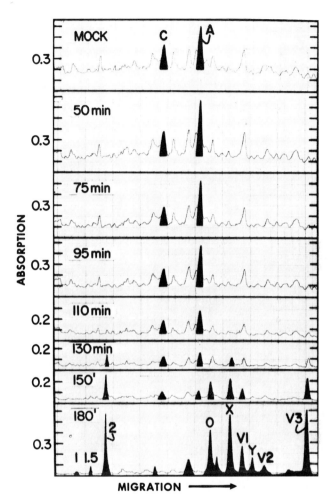

Figure 2 B

FIGURE 2. Alterations in protein synthesis following poliovirus infection of HeLa cells. a) Autoradiograph of extracts from HeLa cells pulse labeled with ^{35}S-methionine for 15 mock-infected and at different times after infection with 50 PFU/cell. b) Densitometer tracings of the autoradiograph (Fig. 2a). The shaded areas indicate the relative amount of A (actin) and C (a chosen cellular protein).

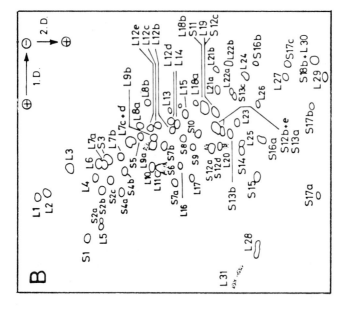
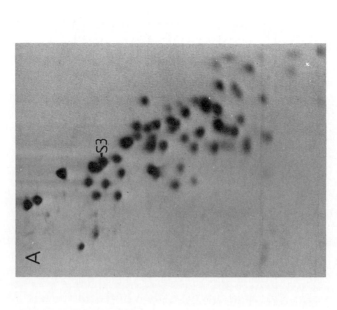

FIGURE 3. Two-dimensional separation of ribosomal proteins of polysomes. Polysomes were prepared from HIB-treated MPC-11 cells and centrifuged through a buffer of high ionic strength. The proteins were extracted with 66% acetic acid, 66 mM Mg (Kaltschmidt & Wittmann, 1972) and analyzed by two-dimensional polyacrylamide gel electrophoresis. The original stained protein pattern is shown in (A) and the nomenclature is given in (B) according to Martini & Gould (1975). Particularly faint spots are represented by dotted areas. About 400 µg of protein was used for this separation.

was almost completely dephosphorylated after glucose starvation of Ehrlich Ascites cells (Kruppa, 1979). It is tempting to speculate that the massive dephosphorylation of a specific small subunit protein accompanying starvation and a tonicity shift specifically affects the initiation process of translation. Infection of HeLa cells by poliovirus leads to quantitative changes in the phosphorylation state of ribosomal proteins of the large subunit (Kruppa, 1979). Whether the observed quantitative changes in the ribosomal protein phosphorylation are cause or consequence of virus induced protein synthesis inhibition remains to be determined.

Membrane-Mediated Release of an Inhibitor of Polypeptide Chain Initiation

Changes in membrane potential (McDonald et al., 1972) and other membrane alterations (Herzberg et al., 1974; Fox et al., 1971) have been considered as signals for the regulation of macromolecular synthesis in tissue culture cells. We have suggested that translational control can be exerted by membrane mediated events. Several conditions that inhibit protein synthesis *in vivo* have little or no effect on *in vitro* protein synthesis (Koch et al., 1976). Starvation, incubation of cells at high densities, arrest in G1 phase, hyperosmolarity of the growth medium, hypertonic salt, DMSO, and ethanol all induce a rapid increase in the number of 80S ribosomes and a concomitant decrease in the size and number of polysomes indicating that protein synthesis is affected at the level of initiation. That amino acid starvation causes a specific inhibition of polypeptide chain initiation was already reported in 1971 by Vaughan et al. The conditions mentioned above which inhibit protein synthesis at the level of initiation all lead to a reduced uptake of amino acids (Koch et al., 1979a). Presently we do not know whether this inhibition of amino acid transport is the cause (as in the case of amino acid starvation) or the consequence of reduced protein synthesis (via a feedback control of amino acid uptake - Ring & Heinz, 1966). The mechanism by which polypeptide chain initiation is inhibited by either HIB, amino acid starvation or upon virus infection appears to be similar. All three processes may involve the activation of a normal host cell mechanism used to regulate protein synthesis at the translational level. We have shown that crude extracts prepared from cells two to three hours after infection with poliovirus, 45 minutes after transfer to amino acid deficient medium, or from cells exposed to HIB for 15 minutes have no detectable endogenous protein synthesis *in vitro*. Protein synthesis in

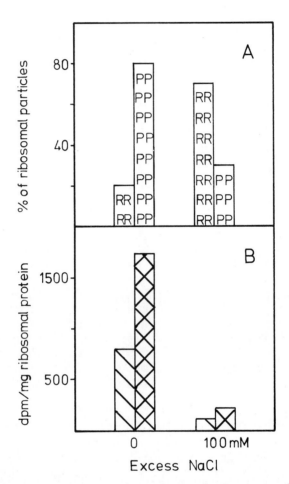

FIGURE 4. Phosphorylation of S3 of messenger-free ribosomes and polysomes from control and salt-treated cells. (A) The proportion of messenger-free ribosomes (R) and polysomes (P) were determined from sucrose gradients in high salt buffer. Single ribosomes bound to mRNA have been included in the polysome fraction. (B) After autoradiography the protein spot S3 was cut out of the gels, and the radioactivity was measured by Cerenkov counting in a liquid scintillation spectrometer. 700-900 µg of ribosomal protein from each of the four preparations were used for the two-dimensional gels. The dpm-values were normalized for reasons of comparison. Shaded bars and cross-hatched bars correspond to $^{32}P_i$-incorporation of S3 of messenger-free ribosomes and polysomes respectively. Results on the left side in A and B are obtained from control cells, those on the right side after raising the NaCl concentration by 100 mM.

these extracts can be partially restored upon Sephadex G 25 gel filtration of the extracts indicating that neither ribosomes nor mRNA are irreversibly inactivated by exposure of cells to these three conditions. Late eluting fractions from the Sephadex G 25 column contain low molecular weight substances (including amino acids) which inhibit protein synthesis in cell-free extracts from untreated cells.

Some characteristic properties of the inhibitor of protein synthesis are listed as follows: Upon dialysis of the cell extract the inhibitor is found in the dialysate. Gel filtration over G 15 and G 10 Sephadex reveals a molecular weight below 1,000. The inhibitor does not give the ninhydrin reaction indicating absence of free primary or secondary amino groups. During high voltage paper electrophoresis at pH 3.5, the inhibitor remains at the origin, indicating the absence of anionic groups e.g. carboxyl or phosphate residues. The compound is quite hydrophobic because it migrates on cellulose thin layer plates using butanol as solvent and is bound to XAD2 columns at pH 2.0 in the presence of 2 M $CaCl_2$. It also binds strongly to glass surfaces, thus hampering the isolation and purification. By adding the compound to an *in vitro* protein synthesizing system its strong inhibitory activity is apparent. The inhibitor binds to ribosomes in equilibrium dialysis and sediments with the ribosomes during sucrose gradient centrifugation. We propose that the inhibitor which acts at the level of initiation is released or activated by a membrane mediated event and is reversibly bound to ribosomes Ribosomes with bound inhibitor are less active or inactive in the formation of initiation complexes.

CONCLUSION

Infection by several viruses (Koch *et al.*, 1976), amino acid starvation (Vaugham *et al.*, 1971), and incubation of cells in hypertonic medium (Saborio *et al.*, 1974) all lead to a specific inhibition of polypeptide chain initation. This inhibition is accompanied by the following events for which a clear causal relationship has not yet been established: a) reduced transport of nutrient amino acids and glucose; b) release or activation of a low molecular weight mediator which binds to ribosomes such that the binding affinity of mRNA for ribosomes and/or initiation factors is reduced; c) changes in the phosphorylation state of ribosomal proteins. We propose that the inhibition of polypeptide chain initiation is membrane mediated. The reduction in overall protein synthesis results in dramatic alterations in the synthesis of individual proteins. Whereas the translation of several mRNAs (viral mRNAs, mRNAs for specialized proteins like immunoglobulins) (Nuss & Koch, 1976b) and globin (Racevskis & Koch, 1976) is hardly affected by this inhibition, the translation of other cellular mRNAs (actin, myosin) and certain RNA tumor virus mRNAs (Koch et al., 1979b) is inhibited by a severalfold higher extent than overall protein synthesis.

Since the inhibition of protein synthesis at the level of initiation effects the intracellular accumulation of many or all proteins to a different degree it will exert a pleiotropic effect on cellular metabolism. The implications of these findings do not appear to be restricted to virus infection or the cellular response to adverse growth conditions. Studies of differentiating cell systems indicate that translational control may be involved in a number of cases where a cell directs its translational activity from a proliferative to a differentiative mode. Translational control mechanisms have been most clearly demonstrated in oocytes where control is mediated by sequestering mRNAs away from the translational apparatus (Gross et al., 1973). We have also been able to demonstrate that inducers of Friend erythroleukemia cell differentiation decrease membrane transport as well as the initiation of protein synthesis (Bilello et al., 1979; Koch et al., 1979b). Globin mRNA has been shown to be favorably translated under conditions of reduced initiation *in vitro* (Lodish, 1974) and *in vivo* (Racevskis & Koch, 1976). Thus production of the major differentiation specific protein is potentiated by inhibition of initiation. The primary importance of translational control based upon reduced initiation is that overt changes in the spectrum of proteins synthesized can occur rapidly and reversibly without the necessity to elaborate a series of specific factors which amplify the translation of specific mRNAs. Although transcriptional control is certainly the primary regulatory event in gene expression, a membrane mediated amplification of control at the level of polypeptide chain initiation will allow the cell to respond rapidly to environmental signals. The further elucidation of these mechanisms appears to be an interesting and fruitful endeavour.

REFERENCES

Bilello, J.A., Warnecke, G. & Koch, G. (1979), Inhibition of polypeptide chain initiation by inducers of erythroid differentiation in Friend erythroleukemic cells. *in* Modern Trends in Human Leukemia III, Eds. R. Neth, P.H. Hofschneider &

K. Mannweiler. Springer Verlag, in press.
Carrasco, L. & Smith, A.E. (1976), Sodium ions and the shut-off of host cell protein synthesis by picorna viruses. *Nature 264*, 807.
Fox, T.O., Sheppard, J.R. & Burger, M.M. (1971), Cyclic membrane changes in animal cells: Transformed cells permanently display a surface architecture detected in normal cells only during mitosis. *Proc. Natl. Acad. Sci. USA 68*, 244.
Gross, K., Ruderman, I., Jacobs-Lorena, M., Baglioni, C. & Gross, P.R. (1973), Cell free synthesis directed by mRNA from sea-urchin embryos. *Nature New Biol. 241*, 274.
Herzberg, M., Breitbart, H. & Atlan, H. (1974), Interaction between membrane functions and protein synthesis in reticulocytes. Effect of valinomycin and dicyclohexyl-18-crown-6. *Eur. J. Biochem. 45*, 161.
Kaltschmidt, E. & Wittmann, H.G. (1972), Ribosomal proteins. XXXII: Comparison of several extraction methods for proteins from Escherichia coli ribosomes. *Biochimie 54*, 167.
Koch, G., Oppermann, H., Bilello, P., Koch, F. & Nuss, D. (1976), Control of peptide chain initiation in uninfected and virus infected cells by membrane mediated events. *in* Modern Trends in Human Leukemia II,(1976), Eds. R. Neth, R.C. Gallo, K. Mannweiler & W.C. Moloney. p. 541, J.F. Lehmanns Verlag München.
Koch, G., Bilello, J.A., Kruppa, J., Koch, F. & Oppermann, H. (1979a), Amplification of translational control by membrane mediated events: A pleiotropic effect on cellular and viral gene expression. *in* Ann. N.Y. Acad. Sci., in press.
Koch, G., Weber, C., Bilello, J.A. (1979b), Alterations in translational control mechanisms in Friend erythroleukemic cells during DMSO induced differentiation. *in* Modern Trends in Human Leukemia III, Eds. R. Neth, P.H. Hofschneider & K. Mannweiler. Springer Verlag, in press.
Kruppa, J. & Martini, O.H.W. (1978), Dephosphorylation of one 40S ribosomal protein in MPC-11 cells induced by hypertonic medium. *Biochem. Biophys. Res. Commun. 85*, 428.
Kruppa, J. (1979), Membrangebundene Biosynthese zellulärer and viraler Proteine in Säugetierzellen. Habilitationsschrift, Universität Hamburg.
Lawrence, C. & Thach, R.E. (1974), Encephalomyocarditis virus infection of mouse plasmacytoma cells. I. Inhibition of cellular protein synthesis. *J. Virol. 14*, 598.
Laskov, R. & Scharff, M.D. (1970), Synthesis, assembly and secretion of gamma globulin by mouse myeloma cells. *J. Exp. Med. 131*, 515.
Leibowitz, R. & Penman, S. (1971), Regulation of protein synthesis in HeLa cells. III. Inhibition during poliovirus infection. *J. Virol. 8*, 661.

Lodish, H.F. (1974), Model for the regulation of mRNA translation applied to hemoglobin synthesis. *Nature 215*, 385.
Lodish, H.F. (1976), Translational control of protein synthesis. *Ann. Rev. Biochem. 45*, 39.
Martini, O.H.W. & Gould (1975), Characterisation of eukaryotic ribosomal proteins. *Molec. Gen. Genet. 142*, 299.
Martini, O.H.W. & Kruppa, J. (1979), Ribosomal phosphoproteins of mouse myeloma cells. *Eur. J. Biochem. 95*, 349.
Matthews, T.J., Butterworth, B.E., Chaffin, L. & Rueckert, R.R. (1973), Encephalomyelocarditis (EMC) virus and rhinovirus 1A (HRV-1A) peptides associated with the infected cell ribosomes. *Fed. Proc. 32*, 461.
McDonald, T.E., Sachs, H.G., Orn, C.W. & Ebert, J.D. (1972), External potassium and baby hamster kidney cells: Intracellular ions, ATP, growth, DNA synthesis and membrane potential. *Developm. Biol. 28*, 290.
Nuss, D.L., Oppermann, H. & Koch, G. (1975), Selective blockage of initiation of host protein synthesis in RNA-virus-infected cells. *Proc. Natl. Acad. Sci. USA 72*, 1258.
Nuss, D.L. & Koch, G. (1976a), Differential inhibition of vesicular stomatitis virus polypeptide synthesis by hypertonic initiation block. *J. Virol. 17*, 283.
Nuss, D.L. & Koch, G. (1976b), Variation in the relative synthesis if IgG and non-IgG proteins in cultured MPC-11 cells with changes in the overall rate of polypeptide chain initiation and elongation. *J. Mol. Biol. 102*, 601.
Oppermann, H., Saborio, J., Zarucki, T. & Koch, G. (1973), Sensitization of cells for viral RNA infection by inhibiton of macromolecular synthesis. *Fed. Proc. 32*, 53.
Oppermann, H. & Koch, G. - unpublished
Racevskis, J. & Koch, G. (1976), Resistance of globin synthesis to inhibition by medium hypertonicity. *Fed. Proc. 35*, 1515.
Racevskis, J., Kewar, S. & Koch, G. (1976), Inhibition of protein synthesis in reticulocyte lysate by poliovirus. *J. Gen. Virol. 31*, 135.
Revel, M. & Groner, Y. (1978), Posttranscriptional and translational controls of gene expression in eukaryotes. *Ann. Rev. Biochem. 47*, 1079.
Ring, K. & Heinz, E. (1966), Active amio acid transport in *Streptomyces hydrogenans*. I. Kinetics and uptake of α-aminoisobutyric acid. *Biochem. Z. 344*, 446.
Saborio, J.L., Pong, S.-S. & Koch, G. (1974), Selective and reversible inhibition of initiation of protein synthesis in mammalian cells. *J. Mol. Biol. 85*, 195.
Schweiger, M., Wagner, E.F., Hirsch-Kaufmann, M., Ponta, H. & Herrlich, P. (1978), Biochemistry of development of E. coli virus T7 and T1. *in* Gene Expression, Eds. B.F.C. Clark, H. Klenow & J. Zeuthen. p. 171, Pergamon Press, New York.

Sonenshein, G.E. & Brawerman, G. (1976), Regulation of immunoglobulin synthesis in mouse myeloma cells. *Biochemistry 15*, 5497.

Vaughan, M.H., Pawlowski, P.J. & Forchhammer, J. (1971), Regulation of protein synthesis initiation in HeLa cells deprived of single essential amino acids. *Proc. Natl. Acad. Sci. USA 68*, 2057.

Wright, P.J. & Cooper, P.D. (1974), Poliovirus proteins associated with ribosomal structures in infected cells. *Virology 59*, 1.

REGULATION OF MACROMOLECULAR SYNTHESIS
BY LOW MOLECULAR WEIGHT MEDIATORS

MODE OF ACTION OF HEMIN-CONTROLLED
TRANSLATIONAL INHIBITOR

César de Haro [1]
Severo Ochoa

Roche Institute of Molecular Biology
Nutley, New Jersey

Despite the finding that the hemin-controlled translational inhibitor (HCI) in reticulocyte lysates is a cyclic AMP-independent protein kinase that phosphorylates the small (38,000 daltons) subunit of the initiation factor eIF-2, the mechanism of inhibition of translation remained unexplained. Whereas treatment of hemin-containing lysates with HCI in the presence of ATP inhibited translation, the same treatment of highly purified eIF-2 did not affect its ability to form a ternary complex with initiator Met-tRNA and GTP or a 40S initiation complex. We have isolated a protein factor (eIF-2 stimulating protein, ESP) that is essential for formation of ternary and 40S initiation complexes by the initiation factor eIF-2, at the low concentrations of eIF-2 present in reticulocyte lysates. The fact that stimulation of complex formation by ESP is virtually abolished when the small subunit of eIF-2 is phosphorylated by ATP in the presence of HCI is consistent with the notion that HCI inhibits translation in lysates by blocking the interaction of eIF-2 with ESP. We have additionally established that, unlike phosphorylation of the small subunit, phosphorylation of the middle (52,000 daltons) subunit of eIF-2, which does not lead to translational inhibition in lysates does not affect eIF-2-ESP interaction. This provides further support for our model of translational inhibition of HCI. Present evidence indicates that formation of the ternary complex is preceded by that of the binary complex eIF-2·GTP and that ESP acts at the level of binary complex formation.

[1] Present address: Centro di Biología Molecular, Universidad Autónoma de Madrid, Madrid, Spain.

INTRODUCTION

The hemin-controlled inhibitor (HCI) of protein synthesis in reticulocyte lysates is a cyclic AMP-independent protein kinase that catalyzes the phosphorylation of the small subunit of the initiation factor eIF-2 (Kramer et al., 1976; Levin et al., 1976; Farrell et al.,1977; Gross and Mendelevski, 1977). eIF-2 forms a ternary complex with Met-tRNA$_i$ and GTP which, upon binding to a 40S ribosomal subunit, give rise to a 40S initiation complex.

In the presence of HCI and ATP, the rate of protein synthesis in hemin-containing reticulocyte lysates soon shows a sharp decline (Clemens et al., 1974). Although this effect would appear to be related to the phosphorylation of eIF-2, treatment of purified eIF-2 with inhibitor plus ATP does not interfere with its ability to form ternary or 40S initiation complexes (Clemens et al.,1976; Datta et al.,1977; Trachsel and Staehelin, 1978). On the other hand, formation of the complex is impaired when partially purified eIF-2 is used (Clemens et al.,1976, Datta et al., 1977). This suggests that inhibition of translation involves another factor(s) present in lysates and in partially purified eIF-2 preparations. Ribosomal salt washes contain, along with eIF-2, a factor necessary for inhibition of ternary complex formation when purified eIF-2 is treated with HCI and ATP (de Haro et al.,1978; de Haro and Ochoa, 1978). The new factor enhances the ability of unphosphorylated eIF-2 to form ternary or 40S initiation complexes but has no effect on phosphorylated eIF-2, it will be referred to as the eIF-2 stimulating protein (ESP). At the low concentrations of eIF-2 in lysates, ESP is essential for eIF-2 function. ESP has no activity in the absence of eIF-2. It would then appear that, at physiological concentrations of eIF-2, phosphorylation of its small subunit inhibits translation by blocking the interaction of eIF-2 with ESP (de Haro et al.,1978; de Haro and Ochoa, 1978). Our earlier observations have now been confirmed with highly purified eIF-2 and ESP (de Haro and Ochoa, 1979a). Moreover, phosphorylation of the middle subunit of eIF-2, which does not inhibit translation in lysates (Benne et al., 1978; Levin and London, 1978; Tahara et al., 1978), does not affect the eIF-2-ESP interaction (de Haro and Ochoa, 1979a).

Further experiments (de Haro and Ochoa 1979b) indicate that formation of the ternary complex is preceded by formation of the binary complex eIF-2·GTP. ESP stimulated binary complex formation with intact eIF-2 but not with eIF-2 whose small subunit has been phosphorylated.

MATERIALS AND METHODS

Conditions for studying the mode of action of HCI were according to published procedures (de Haro and Ochoa, 1978; de Haro and Ochoa, 1979a; de Haro and Ochoa, 1979b).

RESULTS

Effect of ESP and HCI on Initiation Complex Formation

ESP has no effect in the absence of eIF-2 but markedly stimulates ternary complex formation by eIF-2 (Fig. 1A). Incubation with HCI and ATP is without effect on ternary complex formation with eIF-2 alone (Fig. 1B, lower curve) but markedly inhibits the stimulation normally elicited by ESP.

The time course of ternary complex formation, in the absence or presence of ESP and/or HCI and ATP, shows that as observed with the A. salina factors (de Haro et al., 1978), ESP increases both the extent and the rate of ternary complex formation (see Figure 4, left panel). Again, incubation with HCI and ATP has no effect on complex formation by eIF-2 alone but severely depresses the ability of ESP to enhance this reaction.

Figure 2 shows that what is true of ternary complex is also true of 40S initiation complex formation. Thus, the small amount of complex formed in the absence of ESP is not affected by preincubation with HCI and ATP (bars and samples 1 and 2) whereas such preincubation virtually abolishes the considerable stimulation produced by ESP (bars and samples 3 and 4).

Properties of eIF-2 and ESP

After separation from ESP (Table I, step 3), eIF-2 showed a marked decrease in activity unless assayed in the presence of ESP. Under our assay conditions, ESP (fraction CM 200) increased 3-to 4-

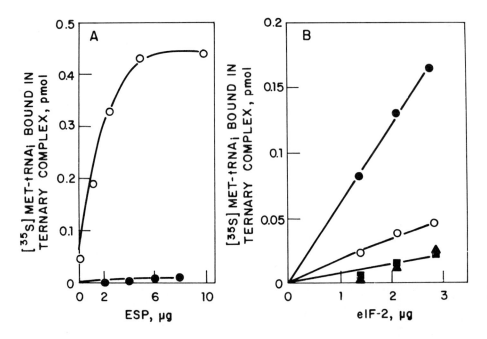

FIGURE 1. (A) Ternary complex formation as a function of the concentration of ESP, without (●) or with (○) eIF-2. (B) Effect of ESP and HCI on ternary complex formation as a function of the concentration of eIF-2. Other addition were as follows: ■ none; ▲ HCI; ● ESP; ○ HCI and ESP (de Haro & Ochoa, 1978).

fold the activity of CM 350 and PC eIF-2 (Table I). On sodium dodecyl sulfate/ polyacrylamide gel electrophoresis PC eIF-2 preparation is estimated to be about 75% pure (de Haro and Ochoa, 1979a).

The estimated molecular weights of the major bands were 38,000, 52,000 and 54,000, corresponding to the α, β and γ subunits of eIF-2. The molecular weight of ESP, estimated from glycerol density gradient centrifugation analyses in the presence of suitable markers, was about 350,000 (de Haro and Ochoa, 1979a). It is evident from Table II that purification of ESP removes proteins that stimulate ternary complex formation nonspecifically. This effect is distinguished from that of ESP in that it is not abolishes by incubation of eIF-2 with HCI and ATP.

Action of Translational Inhibitor

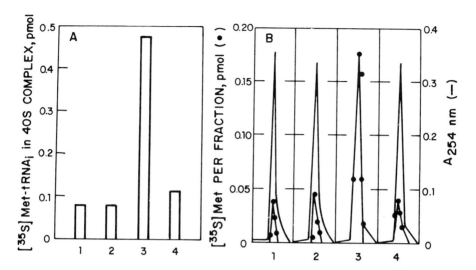

FIGURE 2. Effect of ESP and HCI on 40S complex formation. (A) [^{35}S]Met-tRNA$_i$ bound in the 40S region of the gradient. (B) Sucrose density gradient centrifugation profiles. All samples had ATP and eIF-2. Other additions were as follows: 1, none; 2, HCI; 3, ESP; 4, HCI and ESP (de Haro and Ochoa, 1978).

TABLE I. Purification of eIF-2

Step	Units	Specific activity *	Yield % No ESP	Yield % With ESP
1. High-salt wash, 80% sat(NH_4)$_2SO_4$ppt	6188	4.4	100	100
2. DEAE-cellulose, DE200	7000	70	114	
CM-Sephadex, CM200	135	1.5		
3. CM-Sephadex, CM350	1485	330	24	
CM350 + CM200	5400	1200		88
4. Phosphocellulose, PC	900	1500	15	
PC + CM200	2520	4200		41

Assay conditions were according to published procedures (de Haro & Ochoa, 1979a).

*Units/mg of protein (1 unit is the amount of eIF-2 causing the binding of 1 pmol of Met-tRNA$_i$ under our assay conditions).

Effect of ESP and Protein Kinases in Ternary Complex Formation

Non specific stimulation of ternary complex formation (e.g., by bovine serum albumin, figure 3A) may in large part be due to stabilization of eIF-2, particulary at low concentrations. Figure 3A also shows that ESP produces considerable stimulation beyond that caused by bovine serum albumin. Figure 3B, in the presence of albumin and a small amount (1 pmol) of eIF-2, may mimic conditions obtaining in lysates. Here ESP produced a 17-fold stimulation of ternary complex formation that was largely abolished (88%) by HCI. Figure 3C, with a large amount (11.2 pmol) of eIF-2, clearly shows that, without ESP, HCI does not affect ternary complex formation (3.10 vs. 3.08 pmol) but largely eliminates (86%) stimulation by ESP. Finally, figure 3D demostrates that, in contrast to HCI, reticulocyte casein kinase (that phosphorylates the eIF-2 β but not the α subunit) does not modify the ESP effect. This is further documented in figure 4.

Effect of ESP on Binary Complex Formation

$[^{14}C]$ GTP and $[^{3}H]$ Met-tRNA$_i$ were incubated simultaneously with eIF-2, in the absence or presence of ESP, and the molar binding ratio GTP/Met-tRNA$_i$ was determined as a function of time (Figure 5, inset). The binding of GTP was considerable faster than that of Met-tRNA$_i$. This observation and the well-established fact that eIF-2 binds little or no Met-tRNA$_i$ in the absence of GTP, strongly suggest that ternary complex formation takes place in two steps, the first of which is formation of the eIF-2·GTP binary complex.

Figure 5 shows the binding of $[^{14}C]$ GTP to eIF-2 (in the absence of Met-tRNA$_i$) as a function of the GTP concentration without or with ESP. ESP caused considerable enhancement of binary complex formation. At saturation, about 1 mol of GTP/mol eIF-2 was bound in the presence of ESP but only about one-fifth as much in its absence. There was no binding of GTP by ESP in the absence of eIF-2. The inset of figure 5 also shows that ESP does not alter the GTP/Met-tRNA$_i$ binding ratio. This means that ESP increases the binding of Met-tRNA$_i$ to eIF-2 to the extent that the binding of GTP to eIF-2 is increased.

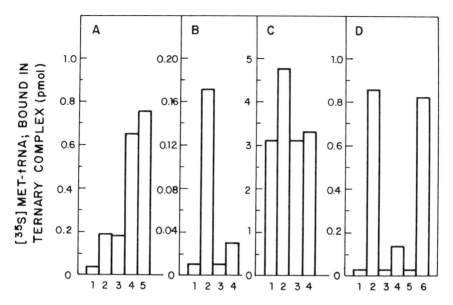

FIGURE 3. Effect of ESP and protein kinase on ternary complex formation. (A) [^{35}S]Met-tRNA$_i$ (1.8 pmol); PC eIF-2 (2 pmol); other additions were: 1, none; 2, BSA (10 µg); 3, BSA (50 µg); 4, ESP; 5, BSA (10 µg) and ESP. (B) [^{35}S]Met-tRNA$_i$ (2 pmol); PC eIF-2 (1 pmol), BSA (25 µg); other additions were: 1, none; 2, ESP; 3, HCI; 4, HCI and ESP. (C) [^{35}S]Met-tRNA$_i$ (8.7 pmol) PC eIF-2 (11.2 pmol); other additions were: 1, none; 2, ESP; 3, HCI; 4, HCI and ESP. (D) [^{35}S]Met-tRNA (2 pmol); CM350 eIF-2 (1.8 pmol); other additions were: 1, none; 2, ESP; 3, HCI; 4, HCI and ESP; 5, casein kinase; 6, casein kinase and ESP (de Haro and Ochoa, 1979a).

DISCUSSION

The work reported in this and earlier papers (de Haro et al., 1978; de Haro and Ochoa, 1978; de Haro and Ochoa, 1979a,b) has disclosed the existence of a new polypeptide chain initiation factor ESP, in eukaryotes. This factor markedly enhances the formation of ternary and 40S initiation complexes by the initiation factor eIF-2. The concentration of eIF-2 in rabbit reticulocyte lysates is of the order of 10-20 nM (de Haro and Ochoa, 1979a). When measured at these concentrations of eIF-2, the formation of ternary complex (eIF-2·GTP·Met-tRNA$_i$) is negligible in the absence of ESP but is increased 20-fold or more by ESP. The data suggest that, at physiological concentrations of eIF-2, ESP is essen-

TABLE II. Purification of ESP

Step	Units	Specific activity *	Sensitivity to eIF-2 kinase %
3. CM-Sephadex, CM200	2700	30	54
4. 25-55% sat(NH_4)$_2SO_4$ fraction	2625	35	70
5. Sucrose gradient	1955	115	70
6. Phosphocellulose	1950	150	87
7. DEAE-cellulose	660	200	88
8. Glycerol gradient	156	520	86

ESP cannot be assayed at steps 1 and 2 prior to separation from eIF-2. Steps 1 and 2 are the same as in Table I. The assay was as reported (de Haro & Ochoa, 1979a).

*Units/mg of protein (1 unit is the amount of ESP promoting the HCI-sensitive binding of 1 pmol of Met-tRNA$_i$ under assay conditions).

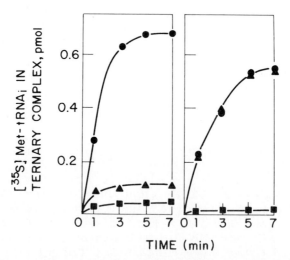

FIGURE 4. Effect of ESP on Met-tRNA$_i$ binding of eIF-2 (ternary complex formation) as a function of time with intact eIF-2 or eIF-2 in the absence of HCI and ATP (phosphorylation of the α subunit) (left) or intact eIF-2 or eIF-2 in the presence of casein kinase and ATP (phosphorylation of the β subunit) (right). (Left) ■ eIF-2 without or with HCI; ● eIF-2 and ESP; ▲ eIF-2, HCI and ESP. (Right) ■ eIF-2 without or with casein kinase; ● eIF-2 and ESP; ▲ eIF-2, casein kinase, and ESP (de Haro and Ochoa, 1979a).

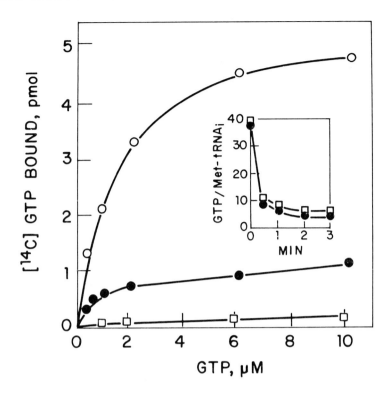

FIGURE 5. Binding of [^{14}C] GTP to eIF-2 as a function of the GTP concentration, in the absence of Met-tRNA$_i$, with 4 pmol of eIF-2 and increasing concentrations of [^{14}C] GTP (900 cpm/pmol) without (●) or with (o) ESP. ☐ Control with ESP but no eIF-2. (Inset) Relative rate of binding of [^{14}C] GTP and [^{3}H] Met-tRNA$_i$ to eIF-2 expressed as the molar ratio, bound GTP/bound Met-tRNA$_i$ ☐ without ESP; ● with ESP. The assay was as reported (de Haro & Ochoa, 1979b).

tial for initiation. ESP has been highly purified and its molecular weight has been estimated to be about 350,000 by glycerol density gradient centrifugation (de Haro and Ochoa, 1979a). The effect of ESP is highly specific because, it is virtually abolished by phosphorylation of the eIF-2 α subunit, and event that leads to translational inhibition in lysates (Kramer et al.,1976; Levin et al., 1976; Farrel et al., 1977; Gross and Mendelewski, 1977) but is unaffected by phosphorylation of the eIF-2 β subunit, which has no effect on translation (Benne et al., 1978; Levin and London, 1978; Tahara et al., 1978). Because phosphorylation of the eIF-2 α subunit is catalyzed by HCI, the heme-controlled inhibitor, our results suggest that translational inhibition by HCI is due to blocking of the interaction between eIF-2 and ESP.

Our present and earlier results (Safer et al., 1975; de Haro and Ochoa, 1979b) suggest that formation of the ternary complex is preceded by formation of the binary complex eIF-2·GTP. ESP acts at the level of binary complex formation. This suggests that ESP may act by binding to the binary complex, displacing the equilibrium toward increased binding of GTP by eIF-2 (reactions a and b). ESP would then be released by Met-tRNA$_i$ to form the ternary complex proper (reaction c)

$$\text{eIF-2} + \text{GTP} \rightleftharpoons \text{eIF-2·GTP} \quad (a)$$

$$\text{eIF-2·GTP} + \text{ESP} \rightleftharpoons \text{eIF-2·GTP·ESP} \quad (b)$$

$$\text{eIF-2·GTP·ESP} + \text{Met-tRNA}_i \rightleftharpoons \text{eIF-2·GTP·Met-tRNA}_i + \text{ESP} \quad (c)$$

We have not been able to detect formation of complexes between eIF-2 and ESP, with or without GTP or GTP and Met-tRNA$_i$, by sucrose or glycerol gradient centrifugation or by gel filtration, in the absence or presence of glutaraldehyde (data not shown), which suggests that the eIF-2-ESP interaction is very weak.

ACKNOWLEDGMENTS

We thank Christa Melcharick for excellent technical assistance.

REFERENCES

Benne, R., Edman, J., Traut, R.S. & Hershey, J.W.B. (1978), Phosphorylation of eukaryotic protein synthesis initiation factors. *Proc. Natl. Acad. Sci. USA* 75, 108.
Clemens, M.J., Henshwa, E.C., Rahamimoff, H. & London, I.M. (1974), Met-tRNA^{Met} binding to 40S ribosomal subunits. A site for the regulation of protein synthesis by hemin. *Proc. Natl. Acad. Sci. USA* 71, 2946.
Clemens, M.J., Pain, V.M. & Henshaw, E.C. (1976), Recent developments in studies of the regulation of Met-tRNA$_f$ binding to 40S ribosomal subunits by translational inhibitor from reticulocytes. Abstracts Cambridge EMBO Workshop on Cytoplasmic Control of Eukaryotic Protein Synthesis.

Datta, A., de Haro, C., Sierra, J.M. & Ochoa, S. (1977), Role of 3':5'-cyclic-AMP-dependent protein kinase in regulation of protein synthesis in reticulocyte lysates. *Proc. Natl. Acad. Sci. USA 74*, 1463.

de Haro, C., Datta, A. & Ochoa, S. (1978), Mode of action of the hemin-controlled inhibitor of protein synthesis. *Proc. Natl. Acad. Sci. USA 75*, 243.

de Haro, C. & Ochoa, S. (1978), Mode of action of the hemin-controlled inhibitor of protein synthesis: Studies with factors from rabbit reticulocytes. *Proc. Natl. Acad. Sci. USA 75*, 2713.

de Haro, C. & Ochoa, S. (1979a). Further studies on the mode of action of the heme-controlled translational inhibitor. *Proc. Natl. Acad. Sci. USA 76*, 1741.

de Haro, C. & Ochoa, S. (1979b). Further studies on the mode of action of the heme-controlled translational inhibitor: Stimulating protein acts at level of binary complex formation. *Proc. Natl. Acad. Sci. USA 76*, 2163.

Farrell, P.J., Balkow, K., Hunt, T., Jackson, R.J. & Trachsel, H. (1977), Phosphorylation of initiation factor eIF-2 and the control of reticulocyte protein synthesis. *Cell 11*, 187.

Gross, M. & Meldelevski, J. (1977), Additional evidence that the hemin-controlled translational repressor from rabbit reticulocytes is a protein kinase. *Biochem. Biophys. Res. Commun. 74*, 559.

Kramer, G., Cimadevilla, J.M. & Hardesty, B. (1976), Specificity of the protein kinase activity associated with the hemin controlled repressor of rabbit reticulocytes. *Proc. Natl. Acad. Sci. USA 73*, 3078.

Levin, D. & London, I.M. (1978), Regulation of protein synthesis: Activation by double-stranded RNA of a protein kinase that phosphorylates eukaryotic initiation factor 2. *Proc. Natl. Acad. Sci. USA 75*, 1121.

Levin, D.H., Ranu, S.R., Ernst, V. & London, I.M. (1976), Regulation of protein synthesis in reticulocyte lysates. Phosphorylation of methionyl-tRNA$_f$ binding factor by protein kinase activity of translational inhibitor isolated from heme-deficient lysates. *Proc. Natl. Acad. Sci. USA 73*, 3112.

Safer, B., Adams, S.L., Anderson, W.F. & Merrick, W.C. (1975), Binding of Met-tRNA$_f$ and GTP to homogenous initiation factor MP. *J. Biol. Chem. 250*, 9076.

Tahara, S.M., Traugh, J.A., Sharp, S.B., Lundak, T.S., Safer, B. & Merrick, W.C. (1978). Effect of hemin on site-specific phosphorylation of eukaryotic initiation factor 2. *Proc. Natl. Acad. Sci. USA 75*, 789.

Trachsel, T. & Staehelin, T. (1978), Binding and release of eukaryotic initiation factor eIF-2 and GTP during protein synthesis initiation. *Proc. Natl. Acad. Sci. USA 75*, 204.

Part V
2',5'-OLIGO A AND INTERFERON

REGULATION OF MACROMOLECULAR SYNTHESIS BY LOW MOLECULAR WEIGHT MEDIATORS

INDUCTION, PURIFICATION, AND PROPERTIES OF 2'5'-OLIGOADENYLATE SYNTHETASE *

L. Andrew Ball [1]
Carol N. White

Microbiology Section
University of Connecticut
Storrs, Connecticut

Treatment of chick embryo cells with interferon causes a dramatic increase in their level of 2'5' oligoadenylate (2-5A) synthetase activity. Upon activation by double-stranded RNA, this enzyme uses ATP to synthesize a series of oligonucleotides with the general structure $(p)ppA^{2'}(p^{5'}A)_n$. Analysis of the polypeptides synthesized in interferon-treated cells reveals the appearance of a new species with an apparent molecular weight of 56,000. This polypeptide is a good candidate for the 2-5A synthetase since it is a major component of partially purified enzyme preparations. Moreover, the 56,000 dalton polypeptide and the synthetase share the following properties: affinity for double-stranded RNA, apparent molecular weight under non-denaturing conditions (50-60,000 MW), elution from DEAE-cellulose colums at 0.05 M KCl, and precipitability with methanol and at pH 5.0.

Preformed shorter oligomers can be further elongated by the activated 2-5A synthetase, indicating that synthesis proceeds by a non-processive mechanism. The direction of chain growth is from the 5' to the 2' terminus. In addition to the shorter oligomeres, a number of other molecules can serve as primers for the synthetase. Among these are: diribonucleoside monophosphates that contain adenosine as their 3' residue

---

* This work was supported by a grant from the National Cancer Institute (CA 14733) and a Research Career Development Award (AI 00273) to L.A.B.

[1] Present address: University of Wisconsin, Madison, Wisconsin

(ApA, CpA, GpA, and UpA, linked either 2'-5' or 3'-5'); diadenosine tetraphosphate (Ap_4A); ADP-ribose and nicotinamide adenine dinucleotide (NAD^+). 2'-adenylated NAD^+ is unable to function as a coenzyme in dehydrogenase reactions. The possible significance of 2'adenylation of cellular constituents for the mechanism of interferon action is discussed.

INTRODUCTION

Treatment of cells with homologous interferon causes several profound changes in their properties. For example, they can acquire resistance to viral infection, a reduced rate of growth, an enhanced sensitivity to the cytotoxic action of double-stranded RNA (dsRNA), and can undergo immunoregulatory changes (Steward, 1977). While it is clear that interferon exerts its effects by causing the derepression of cellular genes, none of its actions is yet understood at the molecular level. Nor is it known if they result from multiple primary effects of represent secondary consequences of a single perturbation of cellular metabolism.

Similarly, extracts of interferon-treated cells shown several altered biochemical properties (Revel & Groner, 1978). The clearest of these is an enhanced sensitivity to the inhibition of cell-free protein synthesis mediated by dsRNA (Kerr et al., 1974). This can be attributed to the presence of two dsRNA-activated enzymes that appear to be induced by interferon: a protein kinase which phosphorylates the small subunit of protein synthesis initiation factor 2 (Zilberstein et al., 1976; Lebleu et al., 1976; Roberts et al., 1976) and 2'5' oligoadenylate (2-5A) synthetase which used ATP to make an oligomeric series of molecules with the general structure $(p)ppA2'(p5'A)_n$ (Kerr & Brown, 1978; Ball & White, 1978; Zilberstein et al.,1978). The trimer and higher oligomers activate a latent endoribonuclease which degrades messenger RNA and thus contributes to the inhibition of cell-free protein synthesis (Clemens & Williams, 1978; Baglioni et al., 1978; Ball & White, 1979; Farrell et al., 1978; Nilsen & Baglioni, 1979).

However, although these enzymes are demonstrable in extracts of interferon-treated cells, their roles in the actions of interferon are not yet clear. For this reason we have undertaken a study of the properties of 2-5A synthetase in an attempt to define the full range of reactions which it can catalyze. Here we report its partial purification and demonstrate its ability to add AMP residues to a number of small molecules, some of which occupy central positions in cellular metabolism.

MATERIALS AND METHODS

The preparation, interferon-treatment, virus infection, isotopic labeling and detergent lysis of primary cultures of chick embryo cells have been described before (Ball, 1979). Post-ribosomal supernatants were prepared by direct centrifugation of the cytoplasmic extracts of 100,000g for 2 hr at 4°; the incubation step described previously was omitted as it caused partial loss of 2-5A synthetase activity.

2-5A synthetase was partially purified by differential precipitation from the post-ribosomal supernatant of interferon-treated cells. Aliquots (0.5 to 3.0 ml) of the post-ribosomal supernantant were subjected to fractional methanol precipitation at -5°. The fraction precipitating between zero and 20% to 30% methanol was 2-3 fold enriched in 2-5A synthetase and depleted of the phosphodiesterase activity responsible for 2-5A degradation. This fraction was redissolved in half its original volume of a buffer containing 20 mM HEPES pH 7.6, 110 mM NH_4Cl, 50 mM Mg acetate, and 1 mM dithiothreitol, and the solution was adjusted to pH 5.0 by the addition of 1.0 M HCl. After 5 min at 0°, the precipitate was removed by centrifugation at 10,000 g for 10 min. Most of the 2-5A synthetase activity remained in the supernatant. However, it was rendered precipitable at pH 5.0 by the addition of polyriboionsinic-polyribocytidylic acid (poly[IC]) to 10 μg/ml, and, after a further 5 min at 0°, it was collected by centrifugation. This fraction contained about 1.4% of the initial protein and 25% of the initial 2-5A synthetase activity, and was thus at least 17-fold purified. Polyacrylamide gel electrophoresis was performed as described before (Ball, 1979).

2-5A synthetase activity was assayed by measuring the dsRNA-dependent incorporation of isotopic label from [^{3}H]ATP into material which bound to DEAE-cellulose paper after digestion with alkaline phosphatase. Reaction mixtures (10 μl) contained 50 mM Mg acetate, 1.5 mM [^{3}H]ATP (16.7 Ci/M), 10 μg/ml poly(IC), and 5 μl of post-ribosomal supernatant extracts. Other components were included as indicated for the individual experiments. Reaction mixtures were incubated at 30° for 2 hr and the reaction was terminated by heating at 95° for 5 min. Precipitated protein was removed by brief centrifugation, and 7.5 μl aliquots of the supernatants were digested with 6 units per ml of bacterial alkaline phosphatase (orthophosphoric monoester phosphohydrolase [alkaline optimum]; EC 3.1.3.1.; [Sigma, type III]) at 37° for 2 hr. Aliquots (6 μl) of the digests were spotted onto 1 cm^2 squares of DEAE-cellulose paper (Whatman DE 81), dried, and washed with several changes of distilled water to remove [^{3}H]adenosine, the phosphatase digestion product of unincorporated [^{3}H]ATP. The incorporated label was eluted from the filters with 0.3 M KOH, and counted in Aqueous Couting Scintillant (Amersham). Reaction mixtures lacking poly(IC) were used as controls, and their incorporation was subtracted from the experimental values.

Large scale syntheses were carried out in 0.1 ml reaction volumes, using [^{3}H]ATP at a specific activity of 167 Ci/M. The products were fractionated by 0.05 - 0.15 M NaCl gradient elution from columns of DEAE-cellulose in 50 mM Tris.HCl, pH 8.0, containing 7 M urea, as described before (Ball] White, 1978). Appropriate fractions were pooled, dialysed for 5 hr against five changes of distilled water, and lyophilized. Alkaline hydrolysis was for 18 hr at room temperature with 0.3 M KOH. Digestion with venom phosphodiesterase was for 30 min at 37° with 50 μg/ml of snake venom phosphodiesterase (oligonucleate 5'-nucleotidehydrolase; EC 3.1.4.1; Boehringer Mannheim). Both digestions were carried out in the presence of 2 mM 3'-AMP to minimize the effects of possible contaminating phosphatase activity.

Reducible NAD^+ and $NADP^+$ were assayed fluorimetrically, in reactions catalysed by yeast alcohol dehydrogenase (EC 1.1.1.1) and glucose-6-phosphate dehydrogenase (EC 1.1.1.49) respectively (Williamson & Corkey, 1969).

RESULTS

Partial Purification of 2-5A Synthetase

Interferon-treatment of primary cultures of chick embryo cells causes a 5,000 to 10,000 fold increase in the level of 2-5A synthetase activity. Moreover, a new 56,000 dalton polypeptide has been detected in the post-ribosomal supernatants of interferon-treated cells (Ball, 1979). Purification of the 2-5A synthetase was undertaken in order to investigate the relationship between the enzyme and the 56K polypeptide. Figure 1 shows the polypeptide profiles of a partially purified enzyme preparation, and the profiles of corresponding fractions from non-interferon-treated chick cells. Although the degree of purification at this stage is less than 20 fold, there is a marked enrichment of the 56K polypeptide. With the exception of cellular actin, the interferon-induced polypeptide is the major labeled component of the partially purified 2-5A synthetase preparations. In addition, the 56K polypeptide and the synthetase activity have similar molecular weights under non-denaturing conditions (50-60,000 MW; Ball, 1979); co-eluate from columns of DEAE-cellulose at 0.05 M KCl (data not shown), and have similar affinities for dsRNA (Ball,1979). These common properties make the 56K polypeptide a strong candidate for a structural component of 2-5A synthetase.

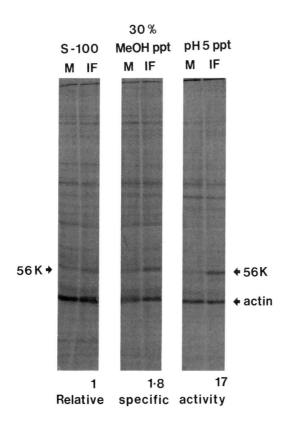

FIGURE 1. Polypeptide analysis of partially purified 2-5A synthetase by SDS-polyacrylamide gel electrophoresis. The fractions shown are the post-ribosomal supernatant (S-100), its 30% methanol precipitate, and the polypeptides that precipitate from this fraction upon subsequent precipitation at pH 5.0 in the presence of poly(IC). The post-ribosomal supernatants were prepared from cells that had been treated with interferon (IF; 250 units/ml) or a mock preparation of interferon (M) for 8 hours, and labeled by incorporation of [^{35}S] methionine from 3 to 8 hr after interferon treatment (Ball, 1979). The specific enzymatic activity of 2-5A synthetase, relative to the post-ribosomal supernatant, is shown. An autoradiograph of the dried polyacrylamide gel is presented, and the positions of the interferon-induced 56K polypeptide and cellular actin are indicated.

Direction of Synthesis of 2-5A

The activated synthetase can further elongate preformed 2-5A moieties. For example, incubation of ^{3}H-labeled trimer with the activated enzyme in the presence of unlabeled ATP for 60 min results in the appearance of labeled tetramer (data not shown). This indicates that the enzyme functions in a nonprocessive manner. Moreover, analysis of the distribution of ^{3}H-label among the four adenosine residues of the tetramer reveals the direction of polymerization of 2-5A. Label distribution was analyzed by digesting the 2-5A with phosphatase, followed by either alkali or venom phosphodiesterase. In the first procedure, the 2' terminus gives adenosine; all other positions give a mixture of 2' and 3' AMP. In the second procedure, the 5' terminus gives adenosine; all other positions give 5' AMP (Kerr & Brown, 1978). These products were distinguished by DEAE-cellulose column chromatography in 7 M urea, and the distribution of ^{3}H-label is shown in Table I. It is clear that elongation of the trimer has occurred by addition of an adenylate residue to the 2' terminus, showing that the direction of 2-5A polymerization is from the 5' end, like that of other polynucleotides.

TABLE I. ^{3}H-Label Distribution Analysis: Trimer Elongation Products [a]

2-5A oligomer	% of ^{3}H-label at:		
	5' terminus	internal position(s)	2' terminus
Trimer (residual)	34.6	32.8	32.6
Tetramer	34.3	28.7 28.7	8.4

[a] Purified ^{3}H-labeled ppA^2'p^5'A^2'p^5'A was incubated with activated 2-5A synthetase in the presence of 6 mM unlabeled ATP for 1 hr at 30°. The products were digested with alkaline phosphatase and fractionaed by chromatography on a DEAE-cellulose-urea column. 30% of the radioactivity eluted in the position expected for the tetramer core and was identified as such by thin-layer chromatography. In the absence of ATP or poly(IC), no tetramer was formed. The tetramer and the residual trimer were analyzed for their distribution of ^{3}H-label as described.

Diribonucleoside Monophosphates as Primers for 2-5A Synthetase

Diribonucleoside monophosphates, linked either 3'-5' or 2'-5', can serve as primers for 2-5A synthetase, but only if they contain an adenosine residue at their 3' terminus. The products of a synthetase reaction carried out in the presence of $C^{3'}P^{5'}A$, for example, consist of a mixture of authentic 2-5A moieties, $pppA^{2'}(P^{5'}A)_n$, and an oligomeric series of primed products, $C^{3'}P^{5'}A^{2'}(P^{5'}A)_n$. The latter can be distinguished from the phosphatase-resistant cores of 2-5A by PEI-cellulose thin-layer chromatography (Fig. 2).

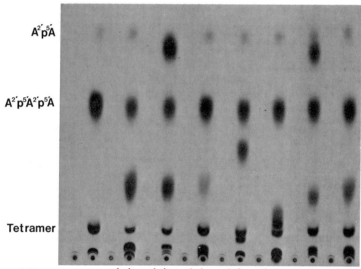

FIGURE 2. Analysis of the 2-5A synthetase products made in the presence of diribonucleoside monophosphate primers. Synthetase reactions [without or with 10 µg/ml poly(IC)] were incubated for 2 hr in the presence of [^3H]ATP (1.5 mM; 167 Ci/M) and the indicated dinucleoside monophosphate (1.5 mM; Sigma). The products were digested with alkaline phosphatase, and 1 µl aliquots were applied to a PEI-cellulose thin-layer plate. Alternate lanes received the products of reactions performed without and with poly(IC). The plate was washed with water to remove [^3H]adenosine derived from the unincorporated ATP, dried, and subjected to chromatography in 1.0 M acetic acid. A fluorograph of the thin-layer plate is shown, and the positions of the unprimed dimer, trimer and tetramer products are indicated.

When the primer has a 3'-5' linkage, the primed products are sensitive to digestion with the appropriate ribonucleases (i.e., GpA-primed products are sensitive to ribonuclease T1; CpA and UpA-primed products are sensitive to ribonuclease A; and all four 3'-5' dinucleoside monophosphate-primed products are sensitive to ribonuclease T2). As shown in Figure 3, digestion with ribonuclease T2 of the trimers made using [^{3}H]ATP in a synthetase reaction primed with unlabeled $A^{3'}p^{5'}A$ results in selective hydrolysis of the primed species to produce $A^{2'}p^{5'}A$. Analysis of the distribution of ^{3}H-label in this product and in the authentic, ribonuclease-resistant, 2-5A trimer gave the results shown in Table II. These data demonstrate the ability of the synthetase to add adenylate residues in 2'-5' linkage to the adenosine residue at the 3' end of dinucleoside monophosphate primers. Whether any of these primed products is able to function as a nuclease activator after the addition of 5' terminal phosphate residues remains to be seen.

Nicotinamide Adenine Dinucleotide as a Primer for 2-5A Synthetase

A number of other molecules were tested for their ability to serve as primers for 2-5A synthetase. Under the standard reaction conditions, no evidence was found for the 2' adenylation of poly(A), poly(A)-containing messenger RNA, transfer RNA, puromycin, S-adenosylmethionine, coenzyme A or adenosine. However, the presence of ADP-ribose, NAD$^+$, or diadenosine tetraphosphate ($A^{5'}ppp^{5'}A$) during the synthetase reaction led

TABLE II. ^{3}H-Label Distribution Analysis: Ribonuclease T2 Digestion Products [a]

2-5A Oligomer	% of ^{3}H-label at:		
	5' terminus	internal position	2' terminus
Dimer	2.0	----	98.0
Trimer	31.3	32.3	36.4

[a] *The products of ribonuclease T2 digestion of the trimer made by [^{3}H]ATP-addition to unlabeled $A^{3'}p^{5'}A$ (Fig. 3, lane e) were resolved by DEAE-cellulose-urea column chromatography and analyzed for their distribution of ^{3}H-label.*

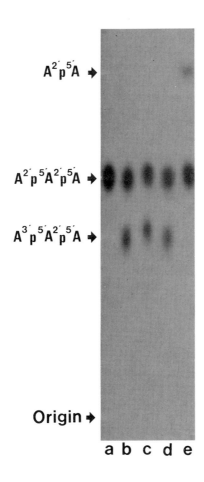

FIGURE 3. Ribonuclease sensitivity of the trimer made by addition to $A^{3'}p^{5'}A$. 2-5A synthetase reaction products were made in the presence of $[^3H]ATP$ and unlabeled $A^{3'}p^{5'}A$, phosphatase-digested, and fractionated by DEAE-cellulose-urea chromatography. The trimer peak, which consisted of a mixture of primed and unprimed products, was digested with ribonucleases and analyzed by PEI-cellulose thin-layer chromatography in 1.0 M acetic acid. A fluorograph of the thin-layer plate is shown. (a) unprimed trimer (marker); (b)-(e) trimer peak from the $A^{2'}p^{5'}A$-primed products, digested with: (b) no ribonuclease; (c) ribonuclease A; (d) ribonuclease T1; (e) ribonuclease T2. The dimer and trimer shown in lane (e) were analyzed for their distribution of ^3H-label (Table 2).

to the appearance of new phosphatase-resistant products. Their structure has yet to be determined but it seems likely that they result from the 2' oligoadenylation of the corresponding primers. Preliminary analysis of the putative NAD^+-primed products shows that their synthesis occurs only in extracts of interferon-treated cells, and only in the presence of dsRNA (Fig. 4). They can be labeled by incorporation of either $[^3H]ATP$ or $[^3H]NAD^+$ and constitute an apparent oligomeric series of products which differ by single negative charges. The major components have net charges which are consistent with their having been formed by the addition of one or two adenylate residues in 2'-5' linkage to the adenosine moiety of NAD^+. Also consistent with this is the observation that $NADP^+$, which carries a phosphate group at the 2' position of the adenosine, is inactive as a primer for 2-5A synthetase.

Incubation of NAD^+ with the activated synthetase results in a decrease in the ability of the coenzyme to be reduced by yeast alcohol dehydrogenase (Fig. 5). Mixing experiments show that this is due to the disappearance of reducible NAD^+ rather than the appearance of an inhibitor. Incubation of $NADP^+$ under the same conditions has no effect on its function as a coenzyme for yeast glucose-6-phosphate dehydrogenase.

DISCUSSION

The induction of 2-5A synthetase by interferon treatment of chick embryo cells can result in a 10,000 fold increase in the level of enzyme activity. The magnitude of this response prompted a search for the structural component(s) of the enzyme, and a candidate 56K polypeptide was identified (Ball, 1979). This 56K polypeptide and 2-5A synthetase co-fractionate during procedures designed to separate proteins on the basis of molecular weight (gel filtration), net charge (ion-exchange chromatography), and ligand binding (affinity chromatography on poly(IC)-agarose). Moreover, a partial purification of the synthetase activity results in a substantial enrichment of the 56K polypeptide. These results show that the polypeptide remains a good candidate for a structural component of the 2-5A synthetase, but much more extensive purification will be required before their relationship can be firmly established. In this respect, it is important to note that the observations of Hovanessian & Kerr (1979) suggest that the murine 2-5A synthetase may be a two (or more) component system.

The ability of activated 2-5A synthetase to elongate preformed 2'5' oligoadenylate moieties shows that the enzyme functions in a non-processive manner like other template-

2'5'Oligoadenylate Synthetase

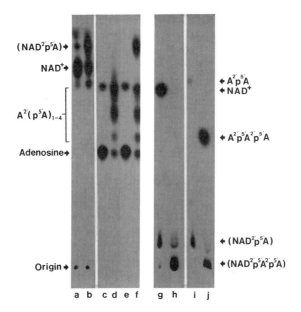

FIGURE 4. Analysis by thin-layer chromatography of the NAD^+-primed products of 2-5A synthetase. Products were labeled with $[^3H]NAD^+$ (adenine-2,8-^3H; 2,960 Ci/M) (lanes a, b, g, h) or with $[^3H]ATP$ (167 Ci/M) (lanes c-f, i, j), and digested with alkaline phosphatase. They were fractionated either by PEI-cellulose thin-layer chromatography on 0.1 M ammonium bicarbonate (lanes a-f), or by initial purification on DEAE-cellulose-urea columns, followed by PEI-cellulose thin-layer chromatography in 1.0 M acetic acid (lanes g-j). Lanes (a) and (b), products labeled with $[^3H]NAD^+$ in the absence (a) and presence (b) of 10 µg/ml of poly(IC). Lanes (c) and (d), products labeled with $[^3H]ATP$ in the absence (c) and presence (d) of 10 µg/ml of poly(IC). Lanes (e) and (f), products labeled with $[^3H]ATP$ in the presence of 2 mM unlabeled NAD^+, and the absence (e) and presence (f) of 10 µg/ml poly(IC). Lanes (g) and (h), products labeled with $[^3H]NAD^+$ and eluting from DEAE-cellulose-urea columns with apparent net charges of -1 (g) and -2 (h). Lanes (i) and (j), products labeled with $[^3H]ATP$ and eluting from DEAE-cellulose-urea columns with apparant net charges of -1 (i) and -2 (j). Fluorographs of the thin-layer plates are shown. Tentative structural identifications are shown in parentheses.

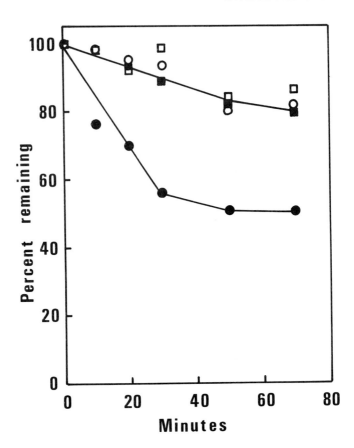

FIGURE 5. Kinetics of disappearance of reducible NAD^+ upon reaction with activated 2-5A synthetase. Reaction mixtures contained 0.2 mM NAD^+, 1.5 mM ATP, and post-ribosomal supernatant extracts of mock interferon- or interferon-treated cells. Incubation was at 30° and 10 µl aliquots were withdrawn at the times shown. They were heated at 95° for 5 min and their supernatants were assayed for NAD^+ by reduction with ethanol in a reaction catalyzed by yeast alcohol dehydrogenase. The results are expressed as percentage of the initial concentration of reducible NAD^+. Extract of mock interferon-treated cells, without (□) and with (■) 10 µg poly(IC) per ml; extract of interferon-treated cells, without (o) and with (●) 10 µg poly(IC) per ml.

independent polymerases. This means that the enzyme can re-attach to its product and add further adenylate residues, a reaction in which the preformed 2-5A acts like a primer. The structural requirements for primer function appear to be the presence, in a small molecule, of a 5'-AMP residue with unsubstituted 2' and 3' hydroxyl groups. The moiety to which the AMP residue is attached seems relatively unimportant, although no macromolecule with primer activity has yet been discovered.

The observation that 2-5A synthetase can adenylate NAD^+ *in vitro* raises the possibility that it can catalyze the same reaction in the intact cell, although we have no evidence for this at present. Since 2'-adenylated NAD^+ is apparently unable to function as a coenzyme, its modification could have profound effects on the redox state of the cell, and hence on the whole of central metabolism. Similarly, the 2'-adenylation of ADP-ribose (and NAD^+) might interfere with the synthesis of poly(ADP)-ribose, the structure of which contains ribose-ribose bonds that involve the 2' position (Hayaishi & Ueda, 1977). Diadenosine tetraphosphate has been suggested as a possible positive "pleiotypic activator" and trigger of DNA synthesis in eukaryotic cells (Rapaport & Zamecnik, 1976; Grummt, 1978). Its 2'-adenylation by an interferon-induced enzyme could render it non-functional, and might, conceivably, account for the reduced growth rate of interferon-treated cells. According to this line of speculation, therefore, some of the multiple actions of interferon might be due to the 2'-adenylation of a number of small molecules that occupy key roles in cellular metabolism. Experiments are in progress to test the validity of these ideas.

REFERENCES

Baglioni, C., Minks, M.A. & Maroney, P.A. (1978), Interferon action may be mediated by activation of a nuclease by $pppA2'p5'A2'p5'$. *Nature (London) 273*, 684.
Ball, L.A. (1979), Induction of 2'5' oligoadenylate synthetase and a new protein by chick interferon. *Virology 94*, 282.
Ball, L.A. & White, C.N. (1978), Oligonucleotide inhibitor of protein synthesis made in extracts of interferon-treated chick embryo cells: Comparison with the mouse low molecular weight inhibitor. *Proc. Natl. Acad. Sci. USA 75*, 1167.
Ball, L.A. & White, C.N. (1979), Nuclease activation by double-stranded RNA and by 2'5' oligoadenylate in extracts of interferon-treated chick cells. *Virology 93*, 384.
Clemens, M.J. & Williams, B.R.G. (1978), Inhibition of cell-

free protein synthesis by $pppA^{2'}p^{5'}A^{2'}p^{5'}$: A novel oligonucleotide synthesized by interferon treated L-cell extracts. *Cell 13*, 565.

Farrell, P.J., Sen, G.C., Dubois, M.F., Ratner, L., Slattery, E. & Lengyel, P. (1978), Interferon action: Two distinct pathways for inhibition of protein synthesis by double-stranded RNA. *Proc. Natl. Acad. Sci. USA 75*, 5893.

Grummt, F. (1978), Diadenosine $5',5''''-p^1p^4$-tetraphosphate triggers initiation of *in vitro* DNA replication in baby hamster kidney cells. *Proc. Natl. Acad. Sci. USA 75*, 371.

Hayaishi, O. & Ueda, K. (1977), Poly(ADP-ribose) and ADP-ribosylation of proteins. *Ann. Rev. Biochem. 46*, 95.

Hovanessian, A.G. & Kerr, I.M. (1979), The 2-5 oligoadenylate ($pppA^{2'}p^{5'}A^{2'}p^{5'}A$) synthetase and protein kinase(s) from interferon-treated cells. *Eur. J. Biochem. 93*, 515.

Kerr, I.M. & Brown, R.E. (1978), $pppA^{2'}p^{5'}A^{2'}p^{5'}A$: an inhibitor of protein synthesis synthesized with an enzyme fraction from interferon-treated cells. *Proc. Natl. Acad. Sci. USA 75*, 256.

Kerr, I.M., Brown, R.E. & Ball, L.A. (1974), Increases sensitivity of cell-free protein synthesis to double-stranded RNA after interferon treatment. *Nature (London) 250*, 57.

Kerr, I.M., Brown, R.E. & Hovanessian, A.G. (1977), Nature of inhibitor of cell-free protein synthesis formed in response to interferon and double-stranded RNA. *Nature (London) 268*, 540.

Lebleu, B., Sen, G.C., Shaila, S., Cabrer, B. & Lengyel, P. (1976), Interferon, double-stranded RNA, and protein phosphorylation. *Proc. Natl. Acad. Sci. USA 73*, 3107.

Martin, E.M., Birdall, N.J.M., Brown, R.E. & Kerr, I.M. (1979), Enzymatic synthesis, characterization, and nuclear magnetic resonance spectra of $pppA^{2'}p^{5'}A^{2'}A$ and related oligonucleotides: Comparison with chemically synthesized material. *Eur. J. Biochem.*, in press.

Nilsen, T.W. & Baglioni, C. (1979), Mechanism for discrimination between viral and host mRNA in interferon-treated cells. *Proc. Natl. Acad. Sci. USA*, in press.

Rapaport, E. & Zamecnik, P.C. (1976), Presence of diadenosine $5',5''''-p^1,p^4$-tetraphosphate (Ap_4A) in mammalian cells in levels varying widely with proliferative activity of the tissue: A possible positive "pleiotypic activator". *Proc. Natl. Acad. Sci. USA 73*, 3984.

Revel, M. & Groner, Y. (1978), Post-transcriptional and translational controls of gene expression in eukaryotes. *Ann. Rev. Biochem. 47*, 1079.

Roberts, W.K., Hovanessian, A.G., Brown, R.E., Clemens, M.J. & Kerr, I.M. (1976), A new double-stranded RNA-dependent protein kinase activity and low-molecular-weight inhibitor

of protein synthesis in extracts from interferon-treated L-cells. *Nature (London) 264*, 477.
Steward, W.E. II (editor) (1977), Interferons and their actions. *CRC Press, Cleveland*.
Williamson, J.R. & Corkey, B.E. (1969), Assays of intermediates of the citric acid cycle and related compounds by fluorometric enzyme methods. *Meth. Enzymol. XIII*, 434.
Zilberstein, A., Federman, P., Shulman, L. & Revel, M. (1976), Specific phosphorylation *in vitro* of a protein associated with ribosomes of interferon-treated mouse L-cells. *FEBS Lett. 68*, 119.
Zilberstein, A., Kimchi, A., Schmidt, A. & Revel, M. (1978), Isolation of two interferon-induced translational inhibitors: A protein kinase and an oligo-isoadenylate synthetase. *Proc. Natl. Acad. Sci. USA 75*, 4734.

REGULATION OF MACROMOLECULAR SYNTHESIS
BY LOW MOLECULAR WEIGHT MEDIATORS

MECHANISM OF ACTION OF 2-5A IN INTACT CELLS

Ara G. Hovanessian, John N. Wood, Eliane Meurs
and Luc Montagnier

Unité d'Oncologie Virale
Département de Virologie
Institut Pasteur, Paris

2-5A (2'-5' linked oligoadenylic acid triphosphate) is a potent inhibitor of protein synthesis in intact cells of different origin, human, hamster and mouse. At a concentration of 10 nM (in AMP equivalents), protein synthesis is inhibited by 50 to 85%. There is also a secondary effect on the total RNA synthesis which becomes evident several hours after inhibition of protein synthesis. All of these effects, however, are transient and after an overnight recovery period, both RNA and protein synthesis resume rates comparable to the appropriate controls. Activation of a nuclease is detected in cells following treatment with 2-5A. The total polyadenylated RNA is much reduced in comparison to that from untreated cells and electrophoretic analysis in polyacrylamide slab gels provides evidence for its degradation. Similarly, there is an apparent degradation of ribosomal RNA. Consistent with these results, extracts from cells pretreated with 2-5A manifest an enhanced nuclease activity in vitro on incubation with exogenous RNA. Here, we propose that as in cell-free systems, the mechanism of action of 2-5A in intact cells involves activation of a nuclease.

INTRODUCTION

Protein synthesis in cell-free systems from interferon-treated cells shows an enhanced sensitivity to inhibition by double-stranded RNA (dsRNA)

(Kerr et al., 1974). A protein kinase(s), 2-5A synthetase (responsible for the formation of a series of 2'-5' linked oligoadenylic acid triphosphate inhibitors of protein synthesis in which the trimer pppA 2'p5'A2'p5'A is predominant) and a nuclease may all be involved in this inhibition (Hovanessian et al., 1977; Kerr et al., 1977; Kerr and Brown, 1978; Ratner et al., 1977). The 2-5A synthetase could be retained on a column of dsRNA bound to a solid support and conveniently used in this form to synthesise large quantities of the inhibitor (Hovanessian et al., 1977; Hovanessian and Kerr, 1979). An identical series of such oligonucleotide inhibitors is also formed on incubation of reticulocyte lysates with dsRNA and ATP (Hovanessian and Kerr, 1978). These oligonucleotide inhibitors are effective at subnanomolar concentrations in cell-free systems prepared from control cells and their effect seems to be mediated at least in part, by a nuclease which degrades mRNA (Clemens and Williams, 1978). The present communication suggests that activation of a nuclease may also be involved in the mechanism of action of 2-5A in intact cells. This is shown by the use of a calcium phosphate-co-precipitation technique to introduce 2-5A into a variety of different cell types.

RESULTS

2-5A was synthesised from ATP by the use of an enzyme fraction from rabbit reticulocyte lysates bound to poly(I).poly(C)-Sepharose (Hovanessian and Kerr, 1978). Its partial purification, and assay by means of inhibition of ^{14}C-labeled amino acid incorporation in rabbit reticulocyte lysate and L cell-free systems were as described (Hovanessian et al., 1977).

Treatment of cells with 2-5A

Growth medium was aspirated and the cell monolayers were incubated in Hepes-buffered saline pH 7.05 (8.0 g/l NaCl, 0.37 g/l KCl, 0.125 g/l Na_2HPO_4 .$2H_2O$, 1.0 g/l glucose, and 5.0 g/l Hepes) (Graham and Van Der Eb, 1973) containing different concentrations of 2-5A as indicated in individual experiments and $CaCl_2$ at a final concentration of

125 mM. After an initial incubation at room temperature for 45 min, the treatment solution was diluted with an equal volume of serum free medium and the cells were incubated at 37°C for 90 min. This liquid medium was then aspirated and the cell monolayers were incubated in growth medium for assay of RNA and protein synthesis.

Inhibition of protein synthesis by 2-5A

2-5A is a potent inhibitor of protein synthesis in intact cells when its uptake is facilitated by incubation in Hepes-buffered saline and $CaCl_2$. Upon addition of $CaCl_2$ (70-125 mM) to solutions of 2-5A (^{3}H-labeled) in the buffered saline a precipitate is formed consisting of 2-5A and calcium phosphate which sediments onto the cells, becomes adsorbed to the cell membrane and then could be taken up by the cell. Inhibition of protein synthesis measured by the incorporation of (^{35}S) methionine into acid insoluble counts was detectable as soon as 20 min after addition of $CaCl_2$ with maximum inhibitory effects (70-95%) at 90 min. In the absence of $CaCl_2$, however, different concentrations of 2-5A had no apparent effect on protein synthesis (Hovanessian et al., 1979). Table I shows the specific inhibitory activity of 2-5A in intact cells of different origin: mouse, human and hamster. At a concentration of 10 nM, 2-5A results in the inhibition of protein synthesis by 50-72%. This is reasonable in comparison with the concentrations required to inhibit protein synthesis in cell-free systems and in intact cells as has been previously reported (Hovanessian et al., 1977; Kerr et al., 1977; Hovanessian and Kerr, 1978).

Activation of a nuclease in cells treated with 2-5A

In vivo RNA and protein synthesis after preincubation in Hepes-buffered saline with $CaCl_2$ was not less than 80% of cells which have been preincubated in normal medium. In the presence of 2-5A, however, both processes were inhibited at different times. Inhibition of protein synthesis was detectable 4-6 hr before any inhibitory effects on RNA synthesis, thus suggesting that the inhibition of RNA synthesis was a secondary consequence of the

Table I. Specific inhibitory activity of 2-5A on protein synthesis in intact cells.

2-5A(M)	% inhibition of protein synthesis		
	MRC5	BHK21	L929
1×10^{-9}	27	28	13
5×10^{-9}	40	47	24
1×10^{-8}	72	69	50
5×10^{-8}	83	82	75
1×10^{-7}	91	84	81

Monolayers of human embryonic fibroblast (MRC5), hamster fibroblast (BHK21) and mouse (L929) cells were treated with different concentrations of 2-5A before incorporation of (^{35}S) methionine into TCA precipitable counts. In the absence of 2-5A incorporation was not less than 50,000 c.p.m.

shut off of protein synthesis (Hovanessian et al., 1979).

Studies on the mechanism of action of 2-5A in cell-free systems have indicated that it activates a nuclease which then degrades mRNA. It was of interest therefore to analyse the RNA extracted from cells after treatment with the inhibitor. For this purpose, cells were incubated in medium containing ^{3}H-uridine for 20 hr before removal of the residual label and treatment with 2-5A. The total and polyadenylated RNA from control cells or cells preincubated in Hepes-buffered saline pH 7.05 containing $CaCl_2$ in the absence and presence of 2-5A were analysed by electrophoresis in polyacrylamide slab gels containing 6M urea. Figs 1 and 2 show an apparent degradation of ribosomal and polyadenylated RNA in extracts from cells treated with 2-5A. In accord with these results, such extracts manifested

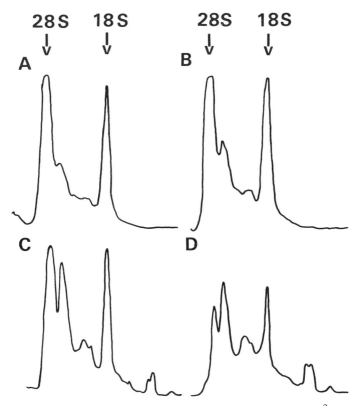

Figure 1. Electrophoretic profiles of (^{3}H) uridine labeled RNA from MRC5 cells. RNA extracted from cultures incubated in medium (A) or treated with Hepes buffered saline and $CaCl_2$ in the absence (B) or presence of 2-5A at a final concentration of 10^{-8} (C) and 10^{-7} (D) M as described. A scan of the electrophoretic pattern of total RNA from the fluorography is shown.

an enhanced level of nuclease activity *in vitro* on incubation with (^{3}H) labeled vesicular stomatitis virus (VSV) mRNA (Fig. 3).

DISCUSSION

The effect of 2-5A in intact cells is dependent on its penetration into the cell. This may occur

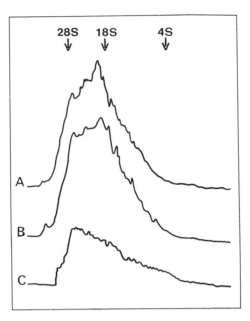

Figure 2. Electrophoretic profiles of (^{3}H) uridine labeled polyadenylated RNA from MRC5 cells. Polyadenylated RNA from cultures incubated in medium (A) or Hepes-buffered saline with $CaCl_2$ in the absence (B) or presence (C) of 2-5A at a final concentration of 5×10^{-8}M were analysed in a 2.6% acrylamide slab gel. A scan of the electrophoretic pattern is shown. Polyadenylated RNA was prepared by poly(U)-Sepharose chromatography.

after treatment of cells in hypertonic medium (Williams and Kerr, 1978) or by adsorption of 2-5A onto the cell membrane. Here, this latter method was used to investigate the mechanism of action of 2-5A in intact cells.

Here, evidence is provided to suggest that activation of a nuclease may be involved in the mechanism of action of 2-5A in intact cells. After treatment with 2-5A both total and polyadenylated RNA measured by ^{3}H-counts were reduced by 50 and 60-70 per cent respectively. In accord with this, electrophoretic profiles of total RNA indicated an apparent degradation of 28S and 18S ribosomal RNA into several small molecular weight components

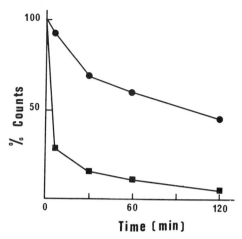

Figure 3. Enhanced nuclease activity in extracts from cells treated with 2-5A. Breakdown of VSV (^{3}H) mRNA on incubation in extracts prepared from MRC5 cells treated in Hepes-buffered saline, in the absence (●) or presence (■) of 2-5A (5 x 10^{-8}M). After 60 min, the cells were washed with Hepes-buffered saline pH 7.5, scraped and lysed with an equal volume of buffer (10 mM Hepes pH 7.5, 10 mM KCl, 2 mM MgOAc and 7 mM 2-mercaptoethanol). The concentration of KCl was then brought to 90 mM, centrifuged at 10,000 X g for 10 min and the supernatant was incubated with VSV (^{3}H)mRNA without any further additions. The VSV (^{3}H)mRNA (2 x 10^5 counts/min per µg) was added to 16 µg/ml and incubated in the cell extract at 30°C. The ordinate gives the residual trichloroacetic insoluble radioactivity per 5 µl-aliquot as a percentage of the zero time value (6200 cpm).

(Fig. 1). Moreover, polyadenylated RNA was also degraded giving rise to a wide spectrum of products which were detectable as a smear on electrophoresis in polyacrylamide gels (Fig. 2). These results are difficult to explain on any other basis than an enhanced level of nuclease activity, since they represent degradation of previously labeled RNA. Accordingly, extracts from 2-5A treated cells manifested an enhanced nuclease activity *in vitro* on viral and cellular RNAs (Fig. 3).

An interesting aspect of the activated nuclease is that its effects are transient (Hovanessian et al., 1979). This is of great advantage since the cell survives after a period when mRNA species are degraded and thus provides a suitable mechanism for the elimination of undesirable messages. In accord with this, preliminary studies indicated that viral RNA synthesis can be reduced to background levels after treatment with 2-5A (unpublished results). It is possible, therefore, that the antiviral action of interferon may at least in part be mediated by the action of 2-5A activated nuclease, i.e., treatment of cells with interferon enhances levels of 2-5A synthetase which in the presence of viral replicative intermediate(dsRNA) and ATP forms 2-5A. The latter in turn activates a nuclease responsible for destroying cytoplasmic mRNA, cellular and viral alike. By this process, viral replication could be eliminated or limited to a certain extent, while the cell may or may not survive. Whatever the situation, however, little virus is produced.

ACKNOWLEDGEMENTS

A.G.H. was the recipient of an European Biology Organisation fellowship. This work was supported in part by a grant from Association for the Development of Pasteur Institute.

REFERENCES

Clemens, M.J. and Williams, B.R.G. (1978). Inhibition of protein synthesis by pppA2'p5'A2'p5'A: a novel oligonucleotide synthesised by interferon-treated L-cell extracts. *Cell 13*, 565-572.

Graham, F.L. and Van Der Eb, A.J. (1973). A new technique for the assay of infectivity of human adenovirus 5 DNA. *Virology 52*, 456-467.

Hovanessian, A.G. and Kerr, I.M. (1978). Synthesis of an oligonucleotide inhibitor of protein synthesis in rabbit reticulocyte lysates analogous to that formed in extracts from interferon-treated cells. *Eur. J. Biochem. 84*, 149-159.

Hovanessian, A.G. and Kerr, I.M. (1979). The (2'-5') oligoadenylate (pppA2'p5'A2'p5'A) synthetase and protein kinase(s) from interferon cells. *Eur. J. Biochem. 93*, 515-526.

Hovanessian, A.G., Brown, R.E. and Kerr, I.M. (1977). Synthesis of low molecular weight inhibitor of protein synthesis with enzyme from interferon-treated cells. *Nature 268*, 537-540.

Hovanessian, A.G., Wood, J.N., Meurs, E. and Montagnier, L. (1979). Increased nuclease activity in cells treated with pppA2'p5'A2'p5'A. *Proc. Natl. Acad. Sci. USA* in press.

Kerr, I.M. and Brown, R.E. (1978). pppA2'p5'A2'p5'A: an inhibitor of protein synthesis synthesised with an enzyme fraction from interferon-treated cells. *Proc. Natl. Acad. Sci. USA 75*, 256-260.

Kerr, I.M., Brown, R.E. and Ball, L.A. (1974). Increased sensitivity of cell-free protein synthesis to dsRNA after interferon treatment. *Nature 250*, 57-59.

Kerr, I.M., Brown, R.E. and Hovanessian, A.G. (1977). Nature of inhibitor of cell-free protein synthesis formed in response to interferon and dsRNA. *Nature 268*, 540-542.

Ratner, L., Sen, G.C., Brown, G.E., Lebleu, B., Kawakita, M., Cabrer, B., Slattery, E. and Lengyel, P. (1977). Interferon, dsRNA and RNA degradation. *Eur. J. Biochem. 79*, 565-577.

REGULATION OF MACROMOLECULAR SYNTHESIS BY LOW MOLECULAR WEIGHT MEDIATORS

2'5'-OLIGO(A): A MEDIATOR OF VIRAL RNA CLEAVAGE IN INTERFERON-TREATED CELLS?

Timothy W. Nilsen
Patricia A. Maroney
Corrado Baglioni

Department of Biological Sciences
State University of New York at Albany
Albany, New York

The replicative complex (RC) of encephalomyocarditis virus has been isolated from infected HeLa cells by isopycnic centrifugation. The RC prepared in this way promotes the synthesis of 2'5'oligo(A) in extracts of interferon-treated HeLa cells. The 2'5'oligo(A) synthesized activates an endonuclease which degrades mRNA. These results are discussed with reference to the proposed localization of the 2'5'oligo(A) polymerase/endonuclease system on the RC formed in interferon-treated cells.

INTRODUCTION

Interferon induces the synthesis of an enzyme, designated 2'5'oligo(A) polymerase in mammalian and avian cells (Baglioni, 1979). This enzyme in the presence of double-stranded RNA (dsRNA) polymerizes ATP into a series of oligonucleotides linked by an unusual 2'5' phosphodiester bond (Kerr & Brown, 1978). The oligonucleotides, designated 2'5'oligo(A) or 2,5A (Baglioni, 1979), activate an endoribonuclease, a constitutive enzyme present in cells treated and not treated with interferon, which specifically cleaves single-stranded RNA (ssRNA) (Baglioni et al., 1978; Clemens & Williams, 1978; Ratner et al., 1978; Zilberstein et al. 1978).

The induction of 2,5A polymerase has recently been correlated with the impaired replication of an RNA virus (Baglioni et al., 1979). In interferon treated HeLa and L cells infected with encephalomyocarditis virus (EMCV), the accumulation of viral RNA is inhibited in direct proportion with the increase in 2,5A polymerase activity (Baglioni et al., 1979. This observation led to the proposal that the 2,5A polymerase could bind to and be activated by double-stranded structures of the viral replicative complex (RC). Synthesis of 2,5A by the polymerase would result in a localized activation of the endonuclease near the viral RC. The endonuclease could preferentially cleave nascent single-stranded viral RNA over cellular RNA because of the rapid catabolism of 2,5A and of the return of the endonuclease to an inactive state in the absence of 2,5A (Williams et al., 1978; Minks et al., 1979a).

This mechanism for preferential degradation of ssRNA covalently linked to dsRNA has recently been demonstrated in vitro (Nilsen & Baglioni, 1979). The replicative intermediate (RI, i.e., the RNA component of RC) of EMCV is degraded to a dsRNA "core" upon incubation with extracts of interferon-treated HeLa cells, but not with extracts of control cells (Nilsen & Baglioni, 1979). This degradation is most likely due to the combined action of the 2,5A polymerase and endonuclease for the following reasons: i) RI promotes synthesis of 2,5A in extracts of interferon-treated cells; ii) when synthesis of 2,5A is prevented by either limiting the ATP concentration or omitting Mg^{2+} from the incubations, the degradation of RI is no longer observed (Nilsen & Baglioni, 1979; unpublished observations). Furthermore, under the experimental conditions described (Nilsen & Baglioni, 1979) RI is preferentially cleaved, since ssRNA present in the incubations is not significantly degraded. This can be explained by a localized activation of the endonuclease near the presumptive site of 2,5A synthesis. In interferon-treated cells infected by RNA viruses localized synthesis of 2,5A may occur at or near the viral RC.

In the present report we show that the RC of EMCV promotes synthesis of 2,5A in extracts of interferon-treated cells. We have devised a procedure for the isolation of "native" RC from infected cells, based on isopycnic centrifugation in gradients of non-ionic silica particles coated with polyvinylpyrrolidone. This procedure may be generally useful to study the association of specific enzymatic activities induced by interferon with viral RC. Furthermore, our results show how the 2,5A polymerase may be activated in interferon-treated cells and suggest that the 2,5A synthesized may mediate the cleavage of viral RNA.

MATERIALS AND METHODS

A. Cell Culture and Virus Infection

HeLa cells were grown in suspension cultures in Joklik's medium with 5% horse serum and infected with EMCV as previously described (Baglioni et al., 1979).

B. Preparation of EMCV Replicative Complex

Cells were labeled 24 h before infection with 10 nCi/ml of [^{14}C]uridine and from 2 to 5 h after EMCV infection with 20 μCi/ml of [^{3}H]uridine (20 Ci/mMol). In this way, prior to infection the cellular rRNA was uniformly labeled and during the infection the viral RNA was specifically labeled in the presence of 5 μg/ml of actinomycin D. The cells were homogenized in 10 mM KCl, 1.5 mM Mg(OAc)$_2$, 0.5 mM dithiothreitol, and 20 mM Hepes-KOH, pH 7.4 (Buffer A), as previously described (Weber et al., 1975). The homogenate was centrifuged 10 min at 30,000 g to pellet membrane associated cellular components. The pellet was resuspended in Buffer A and the detergent Triton X-100 added to 0.5%. The centrifugation was repeated and the supernatant fraction removed and centrifuged 90 min at 150,000 g. The pellet obtained (p-150) was resuspended in Buffer A containing 20 mM EDTA and centrifuged through a 15-30% sucrose gradient made in 10 mM NaCl, 20 mM EDTA, 10 mM Tris-HCl, pH 7.5 (Buffer B). Gradient fractions were collected, an aliquot used to measure labeled RNA, and the remainder combined as indicated in Fig. 1A. Pooled fractions were diluted to 3.5 ml with Buffer B and the final sucrose concentration was adjusted to 30% (w/w). This sample was mixed with 1.5 ml of Percoll (Pharmacia) and centrifuged for 30 min at 30,000 rev/min in a VTi65 vertical rotor (Beckman). The resulting gradients were fractionated by injecting Fluorinert into the centrifuge tubes from the bottom with a motor-driven syringe. The radioactivity in gradient fractions was determined by precipitating the RNA with 3% alkyltrimethylammonium bromide (Sigma) and collecting the precipitate on glass fiber filters. Unlabeled RC were prepared as described above, except that the 150,000 g pellet was resuspended in Buffer A without EDTA and fractionated directly on Percoll gradients containing 3.5 ml of 30% (w/w) sucrose in Buffer B minus EDTA and 1.5 ml of Percoll.

C. *2'5'oligo(A) Polymerase Assay*

The assay to measure synthesis of 2,5A has been described (Minks et al., 1979b). After 75 min incubation at 30°, the oligonucleotides synthesized were isolated by chromatography on DEAE-cellulose (Minks et al., 1979b).

D. *Endonuclease Assay*

The 2,5A-activated endonuclease was assayed in incubations with labeled poly(A)-containing vesicular stomatitis virus (VSV) mRNA as described (Baglioni et al., 1978). The endonucleotylic cleavage of this mRNA was measured by oligo(dT)-cellulose chromatography (Baglioni et al., 1978).

RESULTS

The goal of this investigation was the isolation of the "native" replicative complex of EMCV in order to establish whether this viral structure could promote synthesis of 2,5A in interferon-treated cells. The RC of picornaviruses sediments with the membrane fraction of infected cells (Levintow, 1974). Upon solubilization of the membranes with a mild detergent, the RC can the further purified by centrifugation at 150,000 g (Flanegan & Baltimore, 1979). In the isolation of RC active in the synthesis of poliovirus RNA, precipitation with 2 M LiCl has subsequently been employed (Flanegan & Baltimore, 1979). This treatment followed by centrifugation on sucrose gradients removes all the proteins from poliovirus RC with the exception of the viral replicase (Flanegan & Baltimore, 1979).

We followed the purification scheme devised for poliovirus RC up to the centrifugation step at 150,000 g, but avoided the treatment with LiCl to prevent release of proteins associated with the "native" RC. The 150,000 g pellet was resuspended (see Methods) and an aliquot tested in the 2,5A polymerase assay (Table 1). Viral components in this fraction specifically promoted synthesis of 2,5A, since a corresponding fraction isolated from mock-infected cells showed no activity (Table 1). However, the 150,000 g pellet contains, in addition to the RC, polysomes associated with viral mRNA and other cellular constituents. Further purification of the RC was therefore attempted by equilibrium sedimentation on isopycnic gradients.

TABLE I. Activation of 2'5'oligo(A) Polymerase by a Membrane-Associated Fraction From EMCV-Infected Cells

pMoles of ATP converted into 2'5'oligo(A)	
Infected Cells	Mock-infected Cells
960	0

A culture of 5×10^8 cells was infected with EMCV (Baglioni et al., 1979). 5 h after infection the cells were harvested and a cell extract prepared and fractionated as described in Methods. Identical procedures were followed with a culture of mock-infected cells. The 150,000 g pellet obtained from the membrane fraction was resuspended in 0.5 ml of Buffer A and 5 µl were incubated with 5 µl of cytoplasmic extract from cells treated for 17 h with 100 units/ml of human fibroblast interferon, under the conditions previously described for the 2'5'oligo(A) synthesis assay (Minks et al., 1979): 125 nMoles of $[^3H]ATP$, 25 mM $Mg(OAc)_2$, 4 mM fructose 1,6-biphosphate and 120 mM $KOAc$, in a final vol of 25 µl. Incubations were for 75 min at 30^6 and the 2'5'oligo(A) synthesized was determined by DEAE-cellulose chromatography (Minks et al., 1979b).

Huang and Baltimore (1970) showed that poliovirus RC could be separated from polysomes by centrifugation on sucrose gradients containing EDTA. The polysomes are dissociated into ribosomal subunits by EDTA and viral mRNA complexed with protein (viral mRNP) is released. The ribosomal subunits and mRNP sediment near the top of the gradient whereas the RC sediments as a broad peak in the heavier region of the gradient (Huang & Baltimore, 1970). The gradient fractions containing RC, ribosomal subunits and viral mRNP were further analyzed by Huang and Baltimore by equilibrium sedimentation on CsCl gradients. To prevent dissociation of protein from RNA in the high ionic strength gradient, the fractions were treated with the cross-linking agent glutaraldehyde. The RC of poliovirus was resolved in these isopycnic gradients from other viral and cellular components.

A similar purification scheme was initially adopted for the isolation of labeled RC from EMCV-infected cells. The 150,000 g pellet obtained from the membrane fraction was treated with EDTA and centrifuged on sucrose gradients (Fig. 1A). All the $[^{14}C]$-labeled ribosomes and 90% of the $[^3H]$-labeled viral RNA were found near the top of the gradient (fraction II). The remaining 10% of the viral RNA sedimented in a broad peak toward the bottom of the gradient (fraction I).

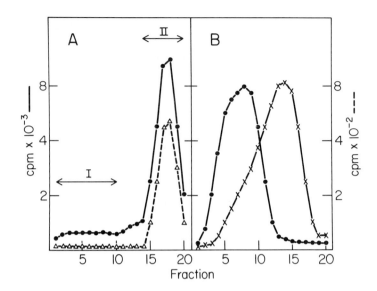

FIGURE 1. Isolation of replicative complex from EMCV-infected cells. HeLa cells were prelabeled with [^{14}C]uridine, infected with EMCV and labeled with [^{3}H]uridine from 2 to 5 h post-infection, as described in Methods.
A. The 150,000 g pellet obtained from the membrane fraction was resuspended in buffer containing 20 mM EDTA and fractionated on 15-30% sucrose gradients. The gradient was centrifuged for 2 h at 27,000 rev/min. Radioactivity in aliquots was determined and the indicated fractions were pooled and centrifuged on the Percoll gradient shown in B, as described in Methods.
B. Equal aliquots of [^{3}H]-radioactivity from peak I and II were loaded on parallel Percoll gradients as described in Methods; 0.25 ml fractions were collected. A. [^{3}H]-radioactivity, ●-●-●; and [^{14}C]-radioactivity, ▲--▲--▲. B. [^{3}H]-radioactivity of peak I from A, x-x-x, and of peak II ●-●-●.

The RNA in these two fractions was further analyzed in gradients containing Na-dodecyl sulphate (SDS). The RNA in fraction II was predominantly 35S viral RNA, whereas the RNA in fraction I sedimented as a heterogenous 30-200S peak, characteristic of the replicative intermediate of picornaviruses (data not shown). This indicated that the RC was effectively separated from the viral components sedimenting near the top of the gradient. This fractionation procedure has the disadvantage, however, that the RC is distributed over several gradient fractions and cannot be directly tested in the 2,5A polymerase assay. The two components resolved by velocity sedimentation were therefore used to develop a new fractionation procedure based on isopycnic equilibrium sedimentation. The use of CsCl gradients did not seem convenient because of the large volumes involved, the need to use a cross-linking agent and the formation of aggregates when concentrated solutions of ribonucleoproteins are cross-linked. Instead, to form isopycnic gradients we used Percoll, a suspension of colloidal silica particles coated with polyvinylpyrrolidone. Upon centrifugation Percoll forms a density gradient because of the size heterogeneity of the silica particles. Percoll is a non-ionic material, and nucleoprotein complexes can be separated in Percoll gradients without the need for cross-linking. Furthermore, it does not interfere with enzymatic assays and forms gradients rapidly. Under the conditions described in Methods, fraction I containing RC was effectively separated from fraction II containing ribosomes and viral mRNPs (Fig. 1B).

The putative RC peak isolated by sedimentation equilibrium in Percoll gradients was characterized by chromatography on colums of Sepharose 2B (Spector & Baltimore,1975), RNAase digestion and sucrose gradient centrifugation before and after RNAase digestion. The labeled material was excluded from Sepharose 2B, was about 25% resistant to RNAase digestion, sedimented as a broad peak in SDS-sucrose gradients before digestion and as a 20S peak after digestion (data not shown). This 20S peak corresponds to the dsRNA "core" of the RI of picornaviruses (Levintow, 1974). We concluded from these observations that the peak isolated by centrifugation on Percoll gradients contained predominantly RC.

This fractionation procedure was scaled up to prepare unlabeled RC from EMCV-infected HeLa cells. Dissociation of polysomes with EDTA was omitted, however, since polysomes have been previously shown to band at a density intermediate between that of 40S and 60S ribosomal subunits (Huang & Baltimore, 1970). The resuspended 150,000 g pellet obtained from the membrane fraction of infected cells was directly fractionated on Percoll gradients (Fig. 2). A sample of labeled RC prepared by the procedure described above was analyzed in parallel (Fig. 2A). Aliquots of the gradient fractions of unlabeled RC were then tested in the 2,5A polymerase assay (Fig. 2B). Fractions corresponding to the labeled RC in the parallel gradient promoted synthesis of 2,5A.

The synthesis of 2,5A was monitored by isolating the oligonucleotides formed by DEAE-cellulose chromatography (see Methods). To further confirm that the oligonucleotides

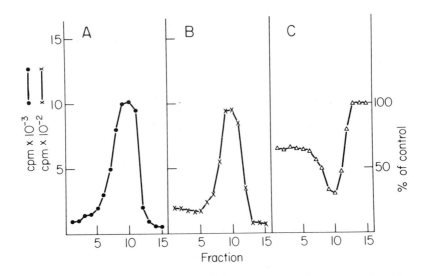

Figure 2. Activation of 2'5'oligo(A) polymerase by the replicative complex of EMCV-infected cells. A. Percoll gradient of fraction I from the gradient shown in Fig.1A. B. Parallel gradient of the 150,000 pellet of the membrane fraction from EMCV-infected cells (0.4 mg of of this fraction corresponding to 2.5×10^6 cells (0.4 mg of nucleic acid in 0.2 ml) was loaded on the gradient. 0.33 ml fractions were collected and 0.05 ml aliquots assayed for activation of 2'5'oligo(A) polymerase. The assay contained 25 µl of cytoplasmic extract of interferon-treated cells (see Table 1), 0.5 mMoles of [^{3}H]ATP and the other components indicated in Table 1 in a final vol of 0.1 ml. The radioactivity of the 2'5'oligo(A) synthesized was determined as described in Methods and is indicated in counts/min. C. Activation of the 2'5'oligo(A)-dependent endonuclease by the oligonucleotides synthesized with gradient fractions. 25 µl aliquots of gradient B fractions were incubated with 10 µl of interferon-treated cell extract, 0.25 mMoles of ATP and the other components indicated in Table 1. After 75 min incubation, the assays were heated to 95° for 5 min, diluted to 1 ml with Buffer A and centrifuged for 2 min at 3,000 g. The supernatant was diluted 10-fold in Buffer A and 10 µl of this dilution used in a 100 µl assay for 2'5'oligo(A)-dependent endonuclease as described in Methods. Each assay contained 10,000 counts/min of poly(A) containing VSV mRNA and 30 µl of HeLa cell extract. After 60 min at 30° the samples were analyzed by oligo(dT)-cellulose chromatography as described (Baglioni, Minks & Maroney, 1978). The percentage of mRNA retained on the oligo(dT)-cellulose, relative to an incubation containing 10 µl of Buffer A, is indicated. In this control incubation, approximately 20% of the mRNA was degraded.

synthesized with RC are authentic 2,5A, we assayed them for their ability to activate the 2,5A-dependent endonuclease of HeLa cells (see Methods). In this assay poly(A)-containing VSV mRNA was used as a substrate and its degradation measured relative to an incubation without added 2,5A. The peak of activation of the endonuclease coincided with that of labeled RC (Fig. 1C). Some activation of the endonuclease was also observed with lighter fractions of the gradient and may reflect the trailing of the RC peak also observed in Fig. 1B. No trailing was observed in the 2,5A polymerase assay shown in Fig. 1B, possibly because this assay is much less sensitive than the biological assay for endonuclease activation.

DISCUSSION

The RC isolated from EMCV-infected cells promotes synthesis of 2,5A in extracts of interferon-treated cells. The oligonucleotides synthesized in this assay have been conclusively identified as 2,5A on the basis of the activation of an endonuclease capable of cleaving mRNA. The component that promotes synthesis of 2,5A has been identified as RC by the following criteria: i) it is found in the membrane fraction of EMCV-infected cells, but not in the corresponding fraction of mock-infected cells; ii) it is found in gradient fractions of density identical to that of labeled RC. These fractions do not contain detectable amounts of 20S viral RNA, the fully double-stranded RNA species found in picornavirus-infected cells (Levintow, 1974). It is therefore unlikely that activation of 2,5A polymerase could be explained by contamination of RC with 20S viral RNA.

The ability of RC to promote synthesis of 2,5A shows that this partially double-stranded viral structure can activate the 2,5A-dependent endonuclease. Localized activation of the endonuclease could account for the specific degradation of viral mRNA in interferon-treated cells.

The participation of the 2,5A polymerase and endonuclease in the degradation of viral RNA in intact cells has not yet been demonstrated. Experiments carried out by several investigators have failed to detect products of degradation of viral RNA in interferon-treated cells infected by a variety of viruses (Baglioni, 1979). It seems possible, however, that once cleaved by an endonuclease viral or cellular RNA may be rapidly degraded to nucleotides by exonucleases. In this case, cleavage of viral RNA would be extremely difficult to detect in intact cells, since no trace of such cleavage could be observed.

The procedure developed in the present study for the isolation of RC by equilibrium centrifugation may allow is to investigate the role of the 2,5A polymerase in interferon-treated cells infected by EMCV. This enzymatic activity may be found associated with the RC isolated from interferon-treated cells as described above. Experiments are presently in progress to test this possibility. This and similar experiments could provide insight into the mechanism of inhibition of viral replication in interferon-treated cells.

ACKNOWLEDGMENTS

This work was supported by Grants AI-11887 and HL-17710 from the National Institutes of Health.

REFERENCES

Baglioni, C. (1979), Interferon induced enzymatic activities and their role in the antiviral state. *Cell 17*, 255.
Baglioni, C., Maroney, P.A. & West, D.K. (1979), 2'5'-oligo(A) polymerase activity and inhibition of viral RNA synthesis in interferon-treated HeLa cells. *Biochem. 18*, 1765.
Baglioni, C., Minks, M.A. & Maroney, P.A. (1978), Interferon action may be mediated ny activation of nuclease by pppA2'p5'A2'p5'A. *Nature 273*, 684.
Clemens, M.J. & Williams, B.R.G. (1978), Inhibition of cell-free protein synthesis by pppA2'p5'A2'p5'A: a novel oligonucleotide synthesized by interferon-treated L cell extracts. *Cell 13*, 565.
Flanegan, J.A. & Baltimore, D. (1979), Poliovirus polyuridylic acid polymerase and RNA replicase have the same viral polypeptide. *J. Virol. 29*, 352.
Huang, A.S. & Baltimore, D. (1970), Initiation of polyribosome formation in poliovirus-infected HeLa cells. *J. Mol. Biol. 47*, 275.
Kerr, I.M. & Brown, R.W. (1978), pppA2'p5'A2'p5'A: an inhibitor of protein synthesis synthesized with an enzyme fraction from interferon-treated cells. *Proc. Natl. Acad. Sci. USA 75*, 256.
Levintow, L. (1974), The reproduction of picornaviruses. in "Comprehensive Virology" (H. Fraenkel-Conrat & R.R. Wagner), Vol. 2, p. 109. Plenum Press, New York and London.
Minks, M.A., Benvin, S., Maroney, P.A. & Baglioni, C. (1979a), Metabolic stability of 2'5'oligo(A) and activity of

2'5'oligo(A)-dependent endonuclease in extracts of control and interferon-treated HeLa cells. *Nucleic Acids Res. 6,* 767.
Minks, M.A., Benvin, S., Maroney, P.A. & Baglioni, C. (1979b), Synthesis of 2'5'oligo(A) in extracts of interferon-treated HeLa cells. *Biol. Chem.* - in press.
Nilsen, T.W. & Baglioni, C. (1979), A mechanism for discrimination between viral and host mRNA in interferon-treated cells. *Proc. Natl. Acad. Sci. USA* - in press.
Ratner, W., Wiegand, R.C., Farell, P.J., Sen, G.C., Cabrer, B. & Lengyel, P. (1978), Interferon, double-stranded RNA and RNA degradation. Fractionation of the endonuclease$_{Int}$ system into two macromolecular components; role of a small molecule in nuclease activation. *Biochem. Biophys. Res. Commun. 81,* 947.
Spector, D.H. & Baltimore, D. (1975), Polyadenylic acid on poliovirus RNA. II. Poly(A) on intracellular RNAs. *J. Virol. 15,* 1418.
Weber. L.A., Feman, E. & Baglioni, C. (1975), A cell-free system from HeLa cells active in initiation of protein synthesis. *Biochem. 14,* 5315.
Williams, B.R.G., Kerr, I.M., Gilbert, C.S., White, C.N. & Ball, L.A. (1979), Synthesis and breakdown of pppA2'p5'A2'p5'A and transient inhibition of protein synthesis in extracts from interferon-treated and control cells. *Eur. J. Biochem. 92,* 455.
Zilbertstein, A., Kimchi, A., Schmidt, A. & Revel, M. (1978), Isolation of two interferon-induced translational inhibitors: a protein kinase and an oligoisoadenylate synthetase. *Proc. Natl. Acad. Sci. USA 75,* 4734.

**REGULATION OF MACROMOLECULAR SYNTHESIS
BY LOW MOLECULAR WEIGHT MEDIATORS**

STUDIES ON INTERFERON ACTION: SYNTHESIS, DEGRADATION AND BIOLOGICAL ACTIVITY OF (2'-5')OLIGO-ISOADENYLATE

Michel Revel, Adi Kimchi, Azriel Schmidt, Lester Shulman, Yuti Chernajovsky

Department of Virology
Weizmann Institute of Science
Rehovot, Israel

Sarah Rapoport, Yehuda Lapidot

Department of Biological Chemistry
Hebrew University
Jerusalem, Israel

Interferon exerts its action on cells by inducing several enzymes. One of the enzymes induced is the (2'-5') oligo-isoadenylate synthetase which, in the presence of double-stranded (ds) RNA polymerizes ATP into $ppp(A2'p)_n5'A$ oligonucleotides. One of the functions of these oligonucleotides is to activate an endonuclease, RNase F, which leads to inhibition of protein synthesis. In cell free systems, mRNA is degraded by RNase F. In SV40-infected monkey cells, in which interferon produces a selective inhibition of viral mRNA translation, mRNAs are not degraded but cleavage of ribosomal RNA is observed.

Interferon inhibits cell proliferation and can block the mitogenic effect of Con A in lymphocytes. The (2'-5') oligo A added to Con A-stimulated lymphocytes also inhibit DNA synthesis if added prior to entry in the S phase. Furthermore, the mitogenic stimulus of Con A induces an inhibitor of (2'-5') oligo A accumulation in lymphocytes. Entry into S phase may, therefore, be controlled in cells by the level of (2'-5') oligo A. The (3'-5') oligo A have no such biological activity.

INTRODUCTION

In extracts of cells treated by interferon, it is possible to identify at least three different mechanisms that reduce the translation of mRNA into proteins (Revel et al.,1978). As schematically shown in Figure 1, these pathways of translation regulation exert their effects on different components of the protein synthesis machinery and each is mediated by a specific enzyme. These enzymes are induced by interferon, since an elevation in the level of the enzymes can be measured a few hours after exposure of the cells to interferon and since this elevation is blocked by inhibitors of RNA and protein synthesis (Kimchi et al., 1979a). The induction of the three enzymes by interferon is illustrated in Table 1. The first pathway of translation regulation results from the induction of a specific protein kinase PK-i, which phosphorylates the small subunit of initiation factor eIF-2. Increased phosphorylation of a 67,000 Mr protein is also associated with this protein kinase activity. The protein kinase PK-i requires an activation step which takes place in the presence of low concentrations of dsRNA and ATP; once activated, the protein kinase is resistant to high concentrations of dsRNA (Kimchi et al.,1979b). A dsRNA-ATP activated preparation of the purified enzyme inhibits mRNA translation (Zilberstein et al.,1978; Farrell et al.,1978) and reduces strongly the formation of met-tRNA-40S ribosome complexes (Chernajovsky et al.,1979) The modification of eIF-2 is most likely responsible for the reduction in initiation complexes, as discussed by De Haro et al., (this volume) for the hemin-regulated eIF-2 protein kinase. We found also a phosphoprotein phosphatase (Revel et al.,1978) which dephosphorylates eIF-2 and the 67,000 Mr protein, and may regulate pathway 1 of Figure 1.

The second interferon-induced enzyme, which is also dsRNA-dependent is the (2'-5') oligo-isoadenylate synthetase E which polymerizes ATP in a series of oligonucleotide of general structure $ppp(A2'p)_n5'A$ (Kerr and Brown, 1978; Ball and White, 1978, Zilberstein et al.,1978). The nucleotide inhibits protein synthesis and enhances mRNA degradation (Baglioni et al.,1978; Ratner et al.,1978) by activating an endonuclease, RNase F, whose activity depends on the presence of the oligonucleotide (Schmidt et al.,1978).

RNase F binds oligo-isoadenylate (Chernajovsky et al., 1979) but is not irreversibly activated. Cells contain a phosphodiesterase which can split the 2' phosphate bond and degrades (2'-5')pppApApA into pA and pppA (Schmidt et al.,1978). This phosphodiesterase 2'-PDi is also increased after interferon treatment of the cells; and probably prevents an excess activation of RNase F. Purified 2'-PDi is however by itself an inhibitor of mRNA translation because it degrades the CCA terminus of tRNA (Schmidt et al.,1979) and is most likely responsible for the tRNA deficiency observed in extracts of interferon-treated cells (Zilberstein et al.,1976) as shown in pathway 3 of Figure 1. The (2'-5') oligo-isoadenylate appears to mediate part of the effects of interferon on protein synthesis, and both its formation and its degradation are subject to regulation by interferon. In this report, we describe some of the biological activities of oligo-isoadenylate as an activator of RNase F and as a mediator of interferon action on cell growth.

MATERIALS AND METHODS

Chemical Synthesis of (2'-5') oligo A. The tri-n-butyl-ammonium salt of adenosine (2',3') cyclic phosphate was chemically polymerized into poly A containing a random distribution of 2'-5' and 3'-5' phosphodiester bonds. 250 mg poly A was exhaustively digested with RNase Pl and the remaining $p(A2'p)_nA$ oligonucleotides separated on a DEAE-Sephadex A25 column with a 0.02-0.5 M gradient of ammonium bicarbonate. The peaks of dimer, trimer and tetramers were lyophilized, dissolved in dimethyl formamide with diphenyl-phosphochloridate and the tri-n-butylammonium salt of pyrophosphoric acid. After 1 hour reaction, the solution was evaporated. The oligonucleotides were precipitated with ether, dissolved in H_2O and chromatographed on a DEAE-Sephadex A25 column with a gradient of 0-0.6 M NaCl. The oligonucleotide peak was desalted on a Biogel P2 column in H_2O, then chromatographed on DEAE-cellulose in 7 M urea pH 7.5 and the peak eluting between charge -5 and -6 was desalted on DEAE-cellulose by elution with ammonium bicarbonate and lyophilized (Lapidot, Rapoport and Revel, unpublished). Enzymatically synthesized $(2'-5')ppp(A2'p)_nA$ dimers, trimers and tetramers were prepared as previously described (Schmidt et al.,1978).

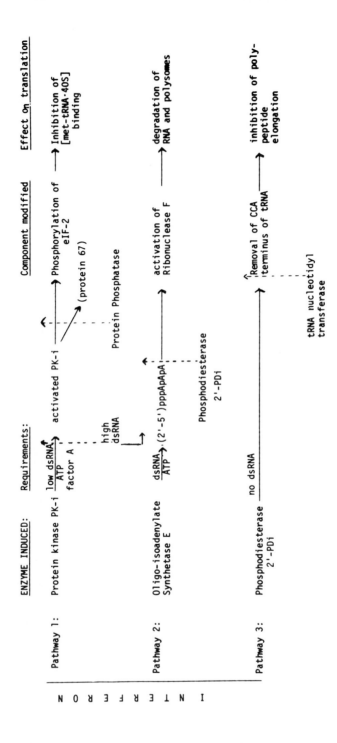

Figure 1. Interferon-induced translation regulations. (see text)

TABLE I. *Interferon-induced Enzymes and Anti-viral State in L Cells*[a]

Interferon U/ml	Expt. I		Expt. II		
	VSV RNA	Protein kinase PK-i	VSV RNA	Oligo A synthetase E	Phosphodiesterase 2'-PDi
0	100	15	100	4	25
2	24	35	12	5	72
10	6	80	5	20	86
50	0	100	2	42	100
250			0	66	
1,250			0	100	100

[a]Results are in per cent of maximum activity. Cells were treated for 8 hours with interferon. Enzymes assays are described by Kimchi et al. (1979a). Measure of vesicular stomatitis virus (VSV) RNA synthesis as in Weissenbach et al. (1979).

Assay of (2'-5') Oligo-isoadenylate Synthetase. The
assay of the enzyme activity in crude cytoplasmic
cell extracts, after binding to poly(rI):(rC)-agarose
beads (AG poly IC, type 6, PL Biochemicals), has
been described (Kimchi et al.,1979a). Purified
enzyme preparations were obtained and assayed as in
Zilberstein et al. (1978). Activation by cellular
RNA was measured in solution by replacing the
poly(rI):(rC) used in the assay reaction,by various
RNA preparations.

*Effect of interferon on the mitogenic stimulation of lymphocytes
by Concanavalin A.* Spleens were removed from 2 month old
Balb/c mice, and cells dispersed by passage through
a metal screen. Erythrocytes were lysed and splen-
ocytes cultures, containing 8×10^6 cells/ml, in RPMI-
1640 with 5% heat inactivated fetal calf serum were
established. Concanavalin A was added at 2.5 µg/ml
and mouse L cell interferon (induced by NDV and
purified by ammonium sulfate precipitation and
carboxymethyl Sephadex chromatography) was added at
the indicated concentrations. Synthetic (2'-5') or
(3'-5')(Ap)$_n$A oligonucleotides were added as indicated. DNA synthesis was measured at 72 hours by a
3 hour pulse of ^{3}H-thymidine (6 µCi/ml; 5 Ci/mmol) in
1 ml cultures. Lymphocyte extracts were prepared at
various times by lysing $0.5-1 \times 10^8$ cells in 0.1 ml of
Hepes buffer pH 7.5, KCl 120 mM, $MgCl_2$ 5 mM, DTT 1 mM
containing 0.5% NP40 and 10% glycerol. Lysates
centrifuged at 8,000 xg for 6 min were used at the
indicated protein concentrations.

RESULTS

RNase F: (2'-5')pppApApA-Dependent Endonuclease

RNase F was first isolated from L cells by its
ability to enhance the translation inhibitory effect
of (2'-5')pppApApA in preincubated and Sephadex-
filtered crude extracts of L cells (line CCL1)(Revel
et al.,1978; Schmidt et al.,1978). A similar procedure was used to isolate this enzyme from mouse
reticulocytes. By filtration on Biogel P-200, the
enzyme behaves as an 60-80,000 Mr protein, when
assayed by the (2'-5') oligo A dependent inhibition
of mRNA translation and we could detect binding of

radioactive (2'-5')pppApApA to the same fraction (Chernajovsky et al.,1979). Figure 2a shows that the degradation of Mengo virus RNA by RNase F (purified 1,000-fold) is completely dependent on addition of (2'-5') oligo A. Mengo RNA fragments of 100,000 to 500,000 Mr, are produced indicating that the cleavage is only at certain sites on the RNA.

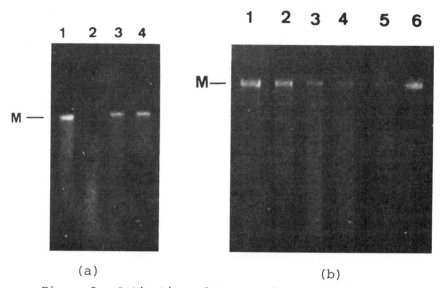

(a)   (b)

Figure 2. Activation of RNase F by enzymatically and chemically synthesized (2'-5')pppApApA.

(a) Reaction (10 µl) in Hepes buffer pH 7.5, 2.5 mM $MgCl_2$, 60 mM KCl, 0.5 mM dithiothreitol contained 1 µg Mengo RNA, with 0.3 µg RNase F (HAP step) alone (slot 1), and plus 0.64 pmoles (2'-5')pppApApA (slot 2). In slot 3, Mengo RNA was incubated without RNase F. In slot 4, Mengo RNA was incubated without RNase F but with the oligonucleotide. After incubation 30 min at 30°C, 6 µl of 33 mM K-phosphate pH 6.5, 50% sucrose, 1.5% SDS were added, the mixture heated to 65°C 5 min and applied to 1.5% agarose gel (horizontal) in 10 mM K-phosphate buffer pH 6.5. After electrophoresis at 6 V/cm, 2 hours, gels were stained with ethidium bromide and photographed under U.V. "M" indicates the position of intact Mengo RNA.

(b) Reactions (10 µl) contained 1 µg Mengo RNA, (slot 1 and 6) plus 0.6 µg RNase F alone (slot 2) or with enzymatically synthesized (2'-5')pppApApA, 70 nM (slot 3). In slot 4 and 5, chemically synthesized (2'-5')pppApApA (see Methods) was added at concentrations of 150 and 300 nM.

The fragments do not show discrete bands, the cleavage being probably at random in certain regions where Mengo RNA has little secondary structure.

Chemically Synthesized (2'-5')pppApApA Activates RNase F.
Figure 2b shows that the chemically synthesized oligonucleotide is biologically active and triggers the degradation of Mengo RNA by a purified preparation of RNase F. The chemically synthesized oligonucleotide was active in the same range of concentrations as the material formed by the enzyme extracted from interferon-treated L cells. In other experiments, activation of RNase F was seen with 30nM of chemically synthesized (2'-5')pppApApA. The 5'-triphosphate of the molecule is essential for activation of RNase F, since (2'-5')pApApA was inactive (Figure 3). The (2'-5')pppApA dimers were also unable to replace the (2'-5')pppApApA trimers in activating RNase F (Schmidt et al.,1978). Both ends of the oligonucleotide activator are, therefore, probably recognized by RNase F.

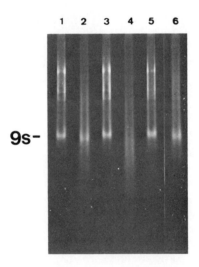

Figure 3. Requirement for 5' tri-phosphate for the activation of RNase F by (2'-5') oligo A.
Reactions (10 µl) as in Figure 2, contained 1 µg globin mRNA (slot 1) plus RNase F alone (slot 2) or plus (2'-5')pppApApA alone (slot 3). In slot 4 RNase F and (2'-5')pppApApA were added. In slot 5 and 6, (2'-5')pApApA was used instead without (slot 5) and with RNase F (slot 6). "9S" designates the position of intact globin mRNA.

The preparation of RNase F used in this work was obtained as outlined in Table 2. The most purified fractions still contained several large polypeptides. RNase F was inactivated at 50°C. It was active at Mg^{++} concentrations of 2 mM as well as 10 mM, and from 80 to 180 mM KCl.

TABLE 2. *Purification of (2'-5') oligo-A stimulated ribonuclease F from L cells*

Step		Activity
		Units per mg protein
I	S150 (cell sap)	10
II	DEAE cellulose pH 7.5, 120 mM KCl	340
III	Phosphocellulose pH 6.7, 430 mM KCl	4000
IV	Phosphocellulose pH 7.9, 200 mM KCl	4000
V	Hydroxylapatite pH 7.2, 100 mM PO_4	11000

One unit is the amount in µg needed to inhibit by 50% Mengo RNA translation in the presence of (2'-5')pppApApA, 70 nM. The SDS-polyacrylamide gel analysis below shows slot 1: DEAE cellulose step; Slot 2 and 3, phosphocellulose steps; slot 4, hydroxylapatite step; slot 5, peak tube of RNase F from hydroxylapatite. Approximate molecular weights are given. On the left, protein markers are shown.

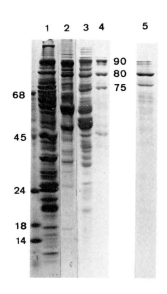

The effect of RNase F is not specific for a given RNA. Nevertheless, Mengo RNA translation was found to be more sensitive to the effect of RNase F than globin mRNA (Schmidt et al.,1978).

Ribosome RNA Degradation in SV40-infected Cells Treated by Interferon

In a previous work (Yakobson et al.,1977) we showed that in BSC-1 cells infected with SV40, the addition of 100 U/ml monkey interferon, 24 hours after infection, produces a marked inhibition of viral protein synthesis (i.e., T-antigen and capsid proteins VP-1 and VP-3). Most host proteins, however, continue to be synthesized with the exception of the histones. SV40 mRNA is made in these cells in normal amounts, but is not found associated with polyribosomes. An increased methylation of internal adenylic acid residues in SV40 mRNAs was found (Yakobson et al., 1978) but no extensive degradation of the viral RNA was detected by sucrose gradient analysis, or by agarose gel and hybridization to SV40 DNA. Recently, we observed that in cytoplasmic extracts of BSC-1 infected with SV40 and treated by interferon during the 24-48 h period after infection, some cleavage of ribosomal RNA became apparent. Figure 4 shows a glyoxal-agarose gel of RNA stained with ethidium bromide in which cleavage products appear at 1.7 and 1.2×10^6 Mr and at 0.32×10^6 Mr in addition to the 2×10^6 Mr and 0.7×10^6 Mr 28S and 18S rRNA bands. Without interferon, or in non-infected cells treated with interferon, the cleavage products are absent. These results suggest that an RNase is activated in the interferon-treated SV40 infected cells. Two experimental facts suggest that this phenomenon is due to the (2'-5') oligo A-dependent RNase F. First, interferon treatment induces the (2'-5') oligo A synthetase in these cells (Kimchi et al., 1979a). Second, RNA from SV40 infected cells can replace a synthetic dsRNA, such as poly(rI):(rC) in the activation of the (2'-5') oligo A synthetase. Table 3 shows that RNA extracted from BSC-1 cells has per µg, about 0.5% of the activity of poly(rI):(rC) in promoting the synthesis of (2'-5') oligo A by enzyme E. Furthermore, (Table 3, Expt.II) poly A-containing cytoplasmic RNA (purified on oligo(dT)-cellulose) had even a higher activity (1-2% of poly(rI):(rC)). Poly A^+-mRNA from human diploid

Studies on Interferon Action

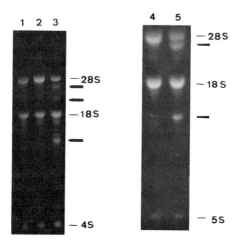

Figure 4. Cleavage of ribosomal RNA in interferon-treated SV40-infected monkey cells.
Monkey kidney cells BSC-1 monolayer cultures infected with SV40 for 24 hours, were treated with 100 U/ml monkey interferon and 24 hours later RNA was phenol-extracted from NP40 cytoplasmic extracts, heated 30 min to 50°C with 13% glyoxal-44% DMSO, and analyzed as in Fig. 2. Slot 1: RNA from uninfected cells treated with interferon; slot 2: RNA from infected cells, untreated; slot 3: RNA from infected cells treated with interferon. RNA from the ribosomal pellet is shown in slot 4, from infected cells; slot 5 infected interferon treated cells.

cells was also active. On a sucrose gradient the active RNA fraction peaked at 14-15S. After RNase A, the RNA was still active but sedimented slower than 4S RNA. Intact polysomes had the same activity as total polysomal RNA. In contrast, poly A^+-RNA from rabbit reticulocytes was devoid of such dsRNA-like activity. Since reticulocytes contain large amounts of the dsRNA-dependent (2'-5') oligo A synthetase and protein kinase (Hovanessian and Kerr, 1978), the lack of dsRNA-like activity in globin mRNA, may be a condition for potent hemoglobin synthesis. Other mRNAs, which contain dsRNA, may have been degraded during the maturation of reticulocytes. Bastos et al.(1977) showed that degradation of non-globin mRNA is required to explain how globin mRNA can accumulate to 90% of the total mRNA in reticulocytes.

TABLE 3. Activation of (2'-5') oligo-isoadenylate Synthetase by RNA from SV40 Infected and Uninfected Cells[a]

Expt.	Cells	Treatment	RNA	Amount	(2'-5') oligo A synthesis $[^{32}P]$-(2'-5')ApA cpm
I	BSC-1	None	Total cell RNA	1 µg	9,000
	"	Interferon	"	"	7,800
	"	SV40	"	"	13,600
	"	SV40+interferon	"	"	18,100
	"	" "	Cytop.RNA	1 µg	1,500
	-	-	poly(rI):(rC)	0.01µg	36,100
	-	-	None	-	150
II	BSC-1	SV40	polyA$^+$ cytop.RNA	0.8 µg	8,800
	Reticulocytes		polyA$^+$ RNA	0.6 µg	145
	-	-	poly(rI):(rC)	0.01µg	9,600

[a] Where indicated, monolayer cultures of BSC-1 cells infected with SV40 for 24 hours were treated by 100 U/ml monkey interferon (Yakobson et al.,1979) and 24 hours later, total RNA was extracted by phenol and 0.5% SDS, and precipitated with 2 M LiCl. From cytoplasmic extracts prepared with 0.5% NP40, cytoplasmic total RNA was obtained by phenol, 0.5% SDS- 5 mM EDTA. PolyA$^+$ RNA was purified on oligo(dT)cellulose. RNAs were added to a 25 µl reaction containing purified (2'-5') oligo-isoadenylate synthetase E (3 µg) and $[^{32}P]$-α-ATP (1 mM; 150 mCi/mmol), incubated 20 h at 30°C and the amount of (2'-5')ApA cores formed measured after alkaline phosphate as in Zilberstein et al. (1978).

The partial degradation of rRNA seen in the BSC-1 extracts, treated with interferon and infected with SV40 may reflect an activation of the interferon-**induced** (2'-5') oligo A synthetase by some RNA present in these cells. Nilsen and Baglioni (1979) have shown that RNAs with double-stranded regions are preferentially degraded in interferon-treated cell extracts. This may result from a local activation of the RNase F system. We have observed a complex between (2'-5') oligo A synthetase E and RNase F (Zilberstein et al., 1978); if the two enzymes function as one entity, the endonucleolytic activity may be functionally localized in the cell. It will be interesting to study whether ribosomes from viral polysomes are preferentially degraded and whether this degradation plays a role in the inhibition of viral mRNA translation seen in the intact SV40-infected cells.

Inhibition of Lymphocyte Mitogenesis by Interferon and by (2'-5') oligo-isoadenylate

Interferon exerts an anti-proliferative effect on cells (for review, Gresser and Tovey, 1978). This phenomena may be observed in a variety of tumor and normal cell cultures and usually requires larger amounts of interferon than the antiviral effect. This anti-growth effect of interferon is particularly clear in cells subjected to a mitogenic stimulus (Lindahl-Magnussen et al., 1972; Balkwill and Taylor-Papadimitriou, 1978) and in human cells is under the control of chromosome 21 like the antiviral effect of interferon (Cupples and Tan, 1977; Gurari-Rotman et al.,1978). We have chosen to study mouse spleen lymphocytes stimulated by Concanavalin A, a system which was shown to be well inhibited by interferon when added prior to the beginning of the S phase (Weinstein et al.,1977; Pacheco et al., 1976). Table 4 shows that the addition of mouse interferon within the first 24 hours after Con A addition inhibits strongly the stimulation of DNA synthesis measured at 72 hours. In this system, we have observed that both protein kinase PK-i and the (2'-5')oligo A synthetase are induced by interferon. We, therefore, assayed whether exogenously added oligonucleotides could mimic the effect of interferon. Since highly charged nucleotides would not penetrate the cells we used the (2'-5')ApApA,(i.e., the alkaline phosphatase resistant cores of the natural (2'-5')pppApApA). Williams and Kerr (1978) showed that these cores were able to inhibit protein synthesis as well as the phosphorylated compound when given to permeabilized cells. Table 4 and 5 show that in intact mouse lymphocytes, (2'-5')ApApA or longer oligomers efficiently inhibit the Con A-stimulated DNA synthesis. No pretreatment of lymphocytes is needed to see this effect and concentrations of 10^{-6} M are effective. In contrast, the (3'-5') isomers are devoid of any inhibitory activity (Table 5), even at

TABLE 4. Inhibition of DNA Synthesis in Con A-stimulated Mouse Spleen Lymphocytes by Interferon and by (2'-5') oligo-isoadenylate

Expt.	Addition to Con A-stimulated cultures	Time of[a] addition	DNA synthesis at 72 h [^{3}H]-Thymidine incorporated	Inhibition
			Counts/min	%
I	None	-	111,020	0
	Interferon, 200 U/ml	0 h	13,175	88
	" "	24 h	66,050	41
	" "	48 h	102,060	8
II	None	-	66,150	0
	(2'-5')ApApApA, 5 μM	0 h	17,665	74
	" " "	24 h	18,800	72
	" " "	48 h	56,765	15

[a] Time of interferon or oligonucleotide addition after Con A. For conditions see Methods.

TABLE 5. Inhibition of DNA Synthesis by (2'-5') oligo-isoadenylate in Con A-Stimulated Mouse Spleen Lymphocytes[a]

Culture	DNA synthesis at 72 h [^{3}H]-Thymidine incorporated
	counts/min
1. Untreated	28,760
2. +Con A 2.5 μg/ml	139,775 (100)
3. +Con A+(2'-5')ApApA, 2 μM	50,175 (19)
5 μM	41,805 (12)
4. +Con A+(3'-5')ApApA, 5 μM	142,665 (102)
10 μM	142,295 (102)

[a] Nucleotides added to lymphocyte cultures 10 min after Con A. Conditions as in Methods.

higher concentrations. The inhibition produced by the (2'-5') oligo A is maximum when the oligo-nucleotides are added together with Con A or 24 hrs later; when added at 48 h the oligonucleotide inhibits only slightly the subsequent DNA synthesis (Table 4). Since stimulation of DNA synthesis by Con A starts at 48 h and is maximum at 72 hours, this indicates that the oligonucleotide has to be added before the cells enter the S phase. A similar result is obtained with interferon itself (Table 4, and Weinstein et al., 1977).

A Mitogenic Stimulus Induces an Inhibitor of (2'-5') oligo-isoadenylate Accumulation

The preceding experiments suggest a role for the (2'-5') oligo A in the interferon-induced inhibition of lymphocyte mitogenesis. We wondered whether the Con A-stimulation may conversely have an effect on the synthesis of (2'-5') oligo A. When we measured the level of oligo-isoadenylate synthetase by binding the enzyme to poly(rI):(rC) Sepharose the Con A treatment neither reduced the basal level of enzymes in lymphocytes nor its induction by interferon (not shown). We, however, found that extracts of Con A-stimulated lymphocytes contain an activity that inhibits the synthesis of the (2'-5') oligo A by purified enzyme E. Thus, the amount of (2'-'5) oligo A formed by a known amount of enzyme E is much smaller in the presence of extract from Con A-stimulated lymphocytes than with those of unstimulated lymphocytes. The differential inhibition of (2'-5') oligo A synthesis is maximum with extracts of lymphocytes prepared 72 hours after stimulation, but is already marked at 24 and 48 h after Con A (Table 6). Fresh lymphocytes also contain the inhibitor. The level of inhibition of (2'-5') oligo A synthesis, is, therefore, high in lymphocytes in which DNA synthesis is active: fresh lymphocytes synthesize DNA and this decreases rapidly unless a mitogen is added to the culture (Sela and Gurari-Rotman, 1978). Similarly, the level of inhibitor of (2'-5') oligo A synthesis decreases when the lymphocytes are maintained in culture, without mitogen, but increase if Con A is added. The nature of the inhibitor is unclear; it is heat labile and nondializable. It is possible that the inhibitor degrades the oligonucleotide formed, as does the 2'-phosphodiesterase that we described previously (Schmidt et al.,1978;1979).

TABLE 6. Inhibitor of the (2'-5') oligo-isoadenylate synthetase in Con A-stimulated mouse spleen lymphocytes

Addition to (2'-5') oligo A synthetase E		(2'-5') oligo A synthesis $[^{32}P]$(2'-5')ApA, cpm		Con A-dependent inhibition of (2'-5') oligo A synthesis %
Lymphocytes extract:		No mitogen added	Con A-stimulated	
Prepared at:	0 h:	795	–	–
	24 h:	1265	475	62
	48 h:	1945	715	63
	72 h:	1890	285	85
No addition		2540		

(2'-5') oligo A synthetase from interferon-treated L cells was bound to poly(rI):(rC)-agarose. Extracts from mouse spleen lymphocytes, prepared at the indicated time from cultures treated with Con A or untreated, were added at a concentration of 0.35 mg protein per ml to the beads and the mixture incubated with $[^{32}P]$-α-ATP for 3 hours at 30°C. After incubation with alkaline phosphatase, the amount of (2'-5')ApA formed was measured by thin layer chromatography on poly-ethyleneimine cellulose with 0.5 N acetic acid. The (2'-5')ApA spots were cut and counted by scintillation.

The Con A mitogenic stimulus is, therefore, accompanied by an increase in an inhibitor of (2'-5') oligo A synthesis which could lower the basal level of oligonucleotide synthesis and allow the entry in S phase. The effect of interferon on Con A-stimulated lymphocytes may be explained by the induction of the (2'-5') oligo A synthetase to a level which exceeds the activity of the inhibitor, allowing accumulation of the oligonucleotide and inhibition of entry in S phase. The (2'-5') oligo A may thus play an important function in cell mitogenesis. Martin (1979) reported inhibition of DNA synthesis in the lymphoblastoid cell line Daudi. Effects on other cells were not observed (unpublished experiments) possibly because the oligonucleotide does not penetrate in non-lymphoid cells. It will be interesting to determine whether the inhibitor of (2'-5') oligo A synthesis, described here, exists in other cells and is increased in fast growing cells. Finally, it will be important to determine whether the antimitogenic effects of (2'-5') oligo A results from phosphorylation of the oligonucleotide in the cell and activation of RNase F or from other functions of these oligonucleotides. The experiments described here, suggest that the (2'-5') oligo A may find some use to regulate the immune response, instead of interferon.

ACKNOWLEDGMENTS

The assistance of Ms Helen Shure, Sara Tzur and Hanna Berissi is gratefully acknowledged. Work supported by a grant from NCRD (Israel) and Gesellschaft für Strahlen- und Umweltforschung (München, Germany).

REFERENCES

Baglioni, C., Minks, M.A., and Maroney, P.A. (1978). *Nature 273*, 684-687.

Balkwill, F.R., and Taylor-Papadimitriou, J. (1978). *Nature 274*, 798-800.

Ball, L.A., and White, C.N. (1978). *Proc.Natl.Acad.Sci. USA 75*, 1167-1171.

Bastos, R.N., Volloch, Z., and Aviv, H. (1977). *J.Mol. Biol. 110*, 191-203.

Chernajovsky, Y., Kimchi, A., Schmidt, A., Zilberstein, A., and Revel, M. (1979). *Eur.J.Biochem,* in press.

Cupples, C.G., and Tan, Y.H. (1977). *Nature 267*, 165-167.

Farrell, P.J., Sen, G.C., Dubois, M.F., Ratner, L., Slattery, E., and Lengel, P. (1978). *Proc.Natl.Acad. Sci. USA 75*, 5893-5897.

Gresser, I., and Tovey, M.G. (1978). *Biochem.Biophys. Acta. 516*, 231-247.

Gurari-Rotman, D., Revel, M., Tartakovsky, B., Segal, S., Hahn, T., Handzel, Z., and Levin, S. (1978). *FEBS Letts.*, *94*, 187-190.

Hovanessian, A.G., and Kerr, I.M. (1978). *Eur.J. Biochem.* *84*, 149-159.

Kerr, I.M., and Brown, R.E. (1978). *Proc.Natl.Acad.Sci. USA* *75*, 256-260.

Kimchi, A., Shulman, L., Schmidt, A., Chernajovsky, Y., Fradin, A., and Revel, M. (1979a). *Proc.Natl.Acad. Sci.* *76*, in press.

Kimchi, A., Zilberstein, A., Schmidt, A., Shulman, L., and Revel, M. (1979b). *J.Biol.Chem.* in press.

Lindahl-Magnusson, P., Leary, P., and Gresser, I. (1972). *Nature New Biol.* *237*, 120-121.

Martin, E.M. (1979). personal communication.

Nilsen, T.W., and Baglioni, C. (1979). *Proc.Natl.Acad. Sci. USA* in press.

Pacheco, D., Falcoff, R., Catinot, L., Floch, F., Werner, E.H., and Falcoff, E. (1976). *Annal Immunology (Inst.Pasteur)* *127C*, 163-171.

Ratner, L., Wiegand, R.C., Farrell, P.J., Sen, G.C., Cabrer, B., and Lengyel, P. (1978). *Biochem.Biophys. Res. Commun.* *81*, 947-953.

Revel, M., Schmidt, A., Shulman, L., Zilberstein, A., and Kimchi, A. (1978). *Fed.Eur.Biochem.Soc.Meet.(Proc.)* *51*, 415-426.

Schmidt, A., Zilberstein, A., Shulman, L., Federman, P., Berissi, H., and Revel, M. (1978). *FEBS Letts.*, *95*, 257-264.

Schmidt, A., Chernajovsky, Y., Shulman, L., Federman, P., Berissi, H., and Revel, M. (1979). *Proc.Natl.Acad. Sci. USA* *76*, in press.

Sela, B., and Gurari-Rotman, D. (1978). *Biochem.Biophys. Res. Commun.* *84*, 550-556.

Weinstein, Y., Brodeur, B.R., Melmon, K.L., and Merigan, T.C. (1977). *Immunology 33*, 313-319.

Weissenbach, J., Zeevi, M., Landau, T., and Revel, M. (1979). *Eur.J.Biochem.* in press.

Williams, B.R.G., and Kerr, I.M.(1978). *Nature 276*, 88-89.

Yakobson, E., Prives, C., Hartman, J.R., Winocour, E., and Revel, M. (1977). *Cell 12*, 73-81.

Yakobson, E., Kahana, H., Revel, M., and Groner, Y. (1978). *Ann.Meet.Israel Bioch.Soc., p.169 (Abstr.)*.

Yakobson, E., Revel, M., and Gurari-Rotman, D. (1979). *Arch.Virol. 59*, 251-255.

Zilberstein, A., Dudock, B., Berissi, H., and Revel, M.(1976). *J.Mol.Biol. 108*, 43-54.

Zilberstein, A., Kimchi, A., Schmidt, A., and Revel, M. (1978). *Proc.Natl.Acad.Sci. USA 75*, 4734-4738.

REGULATION OF MACROMOLECULAR SYNTHESIS BY LOW MOLECULAR WEIGHT MEDIATORS

LOW MOLECULAR WEIGHT MEDIATORS OF MACROMOLECULAR SYNTHESIS - AN OVERVIEW

A.A. Travers

MRC Laboratory of Molecular Biology,
Hills Road, Cambridge, England.

The regulation of macromolecular synthesis by low molecular weight compounds is of wide occurrence in living organisms. The first intimation of a regulatory process of this type has often been the correlation of an *in vivo* change in macromolecular synthesis with an alteration in the levels of a particular compound. The question that we seek to answer is whether these two phenomena are casually related. In particular we aim to unify the biology of the system under study with reactions in a test tube. It is clear that *in vitro* observations can account for some *in vivo* phenomena in qualitative or even quantitative terms but in no case, I think, does there exist rigorous proof that the reactions which we study in the test tube are indeed those responsible for the regulation observed in the living organce. Despite this caveat it is clear that the low molecular weight mediators of macromolecular synthesis are of considerable biological importance. In many cases there is an elaborate enzymatic machinery for controlling their intracellular concentrations at the levels of both synthesis and degradation. However, a further complication has been the realisation that there is redundancy in the regulatory systems controlling the biosynthesis of macromolecules such that the same effect may be elicited by a variety of environmental perturbations, not all of which produce corresponding variations in the level of the putative regulator(s).

I would like to consider ppGpp as the paradigm of an effector whose accumulation is correlated with changes in the selectivity of gene expression. Although a comparative late-comer in the historical perspective the study of the mode of action of this nucleotide has provided the basis for many of our ideas as to how such a pleiotypic effector might work. Although ppGpp is a dispensible metabolite during normal growth (Cashel) it is clear that the cell goes to considerable lengths to control the intracellular concentration of this compound. Both genetic (Atherley, Cashel) and biochemical (Richter, Sy) analysis show that there are two pathways for its synthesis, one dependent on ribosomes and a second whose function is possibly linked to energy metabolism. Similarly the degradation of ppGpp to either ppGp or ppG may be equally complex (Richter).

What is the role of this compound? *In vivo* its accumulation is often, but not invariably, associated with a crash response to sudden environmental stress. One manifestation of this response is a dramatic change in transcriptional selectivity, manifested as a preferential shut off of rRNA synthesis. *In vitro* physiological concentrations of the nucleotide inhibit the initiation of rRNA synthesis with either template bound (Gruber) or free (Travers) RNA polymerase as its target. A second consequence of the imposition of stringency is the unfolding of the bacterial nucleoid. If we take into account this change the 10-20 fold inhibition of rRNA synthesis by ppGpp *in vitro* is quite sufficient to account for the *in vivo* response. Yet we know that changes in gene expression, very similar to those observed during the stringent response, can be elicited by the addition of TPCK to bacteria or by the introduction of a temperature sensitive mutation in the gene encoding fructose 1,6-diphosphate aldolase (Atherley). In these cases ppGpp does not accumulate so it is by no means certain that ppGpp is the immediate effector of the change in macromolecular metabolism occurring during the stringent response although genetic analysis suggests that it is necessary for the response. Indeed ppGp has been suggested as an alternative to ppGpp as the principal regulator of stringency (Gallant). An alternative view to explain this apparent paradox is to suppose that there is no one single effector but that the response of macromolecular synthesis and in particular transcription, to environmental stress depends on the balance between the levels of a number of effectors. Such a mechanism would allow diverse means of eliciting the same response.

In this meeting the talks concentrated on the short term effects of the mediators, but it is my view that the long term effects may be just as important. The accumulation of ppGpp is accompanied by the preferential synthesis of two proteins, B56.5 and stringent starvation protein. Both these proteins bind to RNA polymerase and appear to alter its properties in a similar manner to ppGpp itself. In other words it appears that ppGpp may induce the synthesis of proteins which then act to stabilise the effect of the low molecular weight mediator. This would thus constitute a simple form of differentiation. The B56.5 protein also binds to the ribosome and one might speculate that it would play a role similar to ppGpp in the suppression of translational errors (Gallant).

A second prokaryotic differentiation in which low molecular weight molecules have been implicated as effectors is the process of sporulation in *Bacillus subtilis*. Two mechanisms have been proposed. On nitrogen starvation the GTP pool falls without a concomitant increase in the HPN IV (p_3Ap_3) pool (Freese). By contrast on energy starvation the HPN IV pool increases (Rhaese). In both cases the correlation between the observed phenomena and the physiological response is strong and suggestive of a causal effect. Again it seems plausible that both effects might be mediated through analogous alterations in the balance of effectors, a balance which would suggest the high rate of spontaneous sporulation is comparatively easy to perturb. These results also emphasise that we must not neglect the regulatory importance of the levels of the nucleoside triphosphate pools, an importance also apparent during the shut off of host protein synthesis consequent upon phage T1 infection (Schweiger).

In the search for selective regulators of gene expression in higher eukaryotes we enter murkier waters. The ubiquity of HS3 in eukaryotic cells are the responses of its intracellular concentration to changes in nutritional conditions (Lejohn) are intriguing and suggestive. Yet although the fluctuations of the levels of HS3 and of the phosphorylated compounds appearing after heat shock in *Drosophila* (Travers) correlate with changes in macromolecular synthesis there is really no evidence as to whether they are themselves regulatory molecules or merely intermediates in some unrelated metabolic pathway.

By contrast the regulation of the activity of the eukaryotic protein synthetic machinery can be much more clearly demonstrated. The regulation of activity alone can generate a selective effect on gene expression by a simple sliding scale effect (Koch). In such a control mechanism the translation of different mRNA species, for example, would be differentially affected according to their binding thresholds. One possible regulatory phenomenon of this type is the phosphorylation of r-protein S6 (Gordon, Koch, Plesner) which follows changes in environmental conditions in organisms as diverse as *Tetrahymena* and mice. Here it remains to be established whether this phosphorylation simply alters ribosomal activity or also changes ribosomal discrimination. Similarly, Gp_4G and allied nucleotides in *Artemia*, in addition to their well established storage function can inhibit certain aspects of translation *in vitro* but we do not yet know whether they do so *in vivo* (Warner).

The response of cells to interferon and dsRNA provides an example of another type of redundancy of control mechanisms whereby the same goal of the limitation of protein synthesis may be achieved by a variety of mechanisms. Interferon itself acts directly or indirectly as a regulator of gene expression by inducing the synthesis of eIF-2 kinase and also of two enzymes involved in 2-5 A metabolism (Baglioni, Ball, Revel). It is clear that 2-5 A acts to inhibit protein synthesis *in vivo* probably by activating a nuclease which degrades mRNA (Hovanessian). Of particular interest is the suggestion (Baglioni, Revel) that the selectivity of 2-5 A action with respect to viral RNA degradation may be gained by the compartmentalisation of its synthesis and site of action. The extremely low concentrations of 2-5 A required for endonuclease activation together with rapid degradation of the compound would be fully consistent with such a model (Ball).

This possibility of compartmentalisation in the regulation of macromolecular synthesis was again suggested as a mechanism controlling DNA synthesis in mammalian cell nuclei (Rapaport). Here the effective signal is proposed to be the balance between ADP and ATP. Also established as a putative regulator of DNA replication is Ap_4A, an unusual nucleotide discovered in unusual circumstances by Zamecnik, as an apparently aberrant byproduct of aminoacyl tRNA synthetase activity. One target of this compound is clearly DNA polymerase α (Grammt). The nucleotide binds tightly to a single subunit of the enzyme but again, however, the direct demonstration that Ap_4A is an immediate trigger of replication in eukaryotes is lacking.

What of the future? The recognition of the complexity of the biological control systems regulating macromolecular synthesis in general suggests that there is much to learn and perhaps new magic spots to be identified. I am sure, however, that we shall continue to be surprised by what we find.

INDEX

A

2-5A, 364
　mechanism of action, 322–323
　protein synthesis and, 319–326
2-5A polymerase, induction of, 330
2-5A synthetase, protein synthesis and, 320
Abt, see pppAppp
Adenosine, inhibition of DNA synthesis and, 223–229
Adenosine 5′,3′(2′)-bis-triphosphate, see pppAppp
Adenosine 3′-diphosphate 5′-diphosphate, see ppApp
ADP-ribose, 2′,5′-oligoadenylate synthetase and, 304
Amino acid(s), analogues, TI-host repression and, 253
Aminoacylation, of tRNA, dinucleoside polyphosphates and, 173
Aminoacyl-tRNA synthetase mutants, tRNA charging differences in, 39–40
ANs, membrane fluorescence, TI-infection and, 254–255
Antibody, to ppGpp, 107, 110–112
Ap2A, stacking of, 180
Ap3A, stacking of, 180
Ap4A
　adenylylation of nuclear protein and, 181
　binding site in neuronal cells, 209
　degradation of, 180–181
　formation of, 1, 180
　high metabolic lability of, 182
　mechanism of action, 210, 364
　metabolic role for, 180
　2′,5′-oligoadenylate synthetase and, 304
　as pleiotypic activator, 181–182
　as pleiotypic signal, 235
　pool size, cell cycle and, 181
　replicative DNA synthesis and, 209–219
　stacking of, 180
Ap5A, stacking of, 180
Apparatus, optical rotatory dispersion, aminoacyl tRNA synthetase and, 179
Artemia, cysts, dinucleotide polyphosphates in, 161
Artemia salina, embryos, dinucleoside polyphosphates in, 161–174
ATP
　concentration
　　starvation and, 241
　　TI-infection and, 258–259
　concentration effect of, 234
　mobility of, 186
ATP/ADP ratios, nuclear, 223–229
ATPases, DNS-dependent, ATP/ADP ratios and, 223

B

Bacillus brevis
　ppGpp metabolism in, 85–93
　ppGpp synthesis in, 95–105
　stringent factor of, 95
Bacillus stearothermophilus, ppGpp metabolism in, 85–93
Bacillus subtilis
　ppGpp metabolism in, 85–93
　sporulation in, 127–139, 145–157

Brine shrimp, *see also Artemia*
  embryos, dinucleoside polyphosphates in, 161–174
Brine shrimp cysts, rehydrated, 1

C

Cell division, ATP concentration and, 234, 241
Cell membrane, TI-host repression and, 253–257
Cell proliferation
  Ap4A and, 1
  interferon and, 341
Cell transformation, inhibition of, 2
Chick embryo cells, 2′,5′-oligoadenylate synthetase of, 303–315
Chlamydomonas reinhardi
  ppGpp in, 117–118
  ribosomes, ppGpp synthesis and, 122
Chloroplasts, ppGpp synthesis and, 115, 122
Competition binding, ppGpp radioimmunoassay and, 107–114
Con A, effects, interferon and, 341
Concentration effect, of ATP, 234

D

Decoyinine
  nucleotide concentrations and, 138
  sporulation and, 134
Deoxyribonucleic acid, *see* DNA
Development
  regulation in *Artemia*, 161, 171–174
  source of purine-containing compounds in *Artemia*, 169–171
Diadenosine 5′,5′′′-p¹-p⁴-tetraphosphate, *see* Ap4A
Dictyostelium discoideum, ppGpp in, 118
Differentiation
  control of, 138–139
  initiation, sensor model of, 155, 157
  microbial, 127–141
  prokaryotic, 363
Diguanosine tetraphosphate, *see* Gp4G
Dinitrophenol
  Ap4A levels and, 242
  transcription pattern and, 189
Dinucleoside polyphosphates
  in *Artemia* embryos, 161–174
  HS3 and, 194
Diribonucleoside monophosphates, 2′,5′-oligoadenylate synthetase and, 303

Discrimination specificity
  enhancement of, 11
  of polyU, 5
DNA
  initiation of synthesis, initiation of protein synthesis and, 182
  replication, ATP/ADP ratios and, 223–229
  replicative synthesis, Ap4A and, 209–219
  synthesis
    Ap4A and, 1, 364
    Gp4G metabolism and, 169, 170–171
    sporulation and, 128–130
DNA polymerase, primers for, 2
DNA polymerase α, Ap4A binding activity and, 209–219, 364
DNA-RNA polymerase, interaction, ppGpp and, 73
Downshift, *see also* Shift down
  carbon-source, mutant AA-787 and, 53–64
  de novo nucleotide biosynthesis and, 195
  ppGpp synthesis by mutant AA-787, 62
Drosphila melanogaster, highly phosphorylated compounds in, 185–191, 363
dsRNA, 363
  2′,5′-oligo(A) polymerase and, 329–338, 341–357
  protein kinase activation and, 342

E

Effector
  pleiotypic, ppGpp as, 361–362
  sporulation and, 145–157
eIF-2, 40S initiation complexes and, 292
eIF-2 kinase, synthesis of, 364
eIF-2 stimulating protein, 40S initiation complexes and, 292
Encephalomyocarditis virus, RNA, inhibition of accumulation, 330
Endoribonuclease, activation of, 304, 341
Enzyme, membrane bound, nucleotide synthesis by, 145, 152
Error frequency, UUU codons and, 5–6
Escherichia coli
  ppGpp metabolism in, 85–93
  stringent factor of, 95
Eukaryotes
  lower, ppGpp in, 115, 117–119
  protein synthetic machinery, regulation of 363

Index

G

Gene, for abt synthetase, isolation of, 152–156
Gene expression
 regulation by ppGpp, 35–36
 T1-virus and, 249–261
GMP, synthesis, specific inhibitors of, 141
Gp4G
 in brine shrimp oocytes, 161, 162, 164–165, 363
 as energy source, 1
GTP, mobility of, 186
GTP: GTP guanylyltransferase, localization of, 161, 166–169
Guanosine 3'5'-bispyrophosphate, see ppGpp
Guanosine 3'-diphosphate-5'-diphosphate, see ppGpp
Guanosine 5'-tetraphospho-5'-guanosine, see Gp4G

H

Heat shock
 ppGpp synthesis and, 115, 118–119
 transcription pattern and, 189
Hemin-controlled translational inhibitor, in reticulocyte lysates, 291–300
High pressure liquid chromatography, see HPLC
Host repression, T1-virus and, 250–254
HPLC
 future and, 2
 nucleotide separation by, 235–236
HS3
 nature of, 194
 as pleiotypic regulator of eukaryotes, 193–205
 salvage reactions of nucleotide synthesis and, 196
Hypertonic initiation block, polypeptide chain initiation and, 274

I

Inducer, of sporulation, 134
Inhibition, virus-induced, of protein synthesis, 277–279
Initiation factor eIF-2, phosphorylation of, 342
Initiation complexes, 40S, eIF-2 and, 292
Interferon
 action of, 344–357, 363
 2', 5'-oligoadenylate synthetase and, 303–315
 2',5'-oligo(A) polymerase and, 329–338
 protein synthesis and, 319–320

K

Kinetic amplification, ppGpp and, 20

M

Magic spots I and II, identification of, 116
Mammalian cells, ppGpp in, 115, 119–123
Mediator, low molecular weights, protein synthesis and, 273, 283–285
Membrane, release of inhibitor of polypeptide chain initiation, 283–285
Methanol, ppGpp synthesis and, 103
Methylene-bis-ADP, Ap4A activity and, 210
Methylxanthines, S6 phosphorylation and, 263, 268
Mitochondria, ppGpp synthesis and, 115, 119, 122
Mitogenic effect, interferon and, 341, 346, 353–357
Mouse embryos, ppGpp in, 115, 120–121
Mutants, asporogenous, isolation of
 early-blocked temperature-sensitive, 150–151

N

NAD, 2',5'-oligoadenylate synthetase and, 304
Nuclear compartments, ATP pools of, 223
Nuclease, activation, 2-5A and, 320, 321–323, 364
Nuclei, S phase and, 223
Nucleotides
 de novo biosynthesis, downshift and, 195
 highly phosphorylated, sporulation and, 145–157
 salvage reactions of synthesis, HS3 and, 196
 unusual, 1
Nucleotide pools, changes during initiation of sporulation, 136–139
Nutritional deprivation, sporulation and, 127–130

O

2',5'-Oligo(A), endoribonuclease and, 329
2',5'-Oligoadenylate synthetase, interferon and, 303–315

2',5'-Oligoadenylic acid triphosphate, see
    2-5A
2',5'-Oligo-isoadenylate synthetase interferon
    and, 341-357

P

P3AP3, see pppAppp
pGp, formation of, 85-93
pGpp, formation of, 85-93
Phenotypic suppression, test for, 13-15
Phe-tRNA, ppGppase and, 88-89
Phosphodiesterase, inhibitors, cAMP pools
    and, 268-269
Phosphorylation
    of initiation factor eIF-2, 292
    of protein 56, 238, 263-270, 363
Phosphorylation state, of ribosomal proteins,
    273, 279-283
Pi pool, 263
    internal, compartmentation of, 265
Pleiotropic effect, membrane mediated,
    273-286
Pleiotropic effector, synthesis and
    degradation, 85-93
Pleiotypic response, growth control and, 194
PMF, transport driven by, T1-infection and,
    254-257
Polyphenylalanine, synthesis by mutant
    AA-787, 60-61
Polysome, assembly, methylxanthines and,
    263
ppApp
    as regulator, 77-83
ppApp
    rRNA synthesis and, 82
ppGp
    formation of, 85-93
    stringency and, 362
ppGpp
    accumulation of, 95-96
        in *Drosophila*, 185-191
    basal level, variations in, 49-50
    degradation of, 88-90, 362
    as effector, 361-362
    in eukaryotes
        lower, 115
        mammalian cells, 115
    function of, 1
    level of charged tRNA and, 39
    mutant AA-787 and, 53-64
    [^{32}P]-labeled, 107-114
    production, two pathways for, 25
    radioimmunoassay for, 102-114
    regulation of gene expression and, 35-36

as regulator, 77-83
relaxed mutants and, 25
RNA polymerase and, 82
RNA synthesis and, 67-74
screening assay for, 28
studies in bacteria, 85-93
synthesis of, 361-362
    in *Bacillus brevis*, 95-105
    by mutant AA-787, 61-62
    ribosome-dependent, 87
    translation errors and, 9
ppGpp-ase, ribosomes and, 85-93
ppGpp synthetases, 96
    characteristics of, 95-105
ppp(A2'p)$_n$ 5'A oligonucleotides, formation
    of, 341
pppAppp
    sporulation and, 145-152, 155-157
    synthesis, regulation of, 156
pppAppp synthetase, gene for, 156
pppA2'p5'A2'p5'A
    interferon and, 1-2
    protein synthesis and, 320
Proliferative activities, DNA-dependent
    ATPases and, 223
Prostaglandin E1, S6 phosphorylation and,
    263, 268
Protein
    nuclear, adenylylation of, 181
    ratio to RNA, 49-50
    stringent starvation, 362
Protein B56.5, ribosome and, 362
Protein kinase
    induction of, 342
    initiation factor eIF-2 and, 292
Protein synthesis
    2-5A and, 319-326
    host, phage T1 infection and, 363
    inhibition, virus-induced, 277-279
    initiation, initiation of DNA synthesis and,
        182
    in mutant AA-787, 59-61
    regulation, Gp3G and, 169, 171-174
    restriction, ppGpp concentration and, 39
    sporulation and, 128-130
Proton motive force, see PMF
Pulse-chase experiments, Pi pool and,
    263-270
Purine, deficiency, sporulation and, 130-135
Purine pathway, block in, 130-132
Purine requiring mutant, sporulation of,
    130-135
Prophosphate donor, ppGpp synthesis and,
    85-93

Pyrophosphotransferase, ribosome-dependent, 96

R

Radioimmunoassay, for ppGpp, 107-114, 116
Regulation, of translation, coordination of, 40
Rel A, deletion, rel X and, 53-64
Rel A mutant, ppGpp metabolism in, 85-93
Relative translational efficiency, of each mRNA, 275-277
Relaxed mutants
  ppGpp levels in, 9, 26
  phenotypic suppression in, 17-18
Release, membrane-mediated, of inhibitor of polypeptide chain initiation, 283 -- 285
Rel X
  deletion, rel A and, 53-64
  genetic studies, 58-59
Rel X, rel A, genetic studies, 58-59
Replication, viral, rUGG and, 2
Replicative intermediate, of EMCV, 330
Reticulocyte lysates, hemin-controlled translational inhibitor in, 291-300
Ribonucleic acid, see RNA
Ribonucleic acid, double stranded, see dsRNA
Ribonucleic acid, messenger, see mRNA
Ribonucleic acid, ribosomal, see rRNA
Ribonucleic acid, transfer, see tRNA
Ribosomal mutations, characteristics of, 48
Ribosomal protein F6, phosphorylation, 238, 363
  kinetics of, 263-270
Ribosome(s), ppGpp synthesis and, 87
Ribosome-mRNA-uncharged tRNA-complex, stringent factor and, 85-93
RNA
  poly(A) containing, protein synthesis and, 237,
  ratio to protein, 49-50
  synthesis
    in mutant AA-787, 59
    sporulation and, 128-130
  viral, inhibition of accumulation, 330
mRNA, translation, inhibition, 342
rRNA
  synthesis
    initiation frequency, 67-74
    shut off of, 362
  transcription, restriction of, 116
rRNA promoter(s),
  RNA polymerase and, 67-74, 77-83
tRNA
  charged, level of, 39
  charging, unbalance of, 40
  transcription, restriction of, 116
RNA polymerase
  holoenzyme, different states of, 83
  HS compounds and, 201
  inhibition of, 362
  ppGpp and, 80-82
  restriction fragment and, 67-74
  selectivity of, 77-83
  specificity, subunit and, 82
RNA polymerase-rRNA promoter, stability of, 67-74
RNase F
  activation of, 341, 348-350
  mRNA and, 341

S

Saccharomyces cerevisiae
  ppGpp in, 118-119
  sporulation in, 139-141
Shift down, see also Downshift
  relaxed mutants and, 25, 33
  sporulation and, 145-157
S-100, nuclease-treated, RNA synthesis and, 69-70
S phase
  ATP/ADP ratio and, 223
  2',5'-oligo A and, 341
Sporogenesis, control of, 194
Sporulation
  induction, decoyinine and, 134
  initiation of, 156
  nutritional deprivation and, 127-130
  regulation, by highly phosphorylated nucleotides, 145-157, 363
SpoT mutant, ppGpp metabolism in, 85-93
Stacking, of diadenosine polyphosphates, 180
Starvation
  ppGpp synthesis and, 115
  patterns of protein synthesis and, 236-237
Stringent control, translational accuracy and, 5-20
Stringent factor, ppGpp and, 95
Stringent response, ppGpp and, 1, 25, 362
SV40-infected monkey cells, interferon effects in, 341, 350-353

T

T1-virus
  gene expression and, 249-261
  translation and, 250-253, 363
3T3 mouse fibroblast cells, Pi pools of, 265-268

Temperature shift, nucleotide accumulation in *Drosophila* and, 184–191
Ternary complex, formation of, 292, 293, 296
Tetracycline, ppGpp synthetase and, 95
Tetrahymena pyriformis, regulation of protein synthesis in, 235–246
Thiostrepton, ppGpp synthetase and, 95
Theophylline, cAMP pools and, 268–269
Transcription
  initiation step, its compounds and, 201
  pattern, heat shock and, 189
  restriction of, 116
Translation
  accuracy, stringent control of, 5–20
  control, membrane mediated, 273–286
  coordination of regulation, 40
  errors, 5–7
  inhibition, by phosphorylated eIF-2, 292
  T1-virus and, 250–253

Transport, PMF-driven, T1-infection and, 254–257
Turnover effect, ATP and, 234

U

rUGG, action of, 2
Uptakes, ATP-dependent, T1-infection and, 254–255

V

Virus, replication of, 2

Y

Yeast, mitochondrial ribosomes, ppGpp synthesis and, 122
Yolk platelets, of *Artemia,* dinucleoside polyphosphates in, 161, 165, 168–169